About

JOB'S BODY
A Handbook for Bodywork

A masterful synthesis, copiously illustrated, of body therapies and anatomical-physiological functions.

EAST WEST MAGAZINE

Juhan blends a novelist's sensitivity for language and an academic's respect for accuracy with a bodyworker's respect for the living, sentient flesh.

YOGA JOURNAL

The first comprehensive attempt to explain the physiology of massage is a success.

MASSAGE MAGAZINE

This book should be required reading for every bodyworker. For lay people, it will sharpen appreciation of the human body.

UTNE READER

Despite a highly technical presentation, complete with more than 100 anatomical drawings and photographs, the book is eminently accessible to lay readers. The dangers of touch deprivation are well-documented.

PUBLISHERS WEEKLY

A broad presentation of the rationale of bodywork, including the relevant physiological, anatomical, and neuropsychological information one should know if one practices any form of bodywork.

SOMATICS

This is a very important book indeed, for it is a guide to the anatomy and physiology of a body that is *someone's* body . . . . This is not merely a scientific account of the physical body, but is already a guide to the lived body as well.

STUDY PROJECT IN PHENOMENOLOGY OF THE BODY

Recommended . . . .

LIBRARY JOURNAL

From William Blake's engraved "Illumination" of *The Book of Job* (1825).

JOB'S BODY

A Handbook for Bodywork

Deane Juhan

Foreword by Ken Dychtwald

STATION HILL

BARRYTOWN, LTD.

Published by Barrytown, Ltd. Barrytown, New York 12507, under Station Hill Openings
Web: www.stationhill.org
E-mail: publishers@stationhill.org

Cover by Susan Quasha

Acknowledgments for quoted materials appear on pages 391-398.
Acknowledgments for illustrations appear on pages 399-404.

Library of Congress Cataloging-in-Publication Data

Juhan, Deane.
    Job's body : a handbook for bodywork / Deane Juhan ; foreword by Ken Dychtwald.
    [New, expanded ed.]
        p. cm.
    Previously published in 1987.
    Includes bibliographical references and index.
    ISBN 1-58177-022-7 (alk. paper)
    1. Touch—Therapeutic use. 2. Manipulation (Therapeutics)
3. Human physiology. 4. Mind and body. I. Title.
RZ999.J84   1998
615.8'22-dc21   98-28525
                                                          CIP

Printed in The United States of America

This book is fondly dedicated to Milton Trager, M.D., whose life work is an unequivocal testimony to the marvelous extent to which a pair of sensitive, knowing hands can alleviate the emotional and physical suffering of human beings.

Preface to the Expanded Edition

My compulsive curiosity about the power of touch and its therapeutic significance was not to any agree laid to rest by the publication of *Job's Body*. In fact the kind of clinical evidence that surfaced during my research for the book, and the themes I found developing during the writing, have continued to be restlessly intriguing for me.

What began as an attempt to somehow account for the effectiveness of touch and movement therapies has led to a very different view of the nature and development of the human organism than I could have possibly anticipated. I had thought that bodywork—and all the processes it can animate within us—could be shown to be a viable way out of many painful and pathological conditions. I am now convinced that this is barely the beginning of what touch can do for us and of what we have to learn from it.

Compassionate, nurturing touch has become alarmingly rare in almost all contexts of our modern life. There are far more regulations restricting physical contact than there are educational pursuits that might show us how to develop and use it. The result of this cultural condition is that we are not only "out of touch" with one another, but also out of touch with many of the forces that shape us and misshape us. And as this absence erodes each of us individually in various ways, it also undermines the physical and psychological foundations that could give more daily satisfaction and functional coherence to our social interactions. Friends, lovers, families, communities, schools, corporations, churches—even legislatures— can all benefit in incalculable ways from the grounding, the bonding, and the trust that only honest, actual contact can create. A simple, brief handshake is used to convey that I mean you no harm. If we can manage to take it beyond that, our contact might also convey that I mean you a great deal of good as well.

So effective touch is not just a way out of much of our physical or emotional pain. It is also a way out of the isolation so endemic in our culture, and out of many of the interpersonal, social and philosophical cul-de-sac which that isolation has produced. In the end, touch is far more than therapeutic; it is *communicative*, and has the power to communicate dimensions of ourselves and our intentions that simply cannot be passed on in any other way. This language of touch and all it has to teach us must be added to our social currency before its absence bankrupts us as organisms. The only genuine escape from the labyrinths of solipsism is direct contact, and this escape through contact is one of the chief pleasures of being human.

The chapters that have been added for this second edition are intended to pursue these larger themes: It is not just our bodies that have suffered the lack of contact, but also the very fabric of our common lives. And touch is not just potentially therapeutic; it is a birthright of incalculable value and joy.

D.J., September 1, 1998, Mill Valley

The ancient tradition that the world will be consumed in fire at the end of six thousand years is true, as I have heard from Hell.

For the cherub with his flaming sword is hereby commanded to leave his guard at tree of life, and when he does the whole creation will be consumed, and appear infinite and holy whereas now it appears finite and corrupt.

This will come to pass by an improvement in sensual enjoyment.

But first the notion that man has a body distinct from his soul is to be expunged. This I shall do…

William Blake, *The Marriage of Heaven and Hell*

Contents

Introduction

1   Job's Dilemma

2   Skin

3   Connective Tissue

4 Bone

5 Muscle

6 Nerve

7　Muscle As Sense Organ

8　The Sense of Effort

9   Sensory Engrams

10   Movements Toward Disease and Movements Toward Health

11 Further Views at the Turn of the Century

Acknowledgements

I want to thank the following individuals for their invaluable assistance and guidance through the long process of researching and assembling this book:

I give thanks to Richard Price and Michael Murphy for their friendship and their encouragement of my efforts at Esalen Institute;

to Ken LeBlanc, who gave me my first experience of bodywork, and to Bill Lyles and George King, who first showed me how to give this experience to others;

to Dr. Milton Trager, whose teaching awakened in me many of the ideas I have tried to articulate, and to Betty Fuller, who introduced me to Dr. Trager and helped make me a part of his institute;

to Father Aelrid Squire, whose spiritual and scholarly insights helped me to come to grips with *The Book of Job;*

to Richard Tarnas for his meticulous proof-reading and numerous useful suggestions, and his unflagging moral support;

to George and Susan Quasha, whose tireless editing, art work, and advice began a friendship as well as finished a book;

to my parents, Wyrick and Shirley, who saw to it that I got a good education and the freedom to use it the way I wanted;

to John and Isabel Murphy, who taught me what it is to pursue an idea with gusto;

and last (for which I have saved the best) to my wife Andrea, whose love and faith and support made my task possible.

Foreword

Job's Body is the book we've all been waiting for. Whether you are a health care professional, a part-time student of human potential, or simply interested in achieving the highest levels of mental and physical health for yourself, you will find a cornucopia of information, ideas, and practical healing strategies in this eloquently written text. Deane Juhan introduces us to the inner and outer workings of nature's most extraordinary masterpiece—the human body.

From the first page, we realize that this is not just another lifeless text on human anatomy. Instead, we come to see the psyche and soma through the experience of an individual who has, for the past fifteen years, totally immersed himself in the direct "hands-on" experience of the numerous bodywork therapies at California's famed Esalen Institute. Combining this wealth of practical experience with a decade of anatomical and physiological studies, the author offers in *Job's Body* the most comprehensive examination of the foundations of effective bodywork that has ever been available. And more than that, because of Juhan's prior background as a scholar of philosophy and literature, this book presents a veritable feast of ideas for the more probing and reflective students of mind/body development.

In his unique and challenging fashion, Deane Juhan endeavors to describe and illustrate exactly how bodywork *works*—in what ways our structure, our physiological processes, and our mental development may be positively influenced by touch and manipulation. His discussion of the body and its responses to touch reveals the common ground shared by many different approaches to bodywork and establishes the basis of a scientific understanding of this most ancient of the healing arts.

Unlike most other textbooks on anatomy and physiology, in *Job's Body* we are encouraged to delve into the intellectual and emotional contents of the body's organs, the history of the circulatory system, the aesthetics of muscle, the philosophy of our bones. This study of the human organism is infused with life and movement, and we become fascinated not only with the intricate mechanics of our wondrous bodies, but also with the miraculous play of psyche and spirit within the totality we are.

Notwithstanding the depth and breadth of supportive detail, Juhan's philosophy of mind/body unity is remarkably straightforward. His primary themes are poignant and simple:

> —Without adequate tactile input, the human organism will die. Touch is one of the principal elements necessary for the successful development and functional organization of the central nervous system, and is as vital to our existence as food, water, and breath.

> —Our genetic blueprints are only the starting points of our individual development. The kind of conditioning we receive and the kinds of conscious choices we make play tremendous roles in our physical growth, our acquisition of skills, our health and maturity, and our aging.

—All of the body's tissues are, then, a great deal more "plastic" and responsive to change and improvement throughout our lifetimes than we normally assume. Far from being "fixed" and "determined" by our biological inheritance, we are all still "works in progress."

—There is no sensation or emotion that is not translated into a muscular response of some kind; these feeling states are the primary bases of our habitual postures and our individual patterns of behavior.

—Bodywork, by using tactile input, can actually re-educate and re-program the organism into becoming more coordinated, more flexible, and more appropriately responsive—literally more "intelligent." A body/mind system that is integrated in this fashion will be more able to resist depression or disease, more able to attend to and repair itself in times of stress or injury.

—Various ancient and contemporary forms of bodywork go far beyond temporary pleasure or relief and actually alter conditioned responses, chemical balances, and structural relationships. That is, bodywork has the potential to deeply change and improve the given state of an individual.

—Nothing is more essential to lasting positive change than self-awareness; it is the prerequisite for self-control. Bodywork is a direct and effective way to increase this awareness within an individual.

This fascinating book offers everything a health practitioner, massage therapist, exercise instructor, or student of human potential could hope for. It is an information-packed reference on the workings of the body and mind; and it offers a broad assortment of strategies for releasing tension, freeing energy, and enhancing health through "hands-on" bodywork. *Job's Body* is an eloquent exploration of the most mysterious and powerful of all human interactions—touch.

December, 1986
Dr. Ken Dychtwald
Berkeley, California

Introduction

Our tendency to live in the world of reality leads us to neglect what is going on in the field of sensations.
—Paul Schilder, *The Image and Appearance of the Human Body*

A Phenomenon Needing Explanation

The genesis of this book has been the attempt to explain to myself a phenomenon that I have observed again and again during the twelve years that I have been a professional bodyworker at Esalen Institute in Big Sur, California.

Here at Esalen I have had the opportunity to see many people arrive suffering from various kinds of mental and physical distress, stay for a brief period, and then go back home relieved of significant amounts of their pain and conflict. Both according to my own observations and by their own accounts, the bodywork they received during their stays was substantially—sometimes it seemed exclusively—responsible for these happy changes. And often these improvements have proven not to be just temporary ones; many individuals have come back for more, but they usually have not returned in their original conditions. It is clear that something set them off in new and more positive directions, and that they were interested in coming back to learn more about these changes, to be able to take them further, make them faster, more complete.

What these people have experienced is not a temporary analgesia or placebo effect, but rather a cumulative process of getting to know their own bodies and their own sensations from a fresh perspective, a process that continues to help them discover who and what they are and to learn to exercise some measure of self-control over many of the vagaries of their physical and emotional symptoms. I have seen stoops straighten, gnarled deformities become more comfortable and functional, injuries heal more quickly and more completely. I have seen dozens of imminent surgeries averted, medications reduced or eliminated, eyeglasses upgraded or even occasionally discarded, chronic pain diminish or disappear, various degenerative conditions slow to a halt and even reverse. And I have heard those that returned to Esalen talk enthusiastically about the positive changes they had experienced at home in their relationships with fellow workers, their employers, their friends, their lovers, spouses, children, parents. And with themselves.

Why should such a few sessions of bodywork, often accompanied by a minimum of verbal dialogue, affect so dramatically these people's symptoms, their relationships with themselves, and their relationships with others? Most of the bodywork techniques I have observed and practiced are neither rigidly systematic nor forceful. The usual impression is that the client is being gentled and pleasured, not being "fixed," or "cured." By what possible mechanisms, then, could something so simple as soothing touch alleviate painful and long-standing physical conditions, quell anxieties, foster more productive attitudes? And if simple touching indeed provided some sort of key, then why were some practitioners so much better at achieving these kinds of results than were others? These questions, and many more that come in their train, have led me to the explorations in this book.

Some of the practitioners who frequently achieve these sorts of results have had a great deal of formal training of various kinds to develop their skills. Some have had much less, and some little or none at all. Almost all of them, including those with the most formal training, insist that regardless of what specific technique they may be pursuing, the actual details of the placement and manipulation of their hands are largely guided by some kind of intuitive process, without which any amount of training is inadequate, and that very often they cannot pinpoint precisely which movements or pressures precipitated a dramatic change in the client. And yet the results are there for anyone to see.

If in fact a thing occurs repeatedly, then it must be in some way explainable. And if it is not explainable in currently acceptable terms, then whatever terms do explain it must be sought out, and the categories of what is currently acceptable will have to adjust themselves accordingly. It is the goal of this book to explore some of the verifiable relationships between the body and the mind in search of data that might contribute toward some concrete understanding of what is happening when skillful touching alters the structure, the chemistry, the feelings, and the behavior of a human being. I would like to find concrete reasons for the obvious successes of a largely intuitive process, a process which seems to be able to use very different specific techniques in order to arrive at the same results of relief from discomfort, dysfunction, and anxiety.

Looking about, I have discovered to my delight that there is indeed already a rich supply of hard experimental data which does support some concrete suggestions as to what is at the bottom of these phenomena. Some of the data, in fact, is so commonly available, so clear and unequivocal, that one can only wonder why medical science has not clamored more enthusiastically for the development and application of these kinds of skills. Therapeutic bodywork is presently relegated almost exclusively to departments of physical therapy, and is often so hedged in there by procedural rules imposed by administrators and physicians—many of whom have no special training in bodywork—that the crucial intuitive element in their success is rendered very nearly inoperable. It is the investigation of available data from the perspective of the bodyworker who is having real success with the intuitive use of his or her hands that particularly interests me.

I would, of course, hope that bodyworkers themselves will find things in this book of interest to them, both to help them understand the scientific concerns upon which the professional acceptance and legitimacy of their work turns, and to appreciate the wealth of information that science does already have about the body and the mind which can effectively underpin and guide their efforts. I would also hope that medical professionals will find things here of interest to them as well; typically one joins the health professions because one is sincerely motivated to help people who are in pain, and I hope to argue persuasively that bodywork is a mode of treatment which has as impressive a record of success as many medical procedures in current use, which offers distinct advantages in a large variety of situations, and which poses extremely interesting possibilities for further research.

This book is also for those whose principal active interest in health care is their own. For them I would hope that reading it might at the very least lead them toward experiences that can help them establish a more knowing and friendlier relationship with their own bodies. The best that could happen would be that they would learn how to avoid many of the pitfalls with which the wear and tear of life threatens all of us. Or, if they come with specific problems already, perhaps it can help them see how bodywork might contribute to the management of their ills. If it guides a single sufferer to a single practitioner who is able to eliminate or avoid a single problem, it will have served no small purpose.

The Term "Bodywork"

Throughout the book I will use the term "bodywork" to refer generally to a wide variety of manipulative therapies. It is not within the scope of this present project to describe the operating premises of each different approach, compare them, or examine how each individually affects posture, function, and behavior. My intent is to establish some concepts about skillful touch as an overall genre of treatment; practitioners of each particular discipline naturally have a great deal to say about how they see their own kind of work fitting into a larger picture, but that is another book—perhaps several more books.

There are some respects in which the term "bodywork" does not please me. It conveys well enough the idea of the body being touched in a deliberate fashion for specific results, but it does nothing to evoke the powerful emotional responses and shifts in mental attitudes which often accompany these physical manipulations. And it has the additional drawback of sounding like something one does to dented automobiles. However, I confess I am hard put to find a term which, on the whole, serves better.

The word "massage" covers a number of styles—Swedish, Esalen, sport massage, and so on; but it does not include many approaches, such as Trager, Rolfing, Feldenkrais, Alexander Technique, Craniosacral Manipulation, Zero-balancing, Reichean and Neo-Reichean work, to name but a few, which are quite different from "massage" in any of its guises. And the term "massage," alas, still seems to be tainted in many quarters by its common associations with touchy-feely parlors, and even with disguised prostitution. This is an unfortunate situation, and one that is unfair to a large number of legitimate practitioners, but it is a present condition of language usage that probably cannot be effectively erased.

It is not that I sternly deplore touchy-feely. In a culture that is as starved for touch as is ours, I suspect there may be some healthful benefits to pleasant tactile stimulation in almost any form whatever. But in order to discuss the kind of bodywork I mean, I strongly feel the need for a word that in no way implies contact that is *merely* sensual, or that is sexual in any shape or form.

The term "physical therapy" avoids these associations, but it is also too narrow in the scope of its normal usage. It refers to an official medical discipline, one which is licensed only after protracted and highly specific studies, prescribed only by physicians, and applied through fixed procedures. Such academic rigor certainly does not count against it as a responsible therapeutic practice, but it does effectively partition "physical therapy" off from many other useful kinds of touching and manipulation. In particular, it typically eliminates a good deal of the *intuitive* element which seems to be such an important part of other approaches, and which in fact many physical therapists have confessed to me that they wish they could use more freely in their clinical practice. So I have settled upon the term "bodywork," because it seems to include fewer of the elements I wish to avoid and exclude fewer of the elements I wish to consider.

A Brief History

The idea that touch, knowledgeably directed, can have significant therapeutic value is far from being a new one. Perhaps its earliest actual documentation is found in *The Yellow Emperor's Classic of Internal Medicine,* attributed to Huang Ti, the Yellow Emperor, who died in 2598 B.C. In addition to this ancient Chinese text, bodywork is also mentioned in a number of Sanskrit ones; and the temple at Borobudur, India, built about 800 A.D., contains among its reliefs one of the Buddha receiving a bodywork treatment. Bodywork is mentioned in Homer's *Odyssey* as being welcome relief to exhausted heroes. Hippocrates and

Asclepiades, two Greek physicians commonly associated with the founding of physical medicine, both discuss it as one of their principal therapeutic modalities. Galen, the Roman physician whose authority lasted throughout the Middle Ages, is said by some scholars to have written at least sixteen books on the subject; Cicero and Caesar both enjoyed it frequently, and sang its praises highly.

Guy de Chauliac (1300-1368) was the most important physician of his time; in 1363 he published a book on surgery which became a standard text for the next two hundred years, in which he describes methods of bodywork as an adjunct to surgery. Paracelsus (1493-1541) found it an indispensable therapy. Valentine Greatrakes, one of Cromwell's soldiers in Ireland in the middle of the seventeenth century, became widely known for his successes in curing diseases by the use of his hands. Francis Glisson, one of the founders of the Royal Society, wrote a book on rickets, in which he includes bodywork among the prescribed treatments.

The nineteenth century saw major strides in the development of medical theory and practice, including those of bodywork. Per Henrik Ling was responsible for the establishment of the Central Royal Institute of Gymnastics in Stockholm in 1813, where he practiced and taught the Swedish Massage that spread all over Europe and America. Johan Georg Mezger, an enormously successful Dutch physician, relieved hundreds of patients of their symptoms during the latter half of the century by the use of his thumb. And in 1884, Professor Charcot, the renowned French physician—and one of the teachers most influential to the young Sigmund Freud—lamented the fact that the French doctors of his day did not interest themselves in bodywork as much as he had hoped they would.

During the twentieth century there has been a prodigious increase in the number of bodywork techniques and practitioners. Part of this renaissance has been due to the technological development of various therapeutic machines and gadgets, but it has primarily been a multiplying of the ways in which human hands can be used to influence physiology. Indeed, as we learn more and more about the mechanisms of our bodies, we do not seem to be eliminating the uses for bodywork, but rather we are discovering newer and more effective ways in which we might apply it.

Throughout the history of the development of their art, bodyworkers have learned to provide relief for conditions as varied as muscles that are too loose, muscles that are too tight, constipation, high blood pressure, broken bones or sprained ligaments that are healing, depression, anxiety, asthma, muscle strain or fatigue, sluggish lymph flow, poor veinous return, epilepsy, manipulation of the fetus in the womb, headaches, to list but a few of the conditions that have been specifically referred to. And with advances in scientific understanding, the list continues to grow, not decrease.

In spite of these facts, the use of bodywork as a legitimate therapy has in recent times diminished more or less steadily in America, even while the diversity of techniques has proliferated. The reason for this decline is certainly not that bodywork is no longer useful. Herman L. Kamenetz, a contemporary American physician, speculates on the subject as follows:

> The development of the pharmaceutical industry was no doubt a factor. A greater variety of machines together with new inventions (diathermy, microwaves, ultrasound, etc.) have supplanted older means of physical therapy including massage. A third—and related—factor is what has been called the dehumanization in the relationship between the patients and those who treated them. This aloofness is opposed to the intimate contact between physician and patient or teacher and pupil. Physicians do not take the effort to learn—hence to teach—the time-consuming art of manual treatment. Thus, therapists do not learn about its potentials. Yet, possibly as a reaction to mechanization, there seems to be a tendency toward closer "contact" between those who search for better health and those who might help them to obtain it, between mother and child, teacher and pupil, patient and healer. This is not yet history but it might point to a swing of the pendulum in the reverse direction, a renewal of massage.[1]

What Can Bodywork Do?

One of the debates that has accompanied bodywork throughout its history has been about what specific things various manipulations actually accomplish, and to what conditions it can therefore be applied with some expectation of success. Almost all of the physicians and practitioners just mentioned have agreed on one or two fundamental points. One is that most of the body's processes rely upon the appropriate movement of fluids through our systems, and that bodywork can be an effective means of promoting these circulations. Whether it is blood in the arteries, capillaries, and veins, the contents of the digestive tract, lymph in its vessels, secretions in their glands, or the fluids that fill all of the spaces in between our cells, manipulation can move them around much like I can push water back and forth in a rubber tube; and with a clear knowledge of these fluid pathways and some practice, I can become quite sophisticated in the ways in which I can stimulate their flows.

Now these flows, or the lack of them, can have far-reaching consequences upon many tissues and functions. Nutrients, oxygen, hormones, antibodies and other immunizers, and of course water, must be delivered to every single cell continually if it is to survive and respond the way it should, and all kinds of toxic wastes must be borne away. There is no tissue in the body that cannot be weakened and ultimately destroyed by chronic interruptions of these various circulations.

Another argument frequently made for the efficacy of bodywork is that both our musculature and the connective tissues which hold us together often become stiffened or shortened or thickened, distorting our posture and limiting our movements. These tissues can be especially troubling after surgery or any other trauma, when the muscles are either tightening up in order to brace an injured area or are contracting in a general withdrawal reflex, and when the connective tissues are scarring over a wound.

These bracing and healing mechanisms often overdo their functions, and it is very common that individuals never recover their full range of motion or their normal levels of comfort after an operation or a serious injury. And these stiffenings, shortenings, and thickenings can also happen as a result of a wide array of overuse, disuse, spasm, injury, illness, fatigue, aging, poor habits, or the innumerable physical strains that various occupations demand of us. Bodywork has been used for thousands of years to relax muscles, eliminate spasms, diminish fatigue, soften connective tissue to make it more supple, and so free up the joints, restoring a fuller range of painless movement.

These kinds of effects upon our fluids and upon our solids have been rightfully cited as benefits of any number of approaches to bodywork throughout its history. They would certainly be enough to establish its therapeutic value. But it is my feeling that they do not go half far enough in describing the positive changes that can happen as a result of skillful touching. Even though they are accurate identifications of benefits, they reflect almost exclusively the *mechanical* aspects of bodywork and of our own system's responses—the laws governing hydraulics, the elasticity and tensile strength of tissues, and so on.

We are, of course, mechanical in many of our physical aspects, so there is a great deal of justification for focusing upon these sorts of effects and explanations, as far as they go. But we are much more than mechanical. We are a confluence of physics, chemistry, and consciousness, streams and quanta of energies that interpenetrate one another in enormously complex ways, that moment by moment create layers and layers of effects, and in which the subtle and the gross are always inextricably intertwined.

> These chemical substances are not merely three-dimensional material particles floating in an aqueous medium. This is an oversimplified notion, another surface disguise. The underlying reality pattern is once again of energy currents, of energy interplays.... As one substance or nutrient attaches to another in metabolic transfer, or as they detach in a catabolic phase, we call the exchange "chemical." With equal justification we can see it as an energy phenomenon. Application of pressure (energy) through muscular expansions and contractions fosters these transfers. To get more economical flow, we must start at the macrolevels of muscular and fascial systems in order to influence the microlevels of cellular metabolism.[2]

In my experience, the plasticity of the body goes far beyond such matters as the mechanics of the circulation of fluids, the local contractions of muscles, the stiffening of connective tissue here and there, and so on. There is something in the actively organized relationships between all of the body's various tissues which is more interesting—and more relevant to our overall health—than are local tissue changes in and of themselves. The skin, the connective tissues, and the muscles are vital organs, organs with multiple functions which profoundly affect each other and all of the other organs of the body, and which are affected by these other organs in turn. Every part of us is continually undergoing dynamic changes from liquid sol states to solid gel states and back again as we grow, move, learn, and age, and no single part ever changes its state without sending reverberations out to all the other parts. Our organic life is an interconnectedness that goes far beyond the mechanical relationships between fluids, tubes, levers, cables, and springs. If we can genuinely affect one level of this interaction, then through it we can reach many levels, and the more sophisticated our sense of these interpenetrations are, then the more varied and precise the manipulative facilitations used in bodywork can be.

And no matter how hydraulic, mechanical, or chemical we may be in many respects, the physical laws which govern these kinds of relationships among our elements are only a part of the organization that forms and sustains human beings. Our other great principle of organization is neural, mental. And regardless of how hard purely mechanistic theorists try to explain the basic functions and the subtle nuances of our mental lives in terms of elementary physical laws, there is still a great deal about our states of mind and the consequent states of our tissues, about our experience and our consequent behavior, that so far at least has utterly resisted satisfactory explanation in those terms. The belief that such mechanistic explanations will one day prove to be adequate, once we have discovered and quantified all the details, is a philosophical stance, not a scientific one. Given the array of evidence before us at the present time, there is perhaps equal justification for maintaining that a great deal in our mental lives may *inherently* resist mechanical explanation. Indeed, the behavior of organic chemical compounds sometimes suggests that even they have some kinds of "feelings" for one another, rather than that they combine in some complex way to produce "feelings."

Neural activity is the most pervasive organizing principle in the body. There is no cell whose environment is not directly sustained or adjusted by the activities of the nervous system. It ultimately determines the plasticity of all other tissues and systems, and it is itself the most radically plastic of all the systems in the organism. There can be no movement, neither free nor limited, without muscular activity; there can be no muscular activity without neural stimulation; and the specific quality of every muscular action—its timing, duration, style, effectiveness—is a summation of all the activities of both the central and peripheral nervous systems at that moment. These muscular actions, in turn, are absolutely of the essence in our mental and physical development. Ida Rolf observed that

> for the therapist of the psyche, as well as the therapist dealing with the
> physical man, the goal is appropriate *movement*. The psychotherapist senses
> immobility in the dimension of time rather than of space. The individual,

bogged down, unmoving in time, unable to escape from his infantile or adolescent assumptions or traumata, manifests this physically as well as psychologically. His lack of movement, his general or localized rigidity, are unequivocal in their statement.[3]

Or, as Dr. Milton Trager has alternatively expressed it:

The mind is the whole thing. That is all that I am interested in. I am convinced that for every physical non-yielding condition there is a psychic counterpart in the unconscious mind, corresponding exactly to the degree of the physical manifestation.... These patterns often develop in response to adverse circumstances such as accidents, surgery, illness, poor posture, emotional trauma, stresses of daily living, or poor movement habits. The purpose of my work is to break up these sensory and mental patterns which inhibit free movement and cause pain and disruption of normal function.[4]

Organizing Mind

It is the mind that is the organizer of our health and our strength, of our associations and responses, of our thoughts, our feelings, and our tissues. The laws of physics and chemistry dictate the conditions which it has at its disposal, but so far no one has been even remotely successful in identifying any combination of these laws as the motivating factor behind the development of consciousness and behavior. How then might we successfully influence this organizer, without risking a major disturbance of the delicate chemical and physical balances upon which it unquestionably relies for its continuing vitality? With what can we intervene that will not spoil the soup while we are adjusting the recipe?

Why not try the very thing that the nervous system uses to affect its own organization, a thing which is neither chemical nor structural—tactile stimulation? It is the touching of the body's surfaces against external objects and the rubbing of its own parts together which produce the vast majority of sensory information used by the mind to assemble an accurate image of the body and to regulate its activities. "Only contact with the outside world," says Dr. Paul Schilder, "provides sufficient regulating sensations."[5]

We do not feel our body so much when it is at rest; but we get a clearer perception of it when it moves and when new sensations are obtained in contact with reality, that is to say, with objects.[6]

Not only is it true that the nervous system stimulates the body to move in specific ways as a result of specific sensations; it is also the case that all movements flood the nervous system with sensations regarding the structures and functions of the body. Movement is the unifying bond between the mind and the body, and sensations are the substance of that bond.

Friction on the skin, pressure on the deeper tissues, distortion of the tissues surrounding the joints—these are the media through which the organism perceives itself and through which it organizes its internal and external muscular responses. As we develop and mature, most of us build up and reinforce a reliably consistent sense of our selves by carefully selecting and maintaining a specific repertoire of movement habits—which generate a specific repetoire of sensations—and by surrounding ourselves with a stable environment with which to interact. This careful process of selection is largely unconscious, and so as long as we are comfortable we are rarely aware of any limitations or potential dangers our cultivated habits may entail. And even if a disturbing symptom appears, we generally do not suspect that our well-worn, tried-and-true behavior might be its cause. In fact, the very consistency of our normal patterns frequently prevents us from changing our ways long enough to obtain such an insight.

It is exactly this circular relationship between our habitual behaviors and the chronic conditions of our tissues that skillful touching can so usefully penetrate. New frictions, new pressures, and new movements of the limbs necessarily create new sensations, volumes of new data which the mind can scan in search of clues for new habits, new modifications, more constructive conditions. And here we are close to putting our finger on the possible reason why touch therapies can sometimes produce positive results so quickly, almost "miraculously." No matter how much I move myself around, my strongest tendency is to move in the same ways that I have always moved, guided by the same deeply seated postural habits, sensory cues, and mental images of my body; but if I can succeed in surrendering to the movements that another person imposes on my body, without my own system of cues and responses interfering, it is possible to treat my mind to a flood of sensations that are novel in important ways, sensations that may well be able to indicate what things I have been doing that have produced my aches and pains at the same time as they have reinforced my normal sense of self.

And even more important, this moment of surrender and new sensation can demonstrate to me that I am not permanently obliged to continue acting out a habitual compulsion. I can see that the habit is a habit, that I am something else, and that for the moment at any rate I can choose to repeat it or not. And if I can drop a compulsive behavior or attitude for a moment without causing a crisis, then perhaps I can dispense with it altogether. As every physician knows, this kind of insight can often be worth more than any number of drugs or procedures for the reversal of a chronic condition.

In other words, just as the mind organizes the rest of the body's tissues into a life process, sensations to a large degree organize the mind. They do not simply give the mind material to organize; they are themselves a major organizing principle. As we shall see, severe touch deprivation leads inevitably to psychological derangement and death—in a surprisingly short period of time. The sheer *quantity* of tactile stimulation that an organism receives, in addition to any considerations of the quality of that stimulation, bears a direct relation to that organism's success in sorting out both its relationship to the surrounding world and its own physiological processes. As it turns out, I learn about my body in exactly the same way that I learn about any other object, by feeling it. Without this active and continual tactile exploration, the organism literally loses its sense of where it is in all its pieces and parts, and rather quickly begins to fail to regulate appropriately its many complexly interwoven systems.

We encounter in this tactile exploration a great deal of ambiguity between sensory feeling and mental feelings. This is not just because it is sometimes hard to distinguish between the two of them, but because they are not clearly separable things. They constantly condition and influence one another to such a degree that it simply becomes an academic abstraction to contemplate the nature of one without the other. I "feel" things, and I simultaneously have "feelings" about them, and very often we find ourselves reduced to the question of the chicken and the egg when we try to determine which one causes the specific qualities of the other. Our very language is riddled with these ambiguities, and in fact we don't really have a way of talking about much of our experience without the metaphors we use to express them. An object, a touch, and a mental feeling can all be "light" or "heavy," "hard" or "soft." An attitude can be just as tense and unyielding as a muscle. A nagging symptom can be like "a millstone around my neck," and mental anxiety can flesh itself out as a "heavy heart," or "cold feet," or a stomach that is "tied up in knots."

These are not merely poetic ways of expressing ourselves. They spring as often from attempts to be as precise as possible about the actual quality of our experiences as they do from attempts to dress those experiences up aesthetically. They point not to a fuzziness in our discernments and our descriptions, but to a certain fuzziness in reality that is created by the many levels of mutual interpenetration between "feeling" and "feelings," between sensation and response, attitude and behavior, mind and matter.

After all, we may be able to tell exactly when a person is having an experience by carefully measuring electrical differentials, chemical shifts, muscular responses, and the like; but none of these objectively quantifiable data can give us any idea whatever of the actual content of that experience. And of course it is precisely the content—the context, the quality, and the meaning—of any experience that is the most significant to the individual's own internal processes and his relationships to the external world. This is the epistemological blind alley, the unacceptable paradox, to which rigidly materialistic empiricism leads us with regard to our physiology and our psychology. This is why the science of neurology can tell us so much about the functions of cells and so little about what to do with the suffering we commonly witness within and around us.

Objectively verifiable, measurable observations are certainly the cornerstone of the physical sciences. They are serious inquiry's protection against fantasies, projections, prejudices, and chicanery. The only problem with relying upon them exclusively in the investigation of our bodies and minds is that in order to do so we must systematically ignore huge amounts of data, data that is tainted because it is merely "subjective." Thus we rid research of the "human"—that is, the unreliable—element. But when it is human response that we wish to study, no amount of experimental sophistication can conceal the fact that it is the very things that would be most helpful to know which strict empiricism obliges us to disregard. What I am feeling and how I feel about it are the essence of my experience, and as such are also the essence of both my development and my degeneration. An objective event is the same for all persons who witness it; but the perceived content of that event and the personal consequences of it can be strikingly varied. It is these personal feelings that often have the most to do with our comfort and our health.

> It is an important revelation about the state of our being to realize that our culture has so thoroughly debilitated us that we have come to believe that sensing is primary and sensing ourselves is not worth mentioning. Behind that acculturated blindness lies the reason for most of the ailments, diseases, and strokes that typify contemporary culture.[7]

In bodywork, what is being felt is of major importance. This does not oppose it to science; rather, it puts bodywork in a position to add important dimensions of information to those of weight and measure. A painful spasm, or a chronic contraction which limits movement, or a destructive habit are not only specific neuromuscular events. They also include the feelings that preceded them, the feelings that accompany them, and the feelings that follow as consequences and condition future neuromuscular events. These feelings are not extraneous to physiology. On the contrary, in a large number of instances they can be demonstrated to be the precipitating factors of specific physiological conditions. Sensations and mental responses alter our chemistry and our structure just as frequently as it happens the other way around.

Furthermore, the dangers of fantasy and chicanery notwithstanding, it is not inherently impossible to be honest, accurate, and consistent in our assessment of these feelings and to use them systematically for therapeutic ends. What are the sensory and emotional conditions surrounding the generation of a muscular distortion? And what are the sensory and emotional conditions necessary for its lasting release? These are questions that can be of enormous significance to the individual suffering from such distortions, regardless of whether or not these sensory and emotional conditions can be weighed or chemically analysed. These are questions that I would have the study and the practice of bodywork address.

Sensorimotor Education and Self-Awareness

Bodywork, then, is a kind of sensorimotor education, rather than a treatment or a procedure in the sense common to modern medicine. Nothing material is added or taken away, so there are no dosages to be strictly adhered to, no statistical rates of success for particular manipulations. If a student is having learning difficulties in school, I cannot effectively tutor him by simply pouring into his eyes and ears the information that will appear on the exam. The very nature of his problem is that he is not assimilating these things as he should. I must first find out something about what he does know, and then I must find what stops him from acquiring the things that he does not. I must enter into an active relationship with him, feel him out, discover what forms information must take before he can successfully absorb it, sort it out, and apply it appropriately. This process is not an exact one. It is partly objective and partly subjective, and the effective tutor is the one who knows how to find the balance between the two for each individual. This is the manner in which bodywork proceeds, first finding the reflex patterns of response that are presently active, and then searching for the quality of sensory input that will begin to alter those patterns for the better.

A point worth remembering here is that in this educational experience it is not the bodyworker who is "fixing" the client. The bodyworker is not attacking a localized problem with specialized tools, confident of achieving certain results. Instead, he or she is carefully generating a flow of sensory information to the mind of the client, information that is not being generated by the client's own limited repetoire of movements—new information that the mind can use to fill in the gaps and missing links in its appraisal of the body's tissues and physiological processes. It is then the mind of the client that does the "fixing"—the appropriate adjustment of postures, the more efficient and judicious distribution of fluids and gases, the fuller and more flexible relationship between neural and muscular responses.

The bodyworker is not an interventionist; he is a facilitator, a diplomatic intermediary between physiological processes that have lost track of one another's proper functions and goals, between a mind that has forgotten what it needs to know in order to exert harmonious control and a body politic which increasingly utilizes disruptive demonstrations, terrorist tactics, and even the threat of all-out civil war to regain its governor's attention. Touching hands are not like pharmaceuticals or scalpels. They are like flashlights in a darkened room. The medicine they administer is self-awareness. And for many of our painful conditions, this is the aid that is most urgently needed.

There can be no doubt about the effectiveness of modern medicine in handling many life-threatening health crises. My mother would not have lived nearly so long, and I myself would not be alive today without it. I have no wish to underestimate the heroics of medical research and practice, and I for one have had occasions to be grateful for their interventions, even when they have been extreme, abrupt, and mightily painful. And yet there exist broad areas of health and disease in which such crisis-intervention is not applicable, and is even downright destructive. Many surgical operations and drug prescriptions presently administered are useless or worse to the patients who receive them, regardless of the expense and pain involved in their trial.

What is more, they are all too often administered without even a realistic *hope* that they will genuinely help an individual's situation; they are done simply because no one knows what else to do. This is not by any means to suggest that surgery and pharmaceuticals are themselves worthless. But what it does suggest is that both consumers and health professionals have to keep a firmer grip on the fact that these forms of therapy are genuinely helpful only in certain kinds of situations.

There are hundreds and hundreds of conditions that do not threaten our lives, but only our comfort and our productiveness. There are conditions that lead up to crises, and conditions to deal with after the crises. There are functional disorders that are unquestionably created by habits and attitudes, not faulty parts or chemical deficiencies. There are conditions that do not reveal their causes to conventional diagnosis. There are scores of psychosomatic symptoms that refuse to be cleared up by conventional procedures. In short, our minds and our bodies face many bad situations between the womb and the grave which cannot be resolved by surgical or chemical intervention, but which just might be resolved—or even avoided altogether—by stimulating more self-awareness, restoring more of a sense of self-control, igniting a little bit of will power and aiming it toward feelings that are more conducive, attitudes that are more positive, habits that are more productive. To develop these things we need sensory education, not physical intervention.

Self-awareness, self-control, and the active application of the will to the processes of growth and development are the major themes of this education. Bodywork can get us in touch with our present situation, give us a feel for possible alternatives, help us to grasp the forces that form us, put our fingers on the elements that have been missing in our ways of doing things, give us a renewed sense of our organic strength and intelligence. It can be a tremendous catalyst in the assertion of our self-responsibility for much of what we have become and much of what we can become. It can help us recall that we are living, growing systems, and not just genetic blueprints for engines doomed to begin wearing out the moment that they begin running, that we are an interweaving of processes and not just collections of parts, and that those processes are ultimately open-ended and creative, not mechanically deterministic. It can demonstrate to us that we neither have to collapse before the forces of gravity, disease, and decay, nor exhaust ourselves in a blind struggle against them, that it is possible to enter into active relationships with these forces, to match their insidiousness with our own cleverness, and to live a very satisfactory life in the midst of their threatening waves. It can give us clues about how to disentangle ourselves from widening vicious circles that threaten to engulf us because of our own unwitting compliance. It can be instrumental in putting a large measure of our lives back into our own hands.

These things can come about as a result of clarifications and improvements in our body image, through the discovery of the strengths and joys that coexist with our weaknesses and pains. Far from being mere hedonism, "the improvement of sensual pleasure" is self-serving in the broadest, most enlightened and responsible sense of the word. It implies an intimate embrace of the good things that both the physical and mental sides of life have to offer, and a desire to enjoy those goodnesses with others.

Of course, for the purposes of evoking these improvements it is not touch *per se,* but the knowledgeable development of the quality of touch that is the important thing. A rigorous scientific education does not in any way insure this quality. And a merely intuitive grasp of it may be very limited in the uses to which it can be successfully applied. But neither of these objections should obscure for us the tremendous potential for relief that resides in effective touching. Yet this potential is systematically ignored in much of our existing health care. A New Zealand psychologist from Otago University has conducted a world-wide survey of one hundred and sixty-nine medical schools; only twelve of them (and only five of these are in the United States) offer any course that teaches the value and technique of touching.

The purpose of this book is to explore the various avenues through which intuitive and informed touch can positively affect a wide variety of symptoms and help to change many people's lives for the better. Bodyworkers, whose numbers are presently expanding at an unprecedented rate, are practitioners of an old and valuable art that has had good use for many centures, and that is discovering new approaches, refinements, and successes every generation. If these individuals continue to develop their art and to help people, then wide paths will be beaten to their doors, and the health care professions will surely welcome a worthy addition to the means at their disposal for the alleviation of human suffering. How in particular these things might be achieved is my subject.

D. J.
Big Sur, California

1
Job's Dilemma

While unconscious creation — animals, plants, crystals — functions satisfactorily as far as we know, things are constantly going wrong with man.
— C. G. Jung, *An Answer To Job*

The Old Story

How can it be that we have forgotten to such a degree the significance that touching can have for our physical and psychological health? What attitudes and practices have we internalized which eclipse the pleasures and benefits of this age-old form of healing from our awareness?

The answer to this riddle is not a simple one. It is not that we have just forgotten a few important facts, but rather that we have adopted a whole new way of looking at facts, and a whole new sense of what is *admissible* as fact. The reasons for our forgetfulness are entwined in the developments of modern philosophy, science, and technology, and it is not an easy matter to clearly separate the causes of our regrettable amnesia from the compelling realities which these developments have thrust upon us.

How we conceive of the powers that be is always a matter of extreme and immediate importance. The rules which we presently imagine govern the operations of things in general will be the rules to which we try to adjust ourselves, and those things which we perceive as being the givens in any situation will be the limits within which we understand our problems, conceptualize our choices, and search for solutions. On the whole, we will not attempt the "impossible," nor will we forgo the "obvious." And this is especially true in matters that concern our health—a thing so precious that in our attempts to preserve it we are least apt to question our received knowledge, most apt to seek out current authorities and assiduously follow their advice. Hence we develop a strong conservatism with regard to our selves, a conservatism which on one hand helps us to avoid much of the ignorance and many of the catastrophes of the past, but which on the other hand causes us to forget many things in that past which are of lasting value.

The biblical Job was a man whose inherited wisdom had served him very well, establishing firmly a place for himself and his family in the midst of health and prosperity. He regarded the power that ruled him as omnipotent, eternal, and unchanging, the Creator of a stable and predictable world. The rules that Job imagined governing his relations to that world were not easy ones, but any man who would follow them as he did was everlastingly assured of the just rewards which he enjoyed. His industry and his devotion were universally admired, and his adherence to authority as he understood it was complete and exact in every detail. It was his joy to live righteously and to savor the returns.

Then the rules changed.

The God of his fathers exposed to Job an unforeseen side of His omnipotence, harshly rattling the good man's sense of right actions and just rewards. God had a cherished son, Lucifer; this son wagered that he could destroy Job's faith with devilish adversities just as handily as it had been won over with divine generosity. God took up the wager, thereby dooming Job to torment.

Job's strength and health collapsed, and his world deteriorated radically. He found his stabilities invaded by the random, the gratuitous, the perverse, the undiagnosable, the incurable. He found his inherited wisdom and his cultivated habits useless against these new ills. And he found the authority to which he had previously appealed mute, either ignorant of or unconcerned with his plight. His wife, as dismayed as he was, found fault with the world, and with a Creator who would allow such an event to happen. His friends and physicians, on the other hand, insisted that the fault must lie in Job, that he had obviously sinned against the laws, that there had to be a flaw somewhere in his person and his behavior. But not one among them had any suggestion to offer that had the slightest practical value for helping Job to comprehend his misery or find a way out of it.

Job strove mightily in the face of his pain and confusion to preserve his old faith and the world view which supported it. He was conservative in the best sense of the word, and his conservatism was heroic. But as he struggled with his dilemma, he was forced to admit that the old world view simply did not square with his present experience. Even in the best of conservatisms, everything depends upon the veracity of the voice of authority, and Job could not escape the insight that he had confused his idea of "authority" with the actual "powers that be"; the former, he discovered, is merely a human opinion regarding the latter. And opinions, no matter how venerable, no matter how useful in certain circumstances, can be hideously wrong in others. In these other circumstances, there was no choice left to Job but to cast aside opinions and directly confront the powers.

"I desire to reason with God," he cried. To the voices of authority this challenge was impossible, blasphemous, but for Job it did not spring from the spirit of rebellion. On the contrary, it was an impulse towards sincere and well-meaning inquiry prompted by desperation, the first attempt in his life to look beyond the truths that he had inherited, truths that for the first time conflicted irrevocably with his senses and his judgment. Not only his health and prosperity, but his whole sense of reality was in crisis, and it was imperative that he find a new basis for both. "But where shall wisdom be found, and where is the place of understanding?"

New wisdom, it turned out, was not to be found in the accepted view of things, but in the acts of confrontation and introspection themselves. Left with nothing but his own resources, Job discovered that those resources had an authority of their own, and that they spoke to him of a very different God than the old law-maker to whom he had been supplicating. "Who hath put wisdom in the inward parts? or who hath given understanding to the heart?" he asked as he looked inside himself. He found within his own substance and his own sensibilities a relationship with the powers which was more direct, more intimate, and more complete than anything he had known before: "Deep in my skin it is marked, and in my very flesh do I see God. I myself behold him, with my own eyes I see him, not with another's."[1]

For Job this was revelation — the perception that God was in his very flesh, in the throbbing of his heart, in the singing of his nerves, in the coiling of his muscles, to be touched and felt more intimately than an embrace. Not eternal judge and justice, but immanent event and consequence. Not perfect and complete, but growing and changing, absorbing each new development and giving every apparent contradiction its day. Not infinite in material fact, but infinite in possibilities. Most important of all, not remote and imperturbable, but present and responsive; not unapproachable, but in need of an active partner.

Job's friends had asked him incredulously, "Canst thou by searching find out God?" Having ignored the impossibility of his search, he could now step around the impediment of their skepticism and address that God directly: "I have heard of thee by the hearing of the ear. But now mine eye doth see thee." And having confronted and seen, he then knew what his friends did not. Enduring faith is not blind and obedient, it is keenly attentive and responsible; it is not fed by awe, but by quickening interest; prosperity is not the disappearance of problems, but the continual engagement with the process of finding solutions. Wisdom is not given from on high, but must be painstakingly unravelled from the knots in his own guts. No amount of prayer or ritual can supplant for him the ancient dictum, "Know thyself."

Renewed by this revelation, he could then redirect his own footsteps. Lucifer might still play his old tricks, but Job now knew that there was no place within him where evil could lodge that was inaccessible to the light of his own sensibilities, and he knew that the constant exercise of those sensibilities was his only lasting salvation. His prosperity returned ten-fold, and continued for the remainder of his life.

The Dilemma Today

Job hurts. He is confused in his pain, he even feels betrayed by the body that gives it to him, because by and large his affairs have been as well ordered as he could possibly imagine. As far as he understands matters, he has omitted no safeguard to his health and has committed no glaringly self-destructive act; he has done nothing to deserve the retributions he feels descending without warning upon his chest, his bowels, his joints, his back, his feet. To the best of his lights and abilities, he has worked hard for success and tried to be "perfect and upright, and one that feared God and eschewed evil." And yet, at the height of his prosperity, his productivity is curtailed and his rewards are taxed by pains.

Job's appeal to the experts on his condition often yields no improvements, nor even satisfactory answers about what is going wrong. Too frequently they can only mouth the established opinions of the current authorities, declaring that his symptoms are his disease, that his suffering itself proves that he is somewhere, somehow pathologically flawed, and that any questioning of these opinions is irrational. "Miserable comforters are ye all," he broods as he leaves their examinations, contemplating such cures as they can offer.

Paradoxically, our material existence has never been so fruitful, our authorities so learned. The practical application of the physical laws unearthed by modern science has given us a dominion over nature that would confound our great-grandfathers, let alone biblical Job. We now have salves and procedures by the thousands for his aches and pains, and legions of specialists to decide which one to use in every case.

And yet it has to be admitted that our understanding of our bodies and our minds is still a tenuous thing, leaving malfunctions many fronts on which to exercise their disruptive tyranny. Most of us have no more control over our internal bodily processes than did primitive sorcerers over the weather, the crops, or the coming and going of the moon — possibly less. Our scientific skills have indeed helped us to substantially eliminate scores of external threats — parasites, germs, viruses, toxins — only to have them replaced by a growing list of equally catastrophic functional disorders: heart failure, brain stroke, ulcers, high blood pressure, hardening of the arteries, head ache, back pain, weakened immune systems, cancer — diseases generated not by filth and want, but evidently by prosperity itself.

These sorrows appear to visit us as inevitably and capriciously as the rain, and in spite of the fact that an orderly empiricism promises that all will one day be logically explained, their comings and goings look very much to us like the vagaries of a blind Fate. Most of us do not know why it should be us that hurts, and when we try to find out, we are told that we are a frightfully complex and fragile collection of tissues and chemical exchanges whose interlocking ramifications are comprehensible only to the experts. So we adopt the same attitudes toward our physical conditions that mankind has usually adopted toward an inscrutable Fate: paranoia, fond hopes, consultations with the established oracles, and acceptance of the inevitable. And we revere the experts as the sole mediaries between this fate and ourselves.

Job not only hurts physically; Job suffers emotionally as well. He has acute anxieties that compound his pain because it has been made clear to him that his well-being is largely out of his hands, that he is surrounded by arbitrary forces which dwarf his capabilities. He suffers because hard experience forces him to admit that the world does not function in the ways that he has been taught to trust, or at least been led to hope, that it would. One of the major supports of his efforts — the idea that just rewards naturally follow right actions — has been thrown into serious question, and the only verity that modern mechanistic empiricism has left for him to lean upon is frighteningly flat and grey and hard: Nature in general being what it is, and his genetic inheritance in particular being what it is, things could not be other with him than they are.

But Job is not without opportunity. In fact, in his present situation, his opportunity is not separable from his hurt. A perfectly healthy body is one of which we are in many ways blissfully unaware. When all is well with it, it simply does for us, delivering the goods and paying the bills, administrating our desires without troubling us with all of the pedestrian, functional details. And while this is certainly in some ways an ideal to be wished for, it is not the way that we learn certain things about ourselves.

Our pains announce deficiencies, excesses, breakages, displacements, shifting balances within the complexities of our organic lives, and often in order to resolve them we have to learn something about these complicated internal affairs of ours, their rich interconnectedness, their crucial interdependencies. When a thing doesn't work, we must find out how it *should* work before we can make it work again, and herein is the spur for a great deal of human progress, and for that peculiarly potent satisfaction of having found the way out of the maze, of discovering a solution to the problem. For this experience of self-examination and revelation, we *require* situations which the experts cannot resolve, situations which throw us back onto our own resources. This is how we learn that we *have* resources.

"I desire to reason with God," cries Job, dangling at the end of his rope. How are things designed such that I end up in this uncomfortable place? and How can I get out of it? — these are burning questions for anyone who hurts. And when we are forced to find the answers for ourselves, we have the opportunity to feel not gratitude toward an expert who can fix us, but something more like exaltation with our own self-awareness, our own strength of will, our own hard-won self-control, our own direct contact with the powers that be. Shakespeare's Lear must expose his flesh and his fears to the storm; but having done so, he acquires a wisdom that none of his advisors could have given him.

Pain, Anxiety, and Certainty

We do not typically enter into these difficult introspections in a spirit of disinterested contemplation. Usually we are pushed by something inside of us that hurts, and so we are highly motivated by a powerful sense of self-interest as we try to pick our way through our confusion. We are driven in our search just as much by our pain and suffering as we are by our intelligence and curiosity, and we demand from our investigations answers that will allay our fears as well as accurately describe reality, solutions that are bulwarks against anxiety as much as they are complete and faithful examinations of our organisms and the laws that govern them.

On the surface it appears that these two concerns are utterly compatible, even necessary concomitants to one another. Nothing but accurate information can in fact assure me of a reliable means of avoiding personal catastrophe, and my fear for my life and my comfort is surely one of the most compelling motives imaginable for the arduous task of scientific examination. And as our grasp of reality becomes more and more reliable, our anxiety diminishes. Our material comforts are preserved in proportion to our expanding knowledge. This rational proceeding is in the best spirit of science.

But there is the potential for a peculiar conflict in this partnership between our fears and our intellect, particularly because we have trained our intellect to consider only objectively verifiable, consistently measurable facts in its deliberations and conclusions. The problem is this: The sources of anxiety are ever-present and powerfully compelling, while the establishing of certainties which rely upon the assemblage of incontestable facts tends to be slow, methodical, and subject to distressing surprises and reversals. I may well regard objective evidence and strict logic as my only reliable means of assessing the properties of reality and of avoiding those dangers; but my overwhelming need to end my pain and put my fears to rest is very capable of forcing my logic to take the most nonsensical turns, even capable of distorting my perception itself.

Anxiety is neither an argument nor a piece of objective data, but a feeling state, an especially persuasive feeling state which can affect all the operations of the mind in a way that mere arguments and data never can. It cannot be erased by the promise of logical explanations and practical solutions in the future. It may be ignored or suppressed while we force it to wait, but it will still be there creating its distortions. Ultimately, it can only be supplanted by another feeling state, which — if it is strong enough to do the trick — will in turn have its powerfully persuasive effects upon our perceptions and our intellects.

Now there is a specific feeling which has for millennia tantalized us with the promise of doing away with anxiety once and for all, and that is the feeling of certainty. Certainties are warm, dry shelters in the storm. Even certainty of the worst commonly relieves the shapeless dreads of anxiety; there is a peace in not having to wonder and struggle after answers anymore that seems to be able to surpass the fear of doom itself. Anxiety can cloud my thoughts and suspend me in a helpless paralysis, while a certainty — no matter how small or how grim — lends me a basis for decision and action, a concrete relation to my fate.

We have therefore a strong tendency to bend all our capacities toward establishing and guarding certainties, and once we have tasted it we have a compulsion to invest all of our observations, theories, and beliefs with this feeling, so that they will become potent against our anxiety as well as satisfying to our intellectual curiosity. And of course, the more collective agreement there is concerning "certainties," the more emotionally potent they become. It is one of the implicit aims of modern scientific education to establish such collective agreement upon the current certainties with the only means it recognizes as legitimate — objectively verifiable demonstration. This rigorous concreteness, this exacting and exhaustive reduction of the confusion of life's welter to materially definable mechanisms, has become the foundation of our modern version of faith.

Unfortunately, if the history of mankind, or even of modern science, has any lasting certainty to offer us, it is the fact that it is entirely possible for rational individuals to be absolutely certain about notions which later prove to be utterly preposterous. Nothing is certain except that our certainties about the ways of the world will change, and with them our ideas about how to cope with the conditions within and around us. This is the case not because we are hopeless ignoramuses, but simply because the urgent emotional necessities which push us toward establishing a sense of certainty seldom allow us the large amounts of time necessary to assemble all the relevant facts. We must escape our pain, quiet our fears, and we must act, today, now. For this reason we are always tempted to adopt beliefs and to defend them staunchly as truths, because the possibilities which they imply profoundly soothe our anxieties and produce some measure of practical results, rather than because their actualities have been borne out by unequivocal proofs or continue to offer the very best solution to current problems.

This is most clearly the case in much of the magical "science" of primitive cultures, where elaborate rites were often repeated with evident satisfaction, without having any verifiable effect upon the weather, the crops, the moon, or the cycles of the seasons. The number of self-evident and time-honored truths that have gone the way of these earnest magical efforts is soberingly large, and many assumptions of modern science have proved to be no less vulnerable than those of the "naive" ancients. Weight and measure, space and time, linear cause and effect, verifiable observation, immutable laws, predictable events, the conservation of matter and energy — all these indestructible building blocks of modern method are beginning to look less irrefutable the harder we look at them. They have led us into an unsurpassed technology, but that technology itself has given us the tools with which to investigate relativity, field theories, nuclear physics, quantum mechanics, the speed of light, and celestial black holes, where many of our cherished stabilities, our objectively verifiable principles, show every promise of going to the devil. And the passing of a belief is never a happy event for those that hold it, live by it, trust it, behave according to its tenets, and make their livings by following its ramifications, even if it proves to have been an obstacle in the way of something that later seems to be more like the truth.

Forms versus Formation

This yearning for certainty tends to make us simplify our views of reality and of the responses it demands of us. If we can categorize the endless stream of objects and events into a finite number of fixed concepts, then we will have power over them. We describe these objects and events and assign them conceptual forms; these forms then become the bricks and mortar of our reality, the givens upon which future observations and responses are based. And if some bits of data contradict our definitions, or do not fit into our conceptual forms, we try to avoid muddying our present thinking by sweeping them into a large bin filled with all the things that will be explained later.

So we concern ourselves exclusively with concrete forms that obey fixed laws, and we become quite skilled in manipulating those forms by means of those laws. We do it with buildings and bridges, with automobiles and rockets, with money and commodities, with populations and social trends. And we do it with the pieces and parts and chemicals that make up our bodies. Each time we identify a form, define its concrete properties, and are successful in manipulating it in clearly predictable ways, we become more and more convinced of the truth of our theoretical basis — that the world is entirely made up of such fixed forms, forms that can be jiggled around and changed in specific ways by the systematic application of fixed laws — until this proposition ceases to be regarded as a theory, ceases to be regarded even as a unit of truth that is applicable within certain limits, and becomes instead a "self-evident" property of reality in general.

We come to see all of creation as being built up from pretty much the same elements and relationships, and so we come to view ourselves in the same fashion. Because a certain chemical action occurs repeatedly in a test-tube (where variables can be strictly limited), we assume that something like it must happen when the same chemicals are mixed in the body (where the variables are enormous and largely unforeseeable). And when we obtain a reaction in the body that approximates that in the test-tube, then we declare a new "cure." What is all too often forgotten (or worse, simply ignored for the time being) is the fact that what was precise in the test-tube is something more like a shotgun blast in the body; certain targets are hit and successfully altered at the expense of an "acceptable" amount of damage to surrounding tissues and chemical balances. If the thing only works well enough for the moment, then we can sweep these side-effects into the same bin with all the data that do not matter at the present time. If they eventually prove to be catastrophic, well then we will look for new forms and new laws to adjust them later. The frightening fact of the matter is that in this situation we become the test-tubes.

No one could deny that many wonderful practical results have come from this manner of proceeding. Nor would I wish to deny anyone who is in a genuinely life-threatening crisis the right to have access to anything available that might have the slightest chance of helping in any way. And I acknowledge that medical research is arduous, complex, filled with trial, error, and necessary compromises. But what I *do* want to maintain is that when this method of producing practical results becomes a dogmatic world view, and as such becomes our only manner of proceeding, then we are drifting into dangerous territory. If every developmental anomaly, every minor injury or illness, every vicious circle of interwoven pathologies is treated with the same mechanisms that have been developed in the test-tube, the risk becomes greater and greater that the accumulation of side-effects will create internal conditions that are significantly worse than the original complaints that we sought to cure. This is not just paranoid speculation; it occurs with alarming frequency. The problem is how to introduce procedures into this world view that might work in other ways.

Fixed laws and fixed forms are reassuring certainties only as long as they are not brought into too close a contact with the actualities of physical process. When they are, peculiar and unsettling things happen, because nature refuses to be either simple or precisely repetitious. For instance, if we are to grant on one hand that all physical forms are subject to a universal and continuous network of causes and effects, then we are also forced on the other hand to admit that no single event has ever been repeated exactly the same way twice, because the surrounding conditions have always shifted. This in turn has a curious effect upon the notion of reliably repeatable experimental results; many real variables and many real effects must be left out for the sake of consistency. No matter how analytically useful strict logic and precise measure may be, they have very little to do with the furious interactions and the delicate dynamic stabilities of matter, and confining our reasoning to these categories will always confine our understanding of these interactions and stabilities. Simplification and reductionism are just not nature's way; or, as Salvador Dali expresses the principle:

> It is now known, through recent findings in morphology (glory be to Goethe for having invented this word, a word that would have appealed to Leonardo!) that most often it is precisely the heterogeneous and anarchistic tendencies offering the greatest complexity of antagonisms that lead to the triumphant reign of the most rigorous hierarchies of form.[2]

That is to say, stable forms are never fixed and never simple. They are not merely the passive accretion of matter according to its inherent mechanical compulsions, but are rather the results of the fierce confrontation of those compulsions with the surrounding conditions, conditions that are never the same from instant to instant. These compulsions and conditions constantly alter one another in ways that are often unpredictable, and each specific instance of physical form presents a radically creative solution to the antagonisms of the particular forces of the moment.

> We know today that form is always the product of an inquisitorial process of matter—the specific reaction of matter when subjected to the terrible coercion of space choking it on all sides, pressing and squeezing it out, producing the swellings that burst from its life to the exact limits of the rigorous contours of its own originality of action. How many times matter endowed with a too-absolute impulse is annihilated; whereas another bit of matter, which tries to do only what it can and is better adapted to the pleasure of molding itself by contracting in its own way before the tyrannical impact of space, is able to invent its own original form of life. What is lighter, more fanciful and free to all appearances than the arborescent blossoming of agates! Yet they result from the most ferocious constraint of a colloidal environment, imprisoned in the most relentless of inquisitorial structures and subjected to all the tortures of compression and moral asphyxiation, so that their most delicate, airy, and ornamental ramifications are, it seems, but the traces of its hopeless search for escape from its death-agony, the last gasps of a bit of matter that will not give up before it has reached the ultimate vegetations of the mineral dream.[3]

Or to state the case more briefly: "Force must be met by force, and the structure evolves as the forces are balancing."[4] And this is as true for the human as for the mineral.

Our yearning for concrete certainties and for universally reliable solutions has led us into the temptation of simplifying grossly the endless diversification that proceeds around us and within us. We have become so fixated with forms, static and accountable, that we have lost sight of their process of formation, kinetic and elusive. We have arranged our thinking and our conclusions in terms of space, forgetting what is perhaps the more important element of time, in which all things continually undergo their spontaneous alterations in spite of our efforts to fix or control them. But this simplification and elimination on our parts can change neither the reality of these ongoing processes nor the fact that whatever we define as a "form" is merely a specific stage in a constant metamorphosis. Our sensations and our realities are constantly shifting, and they are not best understood by clinging to any fixed notion, but rather by constant attention to our ongoing processes and by the constant adjustment of our conclusions.

Things have not been created once and for all and then sent on their predictable ways. We are in the midst of active and radical creation, in which all data is permissible at all times and the theoretical limitations of which we cannot presently imagine. And this recollection of our real situation need not be dug out of the manuscripts of some arcane pseudo-science, or be glimpsed fleetingly through some mystical vision. It is a recollection that is being forced upon us by the hard data being turned up by the cutting edge of the modern physical sciences themselves. David Bohm, one of the foremost contemporary theoretical physicists, offers the following observations in a recent interview:

> *Bohm:* You could say that creativity is fundamental . . . and what we really have to explain are these processes that are not creative. You see, usually we believe that in life the rule is uncreativity, and occasionally a little burst of creativity comes in that requires explanation. But . . . creativity is the basis, and it is repetition that has to be explained. . . .
>
> *Weber:* So in your view, it's as if the universe were experimenting?
>
> *Bohm:* Yes, you could look at it that way. Trying out various forms. You can say that natural selection explains the way things survive once they emerge or appear, but it doesn't explain why so many forms have appeared. There seems to be a tendency to produce structure and form, which is intrinsically creative, and survival or natural selection is merely the mechanism which selects which forms are going to remain. Any form incompatible with itself or with the environment is not going to last, that is all.
>
> *Weber:* In that view, then, the universe is learning.
>
> *Bohm:* I think so, yes.[5]

Subjective Data

This infatuation with fixed forms and precise measurement has done more than just legislate our view of the world around us; it has legislated our view of ourselves, our development, our structures, our functions, our minds. And it is in the handling of our selves that the logical cul de sacs and double binds, the inconsistencies and the side-effects are the most vicious and insidious, where the errors and the gaps in our theoretical principles cause the most damage and suffering. And yet, if we reconsider our methods for a moment, might it not be a more careful consideration of our direct experiences of these selves which could not only help us to avoid personal injury or manage personal illness, but to expand our theoretical thinking about the world in general as well? It is true, after all, that we ourselves are forms which are continually undergoing formation and reformation. We are objects, and there is every reason to suspect that the processes of matter within us are no different than natural processes anywhere. Might not a refinement of our sensibilities concerning our own bodies offer fresh insights to an inadequate world view, just as surely as adherence to that world view leads us to make mischief on our bodies?

I both have a body and I am a body, and this intimate relation puts my body in a closer juxtaposition with my immediate awareness than any other object that I can possibly contemplate. No piece of laboratory equipment could ever put me closer to a form and its process of formation than can my direct perception of my own body. The neural tentacles of my mind are rooted in the cellular and molecular depths of this formation, where they register every move, every stage in development, every shift in chemical balance, every nuance of posture, structure, and function.

But no. The currently accepted method defines acceptable inquiry. Everything that is a part of this internal awareness is "subjective." I can neither weigh nor measure my bodily sensations, nor even prove to someone else that I did or did not have them. They are a kind of hallucination, to be stringently avoided in my search for "objective," verifiable information. Feelings are worse than useless, they are noisome complications to our sense of reality, complications without which we can produce more solid, predictable results. And so we systematically ferret them out and expunge them from anything "scientific"—that is, anything worth our serious attention—that we have to say on the subject of forms and formation, even as they relate to our own beings.

This manner of proceeding may be good science in some limited sense of the term, but it has to be admitted that it is a method that begs many questions about our relationships to ourselves and to the world. To a large degree it is frankly not interested in these relationships, because they are not quantifiable facts. And just as it has trained us to view other forms with rigorous objectivity, it has trained us to view our personal forms in the same fashion, and to insist that our internal "subjective" sensations, our most direct means of access to information about our selves, are irrelevant.

By granting an exclusive intellectual and scientific status to this strictly empirical method, and by limiting the admissible data to "objective facts," we have forgotten an important truth about the nature of our self awareness:

> Whether it is a question of someone else's body or of my own body, I have no other means of knowing the human body except that of living its life, that is to say of accepting involvement in the action which passes through it and mingling with it. Thus I am my body, at least to the extent to which I have acquired one and conversely my body is, as it were, a natural subject, a provisional sketch for my whole being. Hence the experience of one's own body is the antithesis of that reflexive movement which disentangles object from subject and subject from object which only gives me the thought of the body or the body as an idea and not the experience of the body or the body in reality.[6]

Our method has enmeshed us in a webbing of static forms and linear effects, and it is very difficult to separate our thinking about our own forms from the way we think about forms in general. We are so immersed in objective descriptions of forms that we have only very stunted ways of talking about how they came to be the way they are, and towards what they are changing. We understand that there is some sort of process going on, but we only allow ourselves to infer it from isolated bits of quantifiable evidence, which is a lot like trying to guess the story line or discuss the quality of acting in a motion picture by meticulously examining a series of still shots extracted from the footage.

Our bodies are giving us every moment a wealth of information about forms and formation. Our bodies are not reducible to simple principles; they are exceedingly complex and chimerical, but perhaps it is because of this very condition that they have so much to teach us about reality. We should under no circumstances ignore or degrade what they are telling us; rather, we should be focusing our earnest attention upon the "field of sensations" at least as much as we do upon the "world of reality." Only in this way can we cultivate an accurate appreciation for the extensive interdependence between the two, and pinpoint the moments of union between spontaneous feelings and physiological responses which constitute that interdependence and direct our development.

The certainty which we try to establish by ignoring the chimerical and subjective elements in this dynamic union can never be a secure one, precisely because the parts that we systematically eliminate from our considerations are constantly operative and vital in any biological process. But fortunately, there is an alternative approach to this reductionism and objectification, an approach that is not required to ignore feeling states, personal reactions, and currently unexplainable anomalies. In the simplest of terms, this approach is that of "being there" in the midst of our experience — bringing all of our sensibilities to bear upon a process in actual motion, gathering together all of our impressions without trying to make hard and fast distinctions between what is "real" and what is mere "sensation," until we begin to get a sense for the ways in which each of these categories leads the other around in mutual circles. This broadening of the range of evidence that may be considered relevant and acceptable is a large factor in that faculty which I term "intuition," the kind of perception and association that is so crucial to bodywork, and, indeed, that is so important to fresh scientific insights of all varieties.

Some will object that this is more contemplation than meaningful investigation, belonging more to the practice of daydreaming than to the practice of science, but it seems to me that this is no serious objection if the approach in fact enlarges the scope of our observations and helps in any way to render insights into the subtle and elusive causes and effects that are typical in biological processes. Some of the most important components of our lives are not simply the sum of their material parts, and large tendencies can sometimes be either created or diverted by small shifts in attitude and behavior at the right moment. Quality of awareness and precise timing are always of the essence in these subtleties, and these are the very things that are often impossible to anticipate objectively. In order to gain an appreciation of them — and even more important, to take positive advantage of them — we must be acutely attentive to a wide variety of events in our interiors; and the subjective field of sensations is the only reliable source of such data. This is why bodywork, deliberate stimulation which provides direct experiences of internal sensations and physical responses, can sometimes generate extremely useful insights into our life processes, while interventions which might seem more concrete and substantial sometimes cannot.

Genetic Certainties

One of the primary objective foundations — one might even say an immutable law — of modern biology is the principle of genetic inheritance. Our physical forms will develop into what our chromosomes dictate; no more, no less. This view has the force of centuries of casual observation and common sense behind it, and its basic principle was obvious to farmers and animal breeders long before anyone discovered a means of unravelling its specific intricacies: Offspring do in fact reflect the physical characteristics of their parents to a very high degree.

In the middle of the nineteenth century the experiments of Gregor Mendel developed this old observation into something much more like a scientific theory. Mendel opened the door, and a century later genetic research was crowned by the discovery of the DNA double helix by James Watson and Francis Crick in 1953, a discovery which promised to reveal in clear detail the molecular basis of the genetic coding of all organisms, the secrets of their forms and the time-tables of their formations.

And indeed, current genetic theory and laboratory technology exercise a control that seems almost God-like over variations of life forms. Yet it must be admitted that this theory does have another appeal, one that is more emotional, and that outstrips even these recent — and extremely impressive — practical accomplishments: If biological and evolutionary developments can be successfully reduced to a molecular determinism, then the life sciences might share the same solid foundation of objective certainty long enjoyed by physics and chemistry. The enormous complexity of living forms will yield themselves up to weight, measure, mathematical expression, and precise manipulation, and we can rest assured that our method will eventually assemble all the relevant facts that will fill in the gaps in our knowledge of life. Even that least scientific of all sciences, psychology, would find a footing in the eternal objective verities; thought and feeling could themselves be explainable in terms of the material forms which produce them.

This compelling tendency toward the feeling of certainty is just as operative in our thinking about genetics as it has been in our thinking about objective, measurable, and reliably repeatable effects in general. It structures the questions that we can formulate, and it structures the ways we can answer them. It is this emotional appeal, and not strict objective scrutiny, which often determines which observations are to be included in our reckonings and which shall be banished as irrelevant aberrations. Above all, it keeps us clinging to concepts with which we have associated certainty in the past, and keeps us poised upon the hope that absolute truth may be milked out of theories that have yielded impressive practical results.

Hence biological science typically strives to identify all of the characteristics of an organism with its genetic code, and all deviations from "normal" structure and function come to be regarded as errors in the arrangement of these coded molecular chains which make up the chromosomes. And further, every sort of malady or deformation that is not attributable to a particular germ or trauma comes to be labeled as a "congenital" condition — another great bin into which we can sweep all anomalies while still preserving the sense that we know what we are talking about. To be sure, we give "environmental factors" a nod of recognition, especially when they are dramatically nurturing or abusive, but the gene is the thing. Its molecular arrangements contain the important factors of the individual, which will unfold in predetermined ways according to a fixed time-table.

These genetic principles have become so generally applied in order to account for almost all the internal conditions and behavior patterns of organisms that their power to eventually explain all biological developments has in turn become another of our unshakeable articles of modern faith. The few objective facts that we do have seem so suggestive, so full of promise, that we tend to forget how large the bin must be in order to hold all the observations which the facts do *not* explain. We forget how often the label "congenital" simply means "we don't really know the causes of this condition yet."

Now this dogged faith in a broad theory based upon a relatively small number of verifiable facts may be a useful mindset in the process of pure research. There it motivates, directs, and sustains efforts in constructive ways. But when this faith is applied too rigidly to the conditions in which we actually find ourselves, when it becomes the only light in which we view the extremely complicated elements involved in human development and human dysfunction, then it can exert a very pernicious effect upon our sense of ourselves and our relationship to the world around us. It masks to a catastrophic degree my own everyday, active complicity in my physical condition and my patterns of behavior; it puts the only important formative elements of my development permanently beyond my personal control. It has the effect of making me passive and helpless, unable to confront my situation meaningfully and change it in any substantive way. It makes my cells and my organs and my systems integers in a fixed equation, rather than grammatical elements in an unfinished sentence. What I may or may not do for myself appears to be of little material consequence. When I am suffering, the only course of action open to me is to consult with experts who speak a technical language that I cannot hope to understand, and to allow them to do whatever they will in an attempt to alter my fate. I am forced to submit to the notions that I am ill, I am suffering, because I am fundamentally flawed somehow, and that it is categorically not within my own power to repair that flaw in any way.

It is true that remarkable formative powers lie coiled in the chromosomes of every cell, powers which regulate to a great degree our coherent development and reproduction. And it is true that many weaknesses and predilections to disease appear to be directly inherited through our gene pools. But our desire to have these clues lead us directly to the exclusive secrets of human developmental process and human behavior tempts us to make far too much of them, and leads us far beyond the bounds of reasonable inference.

> Mechanists usually assume that the genetic program can be legitimately regarded as an aspect of the DNA. Now the reason I think it can't is that we actually know, as a result of mechanistic research, what DNA does. DNA provides a sequence of chemical letters of the genetic code which spells out the sequence of amino acids in proteins. Some of the DNA is involved in the control of protein synthesis rather than directly coding the protein itself. And this is what's been shown, and this is all perfectly reasonable and very interesting. DNA, by providing the code for the sequence of amino acids, enables the cell to make particular proteins. And that is all the DNA can do — enable the cell to make particular proteins. . . . DNA helps us to understand how you get the proteins which provide, as it were, the bricks and mortar with which the organism is built, but it doesn't explain how these bricks and mortar assemble into particular patterns and shapes. The idea of DNA shaping the organism or programming its behavior is a quite illegitimate extrapolation from anything we know about what DNA does. . . . So what starts as a rigorous and well-defined theory about the way the DNA codes the RNA and how the RNA codes the proteins, soon turns into a kind of mystical theory in which DNA has unexplained powers and properties which can't be specified in exact molecular terms in any way at all.[7]

From beginning to end, we are shaped by forces that cannot be adequately accounted for in the coding of our genes, or in the sum total of the chemical changes occurring in our developing tissues. Every human being's life is a long and continually active process. The genetic code is only the point of departure. Given any particular arrangement of DNA in the chromosomes, any one of hundreds of potential individuals may actually develop. Height, weight, profile, skin texture, amount of fat, amount of muscle, tone, strength and stamina, facial expression, acuity of eyesight, functionability of internal organs, neural responsiveness, range of motion, degree of coordination, level of intelligence, self-awareness — all of these elements can fluctuate widely from individual to individual, regardless of similar genetic backgrounds, and — even more importantly — can fluctuate widely within the same person at different times. These are elements which are central to the appearance, the attitudes, and the behavior of the individual, and are in no way "secondary" or "accidental." They are the results of the total formative process of living, and play as significant a role in the shape and quality of life as do any potentials coiled in the genes.

Confronting Our Flesh

The modern Job's dilemma stems from the thoroughness with which he has accepted the mechanistic models offered him, and the thoroughness with which he has internalized the concept of the gene as Fate. The utter passivity which these points of view engender with regard to his personal development is precisely the problem, and a decisive way out of this epistemological and physiological paralysis is the very thing for which he desperately searches. No matter how much he has come to rely upon their advice for a wide range of practical decisions, the "experts" — who for the most part have spent their careers building up these mechanistic models — are scarcely the ones who can be expected to show him the way out. It is only by taking "his flesh into his teeth" and his "life into his own hands" that he can resolve the riddle of his pain and return to prosperity. Job's best impulse is toward a tangible acquaintance with the sensory information that can apprise him of his situation. To gain this, he must for a time set aside experts and models of reality and scientific authority, and find some way to feel for himself the forces which move him, to systematically explore his subjective experiences, to find the basis of his personal responsibility for his own fate.

This personal, sensory engagement with the self does not spring from a rebellion against scientific authority, but rather from a realization of the present inadequacy of that authority's conception of reality, a realization that is not contrived for the purposes of debate but which is forced upon him by his own painful circumstances. This coming to grips with his wealth of subjective information is neither an abandonment of sound principles nor a hedonistic avoidance of the issue; it is in the best spirit of scientific inquiry, and it is motivated by the grimmest necessity. When the conceptions of reality that we maintain do not square with the things that we are experiencing, it is not because we are flawed or because our experiences are wrong, but because our conceptions cannot contain all of the facts as we perceive them. And there is no constructive way out of this crisis but to enlarge our sense of reality to include our actual experiences.

What we have the opportunity to discover in this confrontation is that the conscious exercise of our own perceptions and our own will is a decisive factor in our relationship with the laws of nature. It is categorically impossible to passively receive an adequate sense of reality. Any conception that is not constantly rediscovered or reconfirmed by the efforts of our own participation and scrutiny cannot continue to be actively true for us, cannot continue to be the basis for right actions and just rewards. This passiveness is itself the seed of our destruction. Strength and health cannot be pumped into any organism that clings to its own passivity. The forces that mold us cannot be good or just to an individual who is unwilling to struggle toward a first-hand understanding of his relationship to them, who is not actively engaged with expanding his capacities, who is not himself taking a conscious part in the creation of his own circumstances. Goodness and justice *consist* of the ongoing successful resolutions of these relationships and the active choices that they continually present to us.

A successful relationship with reality must be learned as a dance is learned — not by contemplation but by participation, by an exertion of will, by movement, action, and response. In order to derive strength from God and the angels, it is necessary to enter into active contact with them, to push against them with our muscles and feel them with our senses. We are formed not by abstract laws but by the intimacies of a wrestling match, where we struggle with universal forces until we begin to feel our individual forces grow in relation to them. "Force must be met by force, and the structure evolves as the forces are balancing."

The one quality that is outstanding in all our organs and tissues, almost no matter in what shape we find them, is their great plasticity. Flesh is a highly malleable thing, constantly shifting its depository layers between the demands of the internal and the external environments. And this plasticity does not function within narrowly defined limits. It is radical. It has produced millions of species from the same basic protoplasm, and it can turn out any one of hundreds of possible individuals from the same genetic code.

We are not at all like billiard balls, bits of mechanistic matter that react in singular and predictable ways to singular and measurable forces. Living form responds in ways that are often not predictable to forces that are often not measurable. Nature does not simply pile together forms and forces until they add up to "life." Life is a process that assumes form and then assumes alternatives, constantly creating solutions to constantly shifting conditions.

> This primordial molding matter is neither solid nor fluid; it is both. It is just sufficiently solid to keep a form it assumes, and just sufficiently fluid to change to some other form at need.[8]

Act, reaction, activity; respond, response, responsibility. The telling differences in meanings between these verbal forms themselves is indicative of the ways we normally think — and ways we refuse to think — about matter and process, form and formation. But reducing biological events to fixed forms of chemical reactions and then extracting the subjective, responsive elements so that we can examine those events objectively has nothing whatever to do with comprehending the intelligence, the complexity, and the plasticity of life forms.

We must remember, however, that this marvelous ductility does not in itself constitute the cure for any condition. Left to its own organic devices, without the exertion of sensibility and will, protoplasm will simply respond to local forces, bad as well as good. We are sol/gel, semi-solid, fluid crystal, and when we are not actively firming ourselves up into functional structures we are sagging and slipping. Our flesh is like silly putty that distorts when it is ignored. We are constantly obliged to actively participate in its formation, or else it will droop of its own weight and plasticity.

This incessant formation we cannot stop. We can only make the choice to let it go its own way — directed by genetics, gravity, appetites, habits, the accidentals of our surroundings, and so on — or the choice to let our sensory awareness penetrate its processes, to be personally present in the midst of those processes with the full measure of our subjective, internal observations and responses, and to some degree direct the course of that formation. We do not have the option of remaining passively unchanged, and to believe for a moment in this illusion is to invite distortions and dysfunctions. Like putty, we are either shaping ourselves or we are drooping; like clay, we either keep ourselves moist and malleable or we are drying and hardening. We must do one or the other; we may not passively avoid the issue. Like Job, we must learn to take our flesh in our own teeth, put our lives in our own hands, and actively participate with the subtle and awesome forces which weave the web of our existence.

It has been my experience that bodywork is one of the most humane and the most effective means available for generating the streams of full and precise sensory information which compose the largest and most concrete part of this self-awareness. And as such it is an invaluable tool with which we can facilitate the mending not only of our own bodies, but also of those gaps in our objectified world view — a world view which has led us dangerously far away from our sense of this vital participation in our fates.

> *Take your well-disciplined strengths*
> *and stretch them between two*
> *opposing poles. Because inside human beings*
> *is where God learns.*
> — Rainer Maria Rilke *(Translation by Robert Bly)*

> *But he can make no progress with himself unless he becomes very much better acquainted with his own nature.*
> — C. G. Jung, *An Answer to Job*

2
Skin

Because there is something in the touch of flesh with flesh which abrogates, cuts sharp and straight across the devious intricate channels of decorous ordering, which enemies as well as lovers know because it makes them both — touch and touch of that which is the citadel of the central I-am's private own: not spirit, soul; the liquorish and ungirdled mind is anyone's to take in any darkened hallway of this earthly tenement. But let flesh touch with flesh, and watch the fall of all the eggshell shibboleth of caste and color too.
— William Faulkner, *Absalom! Absalom!*

The Open Barrier

The French have an apt expression used to describe the fortunate person who is comfortable with his own being and with his surroundings; they say that he *feels good in his own skin.* This phrase captures with a single stroke the deep significance for every organism of the two major functions of our surface layer: On one hand, the skin is a barrier, effectively containing within its envelope everything that is ourselves and sealing out everything that is not. On the other hand, it is an open window, through which our primary impressions of the world around us enter into our consciousness and structure our experience. The nature of this envelope itself provides many ways for us to feel good in our own skins.

Every unit of living protoplasm — be it plant or animal, single cell or complex aggregate — requires an effective barrier between its internal affairs and the welter of elements and forces which surround it. The colloidal behavior of primitive proteins and other organic molecules, jelling together in active physical and chemical association, cannot really be called "life" until it has established some means of self-containment.

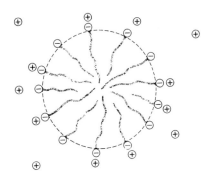

Fig. 2-1: A micelle — a colloidal droplet formed by paraffin-chain ions in a soap solution. The droplet is formed because of the long molecules' reaction to the surrounding water: The negatively charged heads have an affinity for the water (hydrophilic), while the paraffin chains are repulsed (hydrophobic). These two forces orient the molecules in the water as pictured.

Before a membrane is devised, colloidal behavior is only a localized reaction taking place within the context of a larger medium; with the formation of the membrane, it becomes a life form, distinct in profound ways from the medium around it. When a loosely associated colloidal jelly loses its cohesive interaction, we say that it has *dispersed;* but when a membrane is broken and its contents scattered, we must say that the thing has *died.*

This containment is vital to every cell within us and to our organisms as a whole. The enclosing surface is the interface between the life processes and lifeless materials which surround it, the boundary between protoplasm and all other forms of matter, the absolute line between the "me" and the "not me" for every living unit.

This outer surface is all that is normally visible of us; its over-all appearance

Fig. 2-2: In this micrograph of a cell membrane, we can discern the double layer of molecules which form it.

is usually the visual image we have in mind when we refer to "man" or "woman" or "Richard" or "Mary." Our surface defines our sizes, our shapes, our color, texture, odor — nearly everything about us that is readily observable. It is what the mirror shows us, and it is to a very large degree who we are. Much of our beauty is precisely skin deep.

If the only task of this outer surface were to keep what is me always inside and what is not me always outside, its structure and function could be much simpler than they are. But of course organic chemical processes constantly need new raw materials to sustain their activities, and they constantly create by-products that are harmful to those activities. So the barrier must be exquisitely selective. It must freely accept nourishments from without, and these in the proper varieties and amounts, at the same time that it is sealing out toxins; and it must reject metabolic wastes while retaining everything necessary

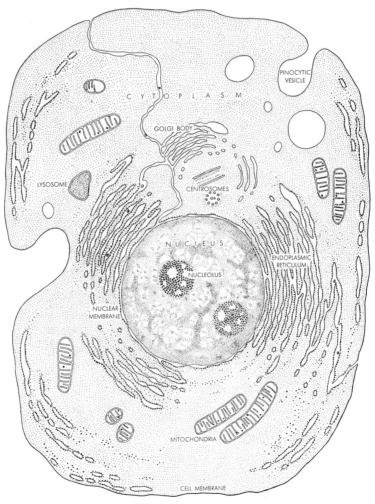

Fig. 2-3: A typical cell. In order to maintain its structure and life forces, it must surround itself with a membrane. All of its internal structures (organelles) are defined by similar membranes.

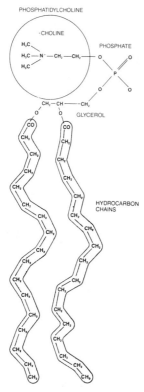

Fig. 2-4: The molecules which form the cell's membrane are phospholipids. Like the molecules in the micelle, the phospholipids have hydrophilic heads (which are attracted by water because of their electrical charge), and long hydrophobic tails (which are repulsed by water because of their oily nature).

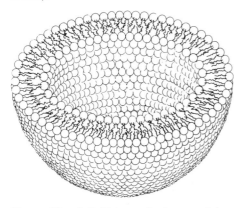

Fig. 2-5: Phospholipids interacting in water. A double layer is formed as all the hydrophobic tails seek to avoid water while all the hydrophilic heads face the water on both sides. The structure tends to form a sphere, so that the tails are not exposed to water along any open edges.

to the internal environment. It is not possible for life merely to hide behind the barrier. Every cell and every organism is forced to somehow explore its surroundings and establish appropriate patterns of acceptance and avoidance. Contacting other things with its outer surface is the only means life has of reaching beyond itself, of tasting the world and differentiating between bitter threats, yeasty necessities, and sweet pleasures.

General Structure and Functions

The skin is one of the largest single organs in the body. For the average adult male it covers eighteen square feet, and weighs about eight pounds — six to eight percent of total body weight. Both the amount and complexity of its activities are enormous. An area of skin the size of a quarter contains some three million cells, one hundred sweat glands, fifty nerve endings, three feet of blood vessels, and nearly as many lymph vessels. The whole skin has approximately six hundred and forty thousand sensory receptors that are connected to the spinal cord by over half a million nerve fibers; tactile points vary from seven to one hundred and thirty five per square centimeter.[1] Hair follicles vary in density from zero on the lips to one hundred and fifty thousand on the scalp.[2] It is its thinnest — about one tenth of a millimeter — on the eyelids, the lower abdomen, and external genitalia, and is its thickest on the palm of the hands and the soles of the feet, where it is three to four millimeters.[3]

Taken together, these elements of the skin orchestrate an impressive array of physiological functions: 1) as a waterproof and puncture-resistant protective covering for all underlying structures, guarding against both foreign invasions and excessive fluid loss; 2) as a temperature regulator (there are far more capillaries in the skin than are necessary for its nourishment; these may either be constricted to drive blood into the interior and preserve body heat, or gorged so that they radiate excessive heat away from the interior; in addition, the sweat glands bathe the skin to cool the over-heated body with evaporation); 3) as

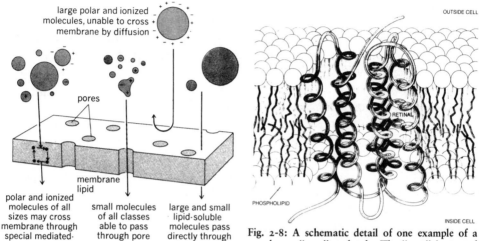

large polar and ionized molecules, unable to cross membrane by diffusion

pores

membrane lipid

polar and ionized molecules of all sizes may cross membrane through special mediated-transport systems

small molecules of all classes able to pass through pore

large and small lipid-soluble molecules pass directly through lipid regions of membrane

Fig. 2-6: The barrier must be selective, so that toxins are locked out, organic nutrients are allowed to pass in, and waste products are expelled.

OUTSIDE CELL

RETINAL

PHOSPHOLIPID

INSIDE CELL

Fig. 2-8: A schematic detail of one example of a membrane "gate" molecule. The "gate" is opened and closed by chemical messengers which alter the shape of the coils, thus expanding or contracting the pore in the phospholipid membrane created by the large imbedded molecule.

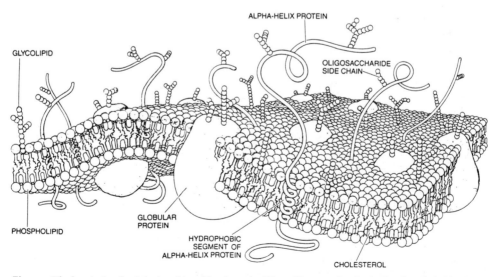

ALPHA-HELIX PROTEIN

GLYCOLIPID

OLIGOSACCHARIDE SIDE CHAIN

PHOSPHOLIPID

GLOBULAR PROTEIN

HYDROPHOBIC SEGMENT OF ALPHA-HELIX PROTEIN

CHOLESTEROL

Fig. 2-7: The barrier's selectivity is achieved by the embedding of large molecules in the phospholipid bilayer. These large molecules penetrate both layers; their shapes and their chemical affinities facilitate the passage of some substances across the membrane, and block others.

an organ of excretion, largely through the sweat glands; 4) as a part of the body's immune system, primarily by means of the rich network of lymph vessels, and also by means of chemical constituents of the skin tissue itself; 5) as a metabolic organ, involved in the metabolism, storage, and catabolism of fat, and in adjusting the body's levels of water and salt through perspiration; 6) as a reservoir for food and water; 7) as the site of vitamin D synthesis; 8) as a facilitator in the two-way passage of gases; 9) as a regulator of blood pressure — the opening and closing of the capillaries can assist in making large shifts in blood flow and peripheral resistance; 10) as a source of moisture and lubrication. This is especially important on the palms and the soles; 11) as a sensory organ.[4]

No tissue in the body is more adaptable than the skin. From the firm velvetiness of the infant to the dry and wrinkled parchment of old age, it is undergoing constant changes in its size, shape, qualities, and functions. According to the demands made upon it, it can become thin and sensitive or calloused and unfeeling. It contains its own mechanisms for healing all manner of traumas and adapting to all sorts of external situations. It is continually shedding and replacing its surface cells; an individual who lives to be seventy years old wears out and replaces approximately eight hundred and fifty skins, each one a little different from the last, reflecting the growth, habits, and over-all health of the organism. As man has increasingly become the bald mammal, his skin has simultaneously become more rugged as it has lost its protective shag, more varied in its roles and textures, and more acute and differentiated in its sensory capabilities.[5] It is almost as though different areas of our skin come from different animals, so widely varied are its qualities.[6]

Human skin derives a good deal of its toughness from being arranged in progressive layers, much like plywood. The outermost covering, the epidermis, generally consists of four distinct layers: The Malpighian layer, which is made up entirely of living cells; the granular and hyaline layers, where these living cells are gradually transformed into dead but very tough plates that make up the final waterproof horny layer, also referred to as squamous cells. It is the epidermis as a whole that is continually being shed and replaced. Constant cell division is occurring in the deepest part of the Malpighian layer, and these new cells then migrate upward, passing through the granular and hyaline phases to become squamous cells and eventually to drop away. The entire cycle takes about twenty-seven days.[7]

None of these epidermal layers have any blood vessels, and they rely upon the underlying dermis for their irrigation. This basement layer of the skin is thicker than any of the others, and contains all of the blood and lymph vessels that serve the epidermis; it also contains the hair follicles, their sebaceous glands and erector muscles, the sweat glands, and most of the skin's nerve endings. The dermis is also heavily laced with areolar or loose connective tissue, which gives the skin both its tensile strength and its pliable and elastic qualities. This connective tissue network is in turn woven into the subcutaneous connective

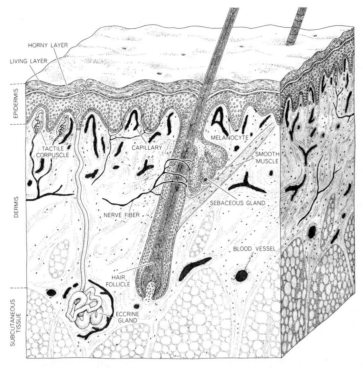

Fig. 2-9: A cross-section of skin. There are three main layers—the epidermis, the dermis, and the subcutaneous. The dermis is interlaced with many connective tissue fibers, which bond the outer layers to the deeper sub-cutaneous layers of fat and muscle.

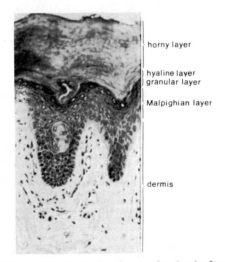

Fig. 2-10: A detail of epidermis, showing its four subdivisions. New skin cells are produced in the Malpighian layer, migrate upward through the granular and hyaline layers, and finally become dead but very tough squamous cells on the surface. These squamous cells are continually sluffed off and replaced from below.

tissue, which in its turn adheres to deeper layers, fixing the skin securely to the underlying structures.

This network of connective tissue in the dermis is not only responsible for the resiliency and adherence of the skin, it also contains the main mechanism for the healing of injuries of all kinds. New fibers of connective tissue are produced by cells called *fibroblasts*. These cells are abundant in the dermis, and when the skin is wounded they are carried to the location by bleeding and by a general inflammatory response. Once at the wound site, they begin manufacturing strands of connective tissue which bridge the gaps created by the injury, filling in the wound. Within this connective tissue matrix, finer repairs of nerves and capillaries proceed at a slower rate; eventually part of the connective tissue is replaced by other regenerated tissues, while some of it remains as supportive scar tissue.[8]

The connective webbing of the dermis is responsible for the changes in appearance of aging skin as well. As a person grows older, an increasing percentage of the moisture in this tissue is lost; as this occurs, the fullness of the skin is literally drained away, leaving it thinner, drier, more easily folded and wrinkled. In addition, with increasing age the connective tissue becomes more densely packed and begins more and more to adhere to itself, markedly reducing its elastic qualities. These signs of aging in the dermis may be more or less advanced in any individual regardless of their chronological age, due to hereditary factors, diet, disease, trauma, or physical and emotional stress. The drying and the self-adhering of the skin's connective tissue may be used to calculate a person's "biological" age, as opposed to their chronological age.

Skin as Sense Organ

The skin is the largest, the most varied, and the most constantly active source of sensations in the body. Most of the other physiological roles of the skin are clearly crucial for our survival; without the protection from foreign invasion, the prevention of excess fluid loss, the metabolic activities, the temperature regulation, the immunological resistances, the healing mechanisms, and the excretory functions which the skin provides, the organism would rapidly expire. We are for the most part unaware of any of these processes. They are either local chemical and cellular responses, or they are orchestrated by normally unconscious levels of our nervous system.

But we are not more aware of anything in our lives than we are of the sensory activities of our skins. From the womb to the grave they flood our conscious minds with constantly changing information, even remaining active as sentinels during our deepest sleep, ready to start us into wakefulness at the first abnormal sensation. And no other function of the skin is more critical to our survival than are these sensations. They contribute more information than any other sensory source to our successful assessment of and appropriate response to our surroundings, and—as we shall see—they help to determine the overall physical and mental health of the organism in powerful ways that are not as obvious as their announcement of this or that sensation.

It was Aristotle who first enumerated the five special senses of sight, smell, hearing, taste, and touch.[9] Of these special senses, only touch involves the entire body; the others are highly localized in their respective organs, and all of these special organs are contained in the head. Our sense of touch, on the other hand, is found almost everywhere on our surfaces to one degree or another. Nor is there just one kind of touch. This sense includes hot, cold, itches, tickles, all degrees of pressure, vibration, and an enormous variety of pains and pleasures.

The variety of touch sensations is further complicated by the fact that different relative intensities and frequencies of any of these single modes may result in very different experiences. Isolated light pressures and tickles, for instance, bear almost no relation to the ecstasies that can be produced by artfully compounding and manipulating these sensations. Nor are the sensations produced by the splashing of a single drop of water on the forehead anything at all like the sensations produced by the thousands of repetitions of single drops in a long, excruciating session of the Chinese "water torture."

The Mother of the Senses

Touch is the chronological and psychological Mother of the Senses. In the evolution of sensation, it was undoubtedly the first to come into being. It is, for instance, rather well developed in the ancient single cell amoeba. All the other special senses are actually exquisite sensitizations of particular neural cells to particular kinds of touch: compressions of air upon the ear drum, chemicals on the nasal membrane and taste buds, photons on the retina. In the human embryo, the sense of touch develops in the sixth week, when we are less than an inch long. Light stroking of the upper lip at this time causes a strong withdrawal, a bending back of the entire neck and trunk.

Touch, more than any other mode of sensation, defines for us our sense of reality. As Bertrand Russell observed, "Not only our geometry and our physics, but our whole conception of what exists outside us, is based upon the sense of touch."[11]

It is a remarkable testimony to this paramount place which it has in our sense of reality that "touch" is the longest single entry in the unabridged dictionaries of many languages, attesting to its wealth of associations in human thought and expression. In the Oxford English Dictionary, for instance, its ramifications continue for fourteen full columns,[12] and if we include the immediately following entries of "touchable," "touching," and "touchy," the entry extends to twenty-one columns. Its metaphorical meanings run the gamut from "pure" to "tainted," from "sensitive" to "deranged," from "contact lightly" to "run through with a sword." Even painting, art for the eyes, and music, art for the ears, can "touch us deeply."

No other sense gives us so much of the world. Helen Keller and Laura Bridgeman, after all, spent their lives blind and deaf; their remarkable educations and achievements were accomplished through no other medium than that of touch.[13] We can compensate for blindfolds, earplugs, or nose plugs with relative ease, but the blockage of all tactile sensations produces a profound and unresolvable disorientation, one that leads rapidly to psychosis.

Types of Sensors

Pain sensors are the most numerous type, reflecting the skin's role as one of the most important sentinels against danger. A variety of pressure sensors (different ones to detect light or heavy, transitory or constant pressure) are next, followed in decreasing order by cold and finally warmth receptors.

There are well over half a million of them altogether, concentrated most densely on the lips and most sparsely on the back. It was once thought that every individual kind of nerve end responded to one and only one type of stimulus. This was called the law of specific sensory modalities. More recent research, however, has suggested that this law may not prevail inflexibly. Heat and cold do both have clearly specific endings, but some endings seem to respond to a wide range of stimuli; some seem to register two or three at once, and all types exist so closely together that it is exceedingly difficult to unravel their particular components in any multiple sensation.[14] At least eight entirely different kinds of tactile nerve endings are known, but there are probably many more variations on these that have not yet been isolated and identified.[15]

One kind found everywhere on the skin are *free nerve endings,* which may have only a few, or may have scores of branches. They can detect light touch and pressure, and it is also postulated that they are the major type of ending that is sensitive to pain, reacting strongly to specific chemicals that are spilled from damaged cells.

Second, there are the acutely sensitive *Meissner's corpuscles,* which are most abundant on the lips and fingers, located just beneath the surface of the epidermis. This kind of ending is particularly sensitive to very light touch moving across the skin. It adapts (that is, it ceases to fire) to a single constant stimulus quite rapidly, so that for this type of ending a light object touching the

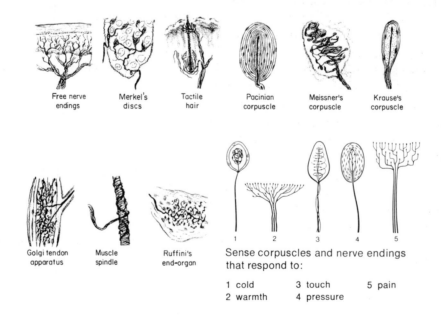

Free nerve endings   Merkel's discs   Tactile hair   Pacinian corpuscle   Meissner's corpuscle   Krause's corpuscle

Golgi tendon apparatus   Muscle spindle   Ruffini's end-organ

1   2   3   4   5

Sense corpuscles and nerve endings that respond to:

1 cold   3 touch   5 pain
2 warmth   4 pressure

Fig. 2-11: Various sensory endings.

A

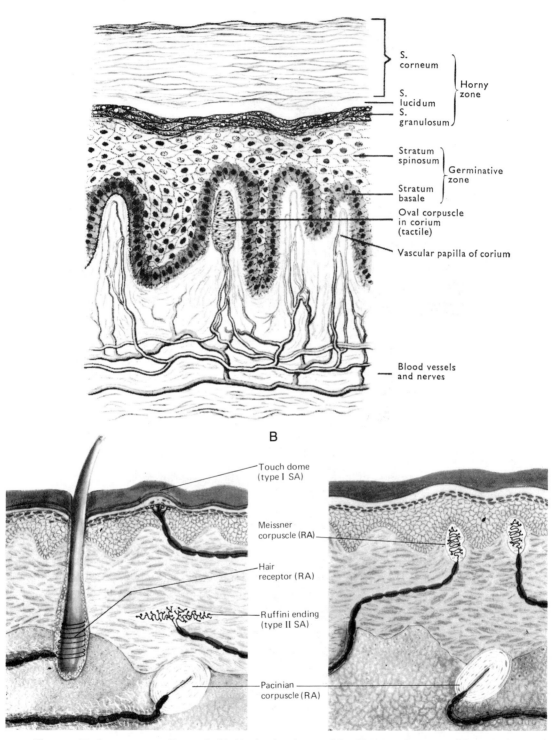

S. corneum

Horny zone

S. lucidum
S. granulosum

Stratum spinosum

Germinative zone

Stratum basale

Oval corpuscle in corium (tactile)

Vascular papilla of corium

Blood vessels and nerves

B

Touch dome (type I SA)

Meissner corpuscle (RA)

Hair receptor (RA)

Ruffini ending (type II SA)

Pacinian corpuscle (RA)

Fig. 2-12: Various sensory endings embedded in the three layers of the skin. Note that pain and light touch endings are closer to the surface, while heat, cold, and various pressure endings are contained within deeper layers.

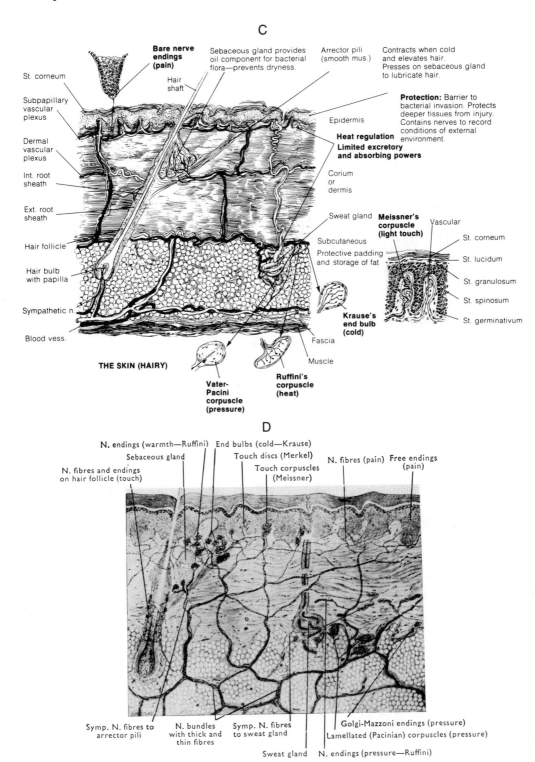

C

Bare nerve endings (pain)

Sebaceous gland provides oil component for bacterial flora—prevents dryness.

Arrector pili (smooth mus.)

Contracts when cold and elevates hair. Presses on sebaceous gland to lubricate hair.

St. corneum

Hair shaft

Subpapillary vascular plexus

Dermal vascular plexus

Int. root sheath

Ext. root sheath

Hair follicle

Hair bulb with papilla

Sympathetic n.

Blood vess.

Epidermis

Protection: Barrier to bacterial invasion. Protects deeper tissues from injury. Contains nerves to record conditions of external environment.

Heat regulation
Limited excretory and absorbing powers

Corium or dermis

Sweat gland

Meissner's corpuscle (light touch)

Vascular

St. corneum

St. lucidum

St. granulosum

St. spinosum

St. germinativum

Subcutaneous

Protective padding and storage of fat

Krause's end bulb (cold)

THE SKIN (HAIRY)

Fascia

Muscle

Vater-Pacini corpuscle (pressure)

Ruffini's corpuscle (heat)

D

N. endings (warmth—Ruffini)

Sebaceous gland

End bulbs (cold—Krause)

Touch discs (Merkel)

Touch corpuscles (Meissner)

N. fibres (pain)

Free endings (pain)

N. fibres and endings on hair follicle (touch)

Symp. N. fibres to arrector pili

N. bundles with thick and thin fibres

Symp. N. fibres to sweat gland

Sweat gland

Golgi-Mazzoni endings (pressure)

Lamellated (Pacinian) corpuscles (pressure)

N. endings (pressure—Ruffini)

2-12: Continued

skin without moving tends to "disappear," while an object that is in motion on the skin can be tracked with great accuracy.

Third, there are associated with Meissner's corpuscles a variation called *expanded tip tactile receptors*. These are similar to Meissner endings, except that they transmit long-continuing signals rather than adapting rapidly, so they can pinpoint an object that is stationary on the skin.

Fourth, every hair on the body has wrapped around its follicle a *hair end organ,* which registers the slightest movement of each hair.

Fifth are pressure sensors that are located in deeper layers of the skin, and also deep within the body, *Ruffini's end organs.* These do not adapt rapidly, and are therefore important for signalling continuous pressures on deep tissues. They are particularly abundant in the joint capsules, where the bones' changing angular relationships create shifting pressures; there the Ruffini organs inform us of the locations and movements of our limbs.

Sixth is another ending located both in the skin and the deeper tissues, the *pacinian corpuscle.* Unlike the Ruffini organs, these endings adapt in a small fraction of a second, and are therefore particularly important for detecting rapid distortions, fleeting pressures, vibrations, and the like.

Although the density of their relative distributions varies from area to area, it appears as though there is no substantial area of skin that is not served by all of these varieties of endings. Their active circles interpenetrate so thoroughly that it is almost never the case that we experience one particular kind of tactile stimulation; even in the laboratory, it is with painstaking difficulty and a certain amount of guesswork that we can map out their separate distributions. Taken together, they tell us most of what we know about the forms and textures of all the material objects around us, and even within us.

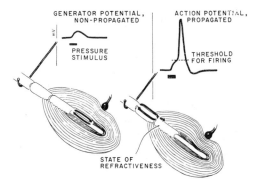

Fig. 2-13: A pacinian corpuscle responding to pressure. As the outer layers are deformed, the buried tip of the sensory axon is stimulated and fires. The more the deformation, the longer the burst of action potentials.

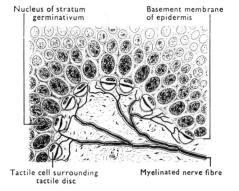

Fig. 2-14: An Iggo dome receptor—several Merkel's discs innervated by a single large nerve fiber. The discs transmit an initially strong signal and then only partially adapt, continuing to send a weaker signal. This both allows them to clearly announce the initial touch of an object on the skin, and then to keep us aware of its extended contact.

Sensing Self

And information about objects is not all that they give us. Every time that I touch something, I am as aware of the part of me that is touching as I am of the thing I touched. Tactile experience tells me as much about myself as it tells me about anything that I contact. I am constantly using the world to explore my reactions just as much as I am using my reactions to assess the world. My sense of my own surface is very vague until I touch; at the moment of contact, two simultaneous streams of information begin to flow: information about an object, announced by my senses, and information about my body announced by the interaction with the object. Thus I learn that I am more cohesive than water, softer than iron, harder than cotton balls, warmer than ice, smoother than tree bark, coarser than fine silk, more moist than flour, and so on.

We could even say that this role of the tactile senses in establishing a fuller and fuller sense of self is their primary function. An infant approaches objects not with an initial idea of research into and manipulation of externals, but with an idea of self-stimulation; and it discovers its own anatomical parts in exactly the same way (and at the same moments) that it discovers other objects. We can never touch just one thing; we always touch two at the same instant, an object and ourselves, and it is in the simultaneous interplay between these two contiguities that the internal sense of self — different from both the collection of body parts and the collections of external objects — is encountered.

> Since [my ideas of] both the body and the world have to be built up, and since the body in this respect is not different from the world, there must be a central function of the personality which is neither world nor body. There must be a more central sphere of the personality. The body is in this respect periphery compared with the central functions of the personality.

That is to say, my tactile surface is not only the interface between my body and the world, it is the interface between my thought processes and my physical existence as well. By rubbing up against the world, I define myself to myself.

This dialectic is life-long, and its formative power can hardly be overstated. It establishes preferences and aversions, habits and departures, becomes the very stuff in which attitudes are ingrained. The "feel" in my skin and the "feelings" in my mind, what I "feel" and how I "feel" about it, become so confounded and ambiguous that my internal "feelings" can alter what my skin "feels" just as powerfully as particular sensations can shift my internal states.

It is not too much to say that the sensory activity of the skin is a major element in the development of disposition and behavior, an element with enough sophistication and plasticity to account for wide divergences of experience and observation:

> The skin itself does not think, but its sensitivity is so great, combined with its ability to pick up and transmit so extraordinarily wide a variety of signals, and make so wide a range of responses, exceeding that of all other sense organs, that for versatility it must be ranked second only to the brain itself.[17]

Skin as Surface of the Brain

Indeed, the associations between the skin and the brain are extremely intimate. This fact is the basis of the "lie detector test"; specific mental states directly influence the electrical properties of the skin, and in regular ways that can readily be measured and correlated. And from blushing to hives, from goose bumps to shingles, the skin demonstrates a reflex expressiveness to scores of mental events.

The Ectoderm

This close association between the skin and the central nervous system could not have more concrete anatomical and physiological connections. All tissues and organs of the body develop from three primitive layers of cells that make up the early embryo: The endoderm produces the internal organs, the mesoderm produces the connective tissues, the bones, and the skeletal muscles, while the ectoderm produces both the skin and the nervous system.

Skin and brain develop from exactly the same primitive cells. Depending upon how you look at it, the skin is the outer surface of the brain, or the brain is the deepest layer of the skin. Surface and innermost core spring from the same mother tissue, and throughout the life of the organism they function as a single unit, divisible only by dissection or analytical abstraction. Every touch initiates

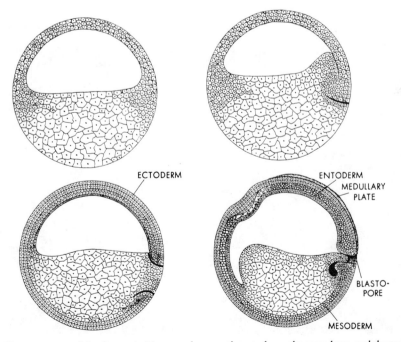

Fig. 2-15: The emergence of the three primitive germ layers—the ectoderm, the mesoderm, and the endoderm. The ectoderm gives rise to the nervous system and the skin, the mesoderm to the muscles and connective tissues, and the endoderm to the internal organs.

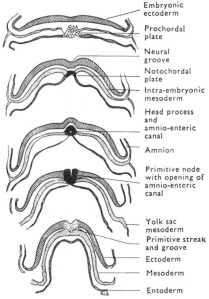

Fig. 2-16: The three germ layers in an 18-day-old embryo.

a variety of mental responses, and nowhere along the line can I draw a sharp distinction between a periphery which purely responds as opposed to a central nervous system which purely thinks. My tactile experience is just as central to my thought processes as are language skills or categories of logic.

Between the third and fourth week of the embryo's life, the ectoderm begins to differentiate into skin tissues and neural tissues. The skin layer separates from the central neural tube and migrates outward, as the developing muscles, bones and organs push it farther and farther away from the core. However, this is far from the end of the close chemical, structural, and functional affinities between the skin and the brain.   In spite of the increasing distance that separates them, properties of the skin continue to play a material role in the development and organization of the central nervous system.

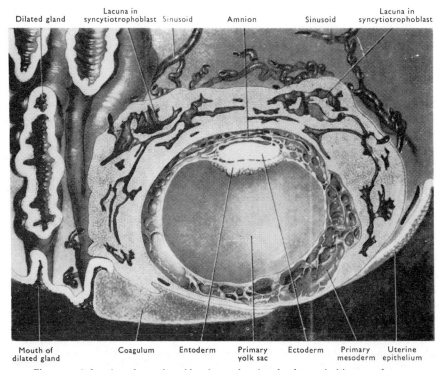

Fig. 2-17: A drawing of a 12-day-old embryo, showing the three primitive germ layers.

Neural Mapping

In order to accurately locate any stimulus on the body's surface, the brain relies upon a precise spatial arrangement of its circuitry, such that specific adjacent nerve endings on the skin transmit their signals through parallel neurons which terminate in specific adjacent cell bodies in the sensory cortex. The principle involved is like that used in modern fiber optics: If the glass fibers at the observer's end of the bundle are arranged as they are at the far end, he will receive a clear, unscrambled image of any object toward which the far end is pointing.

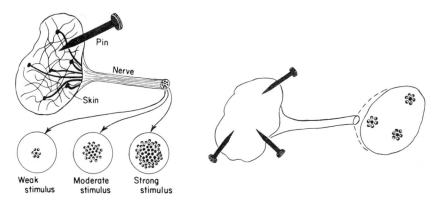

Fig. 2-18: An area of skin, innervated by endings from several axons. The axons are gathered together into a nerve trunk, where they preserve their parallel arrangements throughout their full length. Thus, when three separate but simultaneous pin-pricks touch the skin, the spatial relationship of the stimulated axons in the trunk correspond to the arrangement of the pin-pricks on the surface of the skin. The principle is similar to fiber optics.

The spatial relationships of the various parts of the periphery are, then, projected by their parallel nerve fibers and "mapped" onto corresponding areas of the cortex, where they are arranged as the familiar sensory homunculus. It is in this way that the brain separates functions and pinpoints locations throughout the body.

For instance, in the dorsal column of the spinal cord (where all the sensory tracts are bundled together), the sensory fibers from the feet lie closest to the midline, and as we rise up the body and up the column, successively added sensory trunks add their fibers progressively toward the lateral sides of the dorsal column.

This spatial organization is maintained with precision throughout the sensory pathway all the way to the somesthetic cortex. Likewise, the fiber tracts within the brain and those extending into motor nerves are spatially oriented in the same way.[18]

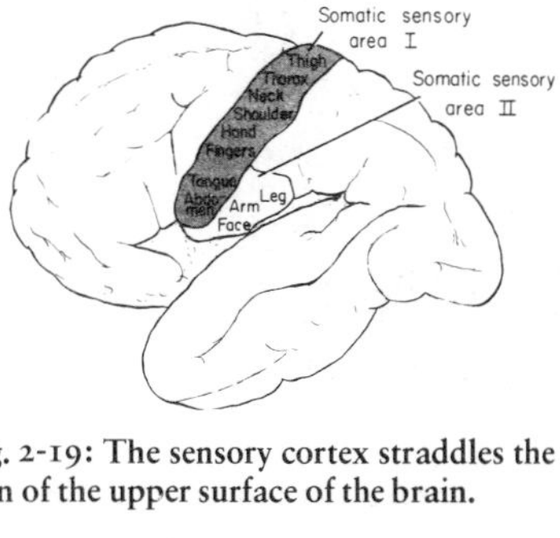

Fig. 2-19: The sensory cortex straddles the midsection of the upper surface of the brain.

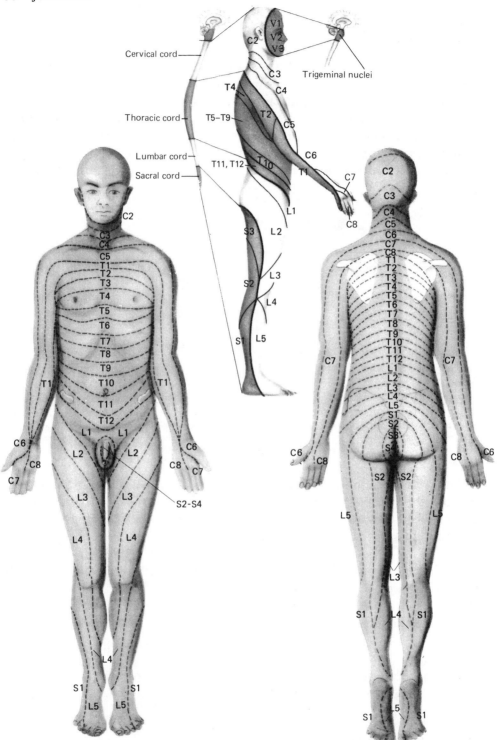

Fig. 2-20: Dermatomes. Each area of skin illustrated is innervated by a specific spinal nerve trunk as indicated—S, sacral; L, lumbar; T, thoracic. Axons coming from each area converge upon the sensory tracts of the spinal cord at specific segmental levels, with each higher segment adding its corresponding spinal tract lateral to those of the segments below it in an orderly fashion.

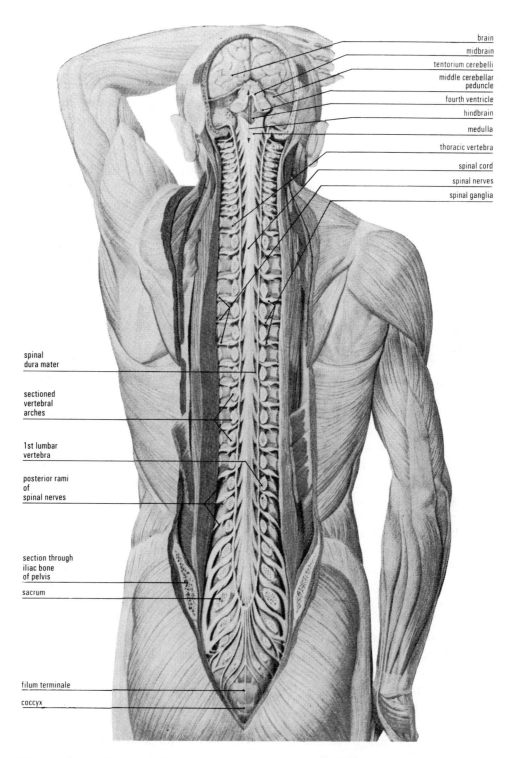

brain
midbrain
tentorium cerebelli
middle cerebellar peduncle
fourth ventricle
hindbrain
medulla
thoracic vertebra
spinal cord
spinal nerves
spinal ganglia

spinal dura mater

sectioned vertebral arches

1st lumbar vertebra

posterior rami of spinal nerves

section through iliac bone of pelvis

sacrum

filum terminale

coccyx

Fig. 2-21: The orderly stacking of nerve trunks from each segment of the body as they converge upon the spinal column.

This same principle of parallel fiber arrangement produces a similar homunculus upon the motor cortex, and another upon the cortex of the cerebellum. These miniature maps of the body correspond not only to the peripheral arrangement, but also to one another, so that parallel circuits not only link my actual hand to my sensory cortex "hand"; they also link this sensory "hand" to my motor cortex "hand" and to my cerebellar "hand" as well. Each map corresponds point for point to all the others, and all are linked together by parallel circuits.

At one time it was suggested by neural anatomists that such parallel pathways developed in the embryo before being assigned to any specific functions, and that particular channels were later created by habitual usage, much like water wears a channel for a stream bed. Conflicting or extraneous paths would be eliminated by atrophy, and new channels could be re-routed in case of damage. Later it was postulated that these parallel circuits were established genetically in the developing brain and spinal cord, which then reached out from the central core with axons and nerve ends to contact the periphery. It was not clear just how the nerve ends knew exactly where to go; perhaps there was a good deal of randomness to their branching which the organism later sorted out by means of trial and error.

More recently, it seems that neither of these earlier views may be the case. It is now appears that the organization of this parallel circuitry is actually *initiated at the periphery.* Local qualities in the skin, the joints, and the deep tissues "tag" the nerve ends which contact them with subtle chemical messages, and these chemical "tags" direct axons growing *inward* toward the appropriate connections in the spinal cord and brain. The process is anything but random or trial and error. It is highly specific, and it is the periphery which helps to organize the connections in the central nervous system, not an organized central nervous system which reaches out to innervate the periphery.

This more recent view that it is peripheral conditions which organize the actual development of neural circuits and guide the process of mapping is of central importance to bodywork. It suggests that the use of touch and sensation to modify our experience of peripheral conditions exerts an active influence upon the organization of reflexes and body image deep within the central nervous system. Such a view seems to be borne out by the following kinds of experiments.

If a patch of skin from the belly of a frog embryo is exchanged with a patch from his back, and the two patches are allowed to re-attach in their new locations and regenerate their nerve ends, a very curious thing happens. When I tickle the patch on the frog's belly, he rubs his back, and when I tickle the one on his back, he rubs his belly. In spite of the fact that the major sensory nerve trunks deep to the skin have not been disturbed, the frog confuses front and back when his transposed patches are stimulated. Nerve ends in the skin taken from the belly still somehow manage to excite the "belly" areas in the frog's sensory and motor cortexes, even though the signals must now go through circuits associated with the back. The neural connections in the brain have somehow shifted to accommodate the shifted areas of skin. These transplants can be repeated with many variations, involving various muscle sets, entire limbs and even the eyes. In every instance, their results are similar — stimulation of transplanted tissue results in reactions that would be appropriate if the tissue were still at its original site, and not at the the new one. "It is necessary to conclude," says R. W. Sperry, the author of a series of these experiments,

> that the sensory fibers that made connections with the grafted tissues must have been modified by the character of these tissues. . . . Growing freely into the nearest areas not yet innervated the fibers established their peripheral terminals at random. Thereafter they must proceed to form

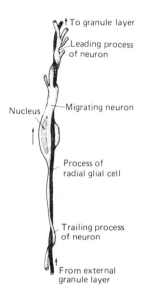

Fig. 2-22: The growing sensory axon in the foetus follows the lead of a glial cell—its supportive and nutritive partner. Presumably, the glial cell carries chemical messengers from the skin which are matched against chemical tags at each ascending level up the spinal column and the brain, identifying the appropriate internal connections. Each axon is then chemically coded to follow its particular glial cell.

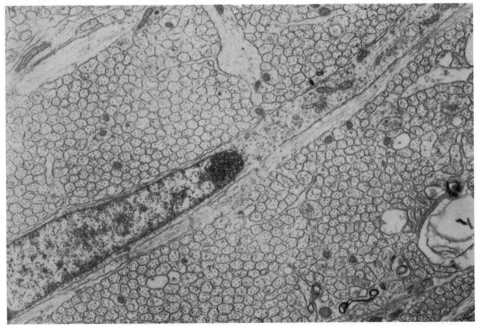

Fig. 2-23: An axon following its glial partner as they grow. The nerve cell's nucleus is the large dark oblong shape. The glial cell is the grey band running from corner to corner diagonally, and the nerve's growing axon is next to it.

central hook-ups appropriate for the particular kind of skin to which they have become attached. It seems clearly to be the same quality in the skin at the outer end of the circuit that determines the pattern of reflex connections established at the center. . . . It is only when contact is made with central nerve cells which have the appropriate chemical specificity that the growing fiber adheres and forms the specialized synaptic ending capable of transmitting the nerve impulse.[19]

There is some kind of specific chemical coding which matches areas in the brain with areas at the periphery, and which guides the growing nerve circuits to their appropriate terminals. And this specificity is strong enough, in the case of the frog and other lower animals, to alter central connections to rematch transplanted peripheral tissues. Local qualities in the skin closely regulate the structuring of the nervous system in the embryo, including the formation of neural connections deep within the developing brain.

In this way the local-sign properties of the skin become stamped secondarily upon the cutaneous nerves and are carried into the sensory centers of the brain and spinal cord. . . . Implicit in this theory is the assumption that in the embryo the cerebral cortex and the lower relay centers also undergo a differentiation that parallels in miniature that of the body surface. In other words, just as from the skin to the first central connection point in the spinal cord, so from relay to relay and finally to the cortex the central linkages arise on the basis of selective chemical affinities. At each of its ascending levels the nervous system forms a map-like projection of the body surface.[20]

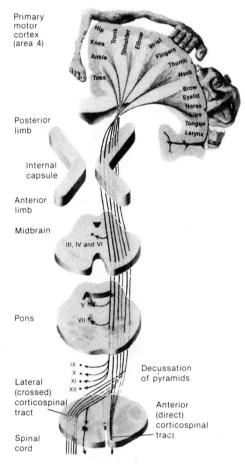

Fig. 2-24: The parallel tracts run up the spinal cord, through the various levels of the pons and the midbrain, and are projected in an orderly fashion upon the cerebral cortex. Each separate group of cells in the cortex corresponds to a separate area on the surface of the body, and the cortical projections form a coherent body map, or homunculus, on the surface of the brain. The homunculus appears distorted because areas serviced by more sensory axons take up more area on the cortex. Thus the cortical "tongue" is larger than the cortical "foot," since the tongue has more sensory elements than does the foot.

The pathways pictured are those of the pyramidal motor system, but the principles of mapping are the same for the sensory tracts.

The "skin" portion of the ectoderm, then, does not really separate from the "neural tube" portion and migrate outward at all. The neural tube expands by means of trunks, axons, and endings, with the skin as its advancing boundary. And it is the chemical and sensory make-up of the skin which provides the "template" for the connections and reflex patterns within the brain, not the other way around.

The skin is no more separated from the brain than the surface of a lake is separate from its depths; the two are different locations in a continuous medium. "Peripheral" and "central" are merely spatial distinctions, distinctions which do more harm than good if they lure us into forgetting that the brain is a single functional unit, from cortex to fingertips to toes. To touch the surface is to stir the depths.

Touch as Food

So the skin, it would seem, offers an excellent means of influencing internal processes. Its sensory pathways unite the surface and the interior of the organism, and its surface does not shield any more than it exposes.

It is also obvious, of course, that internal states of mind and of physical health directly affect the skin. Transitory moods are reflected by paling, flushing, goose flesh, cold shivers, sweating. Chronic anxiety and exhaustion darken the area under the eyes. Healthy circulation makes the surface pink and warm; liver failure makes it yellow and clammy. Faulty diet can make it too dry or too oily. Acute emotional distress can erupt in a plethora of rashes, bumps, pimples, and boils. Neurotic dispositions can render it nearly numb at one extreme, or excruciatingly sensitive at the other.

But what about the other way around? Can conditions and sensations on the skin really have equally potent effects on our organs, our circulation, our moods, our personalities? A soothing hand will calm a frightened animal or an injured child, and a cool cloth can diminish a raging headache — these are effects everyone has experienced. But doesn't the intimate connections between the skin and the central nervous network suggest that the relationships might go further than these familiar palliatives? Evidence from many quarters indicates that indeed it does go much further.

In the Orphanage

> As recently as 1915 James H.M. Knox, Jr. of the Johns Hopkins Hospital noted that, in spite of adequate physical care, 90% of the infants in Baltimore orphanages and foundling homes died within a year of admission.[21]

In the same year a New York pediatrician, Dr. Henry Dwight Chapin, published a report concerning children's institutions in ten different cities, which described a similar situation, except that in the orphanages he studied the infant mortality within one year after admission was much closer to 99%. Such statistics for abandoned infants were in fact almost universal in the nineteenth and early twentieth centuries.

The "disease" was called *merasmus,* a Greek word meaning "wasting away," and it was tacitly assumed that almost every infant in every orphanage would succumb to it, in spite of an adequate diet and professional medical attention. Even those who lived through this initial year were clearly damaged by it.

> The period between the seventh and the twelfth month of life was the time of the highest fatalities. Infants who managed to survive their first year uniformly showed severe physical retardation.[22]

As one would hope, these shocking figures spurred investigations and reforms in orphanages in Europe and America. The largest single factor that emerged from the closer study of these institutions proved to be their sparse staffing. Attendants had just enough time to clean and feed the infants and plop them back into their solitary cribs, where they died of loneliness, of inadequate sensory stimulation. When extra help was added, so that there were enough attendants for each infant to be held, handled, talked to, played with for ample lengths of time every day, infant mortality rates plummeted.

And not only did more infants survive, but the survivors were not marred by the stunted growth and the mental retardations of "deprivation dwarfism." Even those who had been previously retarded showed dramatic increases in their weight, height, energy, and mental acuity. In institution after institution, the mystery of infant merasmus was cleared up: The *tactile stimulation* associated with tender, loving care was absolutely crucial to a baby's development. Without it, no amount of food and no kind of medicine could produce a healthy individual.

In the Hospital

Continued observations have confirmed that this connection between adequate stimulation and healthy physical development is unequivocal. In 1942, Harry Bakwin of New York University began observing infants who were removed from their homes for hospital care. He noted that they soon became listless, apathetic, depressed. Their bowel movements were more frequent, and even though they were well fed, they failed to gain weight at normal rates. They suffered from an increase in respiratory infections, and they persistently developed fevers of unknown origins. "All such abnormalities, however, quickly disappeared when the infants were returned to their home and mother."[23]

A similar study was conducted by Margaret A. Ribble over an eight year period in three New York maternity hospitals. She found that when an infant was isolated from its mother, diarrhea was more prevalent, and muscle tone commonly decreased. The absence of normal mother-infant interaction was not, Ribble concluded, merely a sentimental concern, but was "an actual privation which may result in biological, as well as psychological damage to the infant."[24]

In 1965 a mother in upper New York State gave birth to a daughter with an incomplete esophagus. A feeding tube was surgically inserted into the baby's stomach, and for fifteen months the mother meticulously fed the child with standard daily doses of nutrient formula administered through the tube. However, the mother was afraid of disturbing the tube in any way, and consequently she did not play with or cuddle the child for the entire time. At the end of fifteen months the child was extremely depressed, showed evidence of motor retardation, and her physical development was that of an eight-month-old. At this point she was hospitalized for observation. During this observation period she received a great deal of attention and handling from the doctors and nurses trying to find her problem. She responded dramatically to this added stimulation by gaining her normal weight and making up for lost growth. Her emotional state also improved strikingly.

> Moreover, these changes were demonstrably unrelated to any change in food intake. During her stay in the hospital she received the same standard nutrient dosage she had received at home. It appears to have been the enrichment of her environment, not of her diet, that was responsible for the normalization of her growth.[25]

In this case, the child fared better in the hospital than at home, just the reverse of the cases studied by Bakwin and Ribble. This would seem to further underscore the point that it is adequate *tactile stimulation* — whatever its source — that is necessary for healthy physical development, and neither some more mysterious "bonding" with the mother *per se*, nor unique genetic characteristics inherited from her.

Sensory Malnutrition

These symptoms of sensory deprivation — retarded bone growth, failure to gain weight, poor muscular coordination, immunological weakness, general apathy — are strikingly similar to those of malnutrition, and yet improper feeding must be ruled out as the causative factor in all these instances. Every one of these children's bellies was being well fed. But a bellyful is not enough. The similarities between malnutrition and sensory deprivation are probably not coincidental at all; it is most likely the case that we suffer from the lack of one sort of nutrition to the same degree that we suffer from the lack of the other.

Indeed, some researchers, such as R.H. Barnes and David Levitsky of the Cornell University Graduate School of Nutrition, regard malnutrition and sensory deprivation as being very closely linked in their causes and effects, and have proposed that

> certain effects of malnutrition may actually be secondary effects of environmental impoverishment. That is, since a prominent effect of malnutrition is to make the person or animal apathetic and unresponsive to the environment, the individual then suffers from lack of stimulation, and this may be the direct cause of some of the symptoms usually associated with malnutrition. Current research suggests that some of the effects of malnutrition

may be offset by programs of environmental stimulation or increased by environmental impoverishment.[26]

Tactile stimulation, physical contact with the environment, appears to be a food that is as vital for development as is any protein.

A Mother's Touch

A large number of observations and experiments have been inspired by these revelations about the cause of infant merasmus, and as evidence continually accumulates, the role of touch in the physical and mental health of animals of all kinds broadens and deepens in its significance. It was formerly assumed, for instance, that it was some sort of hormonal secretion in pregnant animals which initiated the physiological and behavioral changes we refer to as the "mothering instincts," so necessary in order for the mother to successfully gestate, deliver, nourish, and instruct her young.

However, if pregnant rats are simply fitted with a wide collar, so that they cannot lick themselves, dramatic things happen to the quality of their motherhood: Their mammary glands attain only 50% of the growth of uncollared pregnant rats; they amass the material for nests, but they scatter them about loosely and do not construct useful nurseries; they are not attentive to the pups as they are born, and do not lick them, do not fondle them, do not clean up the afterbirth; they do not nurse their pups to any extent, and even seem to avoid contact with them.[27]

On the other hand, stereotyped maternal behavior can be induced in virgin female rats by merely putting them into a small cage filled with pups, where they share many close physical contacts. It is evident that adequate tactile experience is at least as important as are hormones for the conditioning of the mothering instincts, and in fact it may well be specific quantities and qualities of tactile experience which trigger the release of the hormones themselves.[28]

This relationship between self-licking and successful motherhood has been confirmed in other animals, and it seems to be closely akin to the almost universal practice of the mother thoroughly licking her young. It has been tacitly assumed that this licking demonstrates an admirable, an almost human, fastidiousness among animal mothers; but for the pups, kittens, kids, and other whelps, something much more significant than a little tidying up is going on. By virtue of the skin's close association with the central nervous system, this cutaneous stimulation is literally awakening organic functions in the newborns' internal organs, and without it their chances of survival are markedly diminished.

Animal breeders, farmers, veterinarians, and zoo-keepers all concur that the new animal must be licked if it is to live. In particular, the perineal region between the genitals and the anus, and the lower abdomen need a good deal of cutaneous stimulation, or else the young animal is very likely to die of a functional failure of either the genitourinary system or the gastrointestinal system: It does not learn to urinate or defecate, and it perishes.

This association of tactile stimulation and proper organ function was force-fully underscored by some accidental findings in a series of experiments having nothing to do with licking. Professor James A. Reyniers and his colleagues of the Lobund Laboratories of Bacteriology of the University of Notre Dame were interested in raising animals in a germ-free environment. During the early days of their experiments, they lost virtually all their animals because they died of early genitourinary and gastrointestinal failure. Only after a former zoo-keeper advised the investigators of her experience with newborn litters, and they began to systematically stroke the animals after each feeding, did their subjects begin to survive.

> Rats, mice, rabbits, and those mammals depending upon the mother for sustenance in the early days of life apparently have to be taught to defecate and urinate. In the early period of this work we did not know this and conse-quently lost all our animals. The unstimulated young die of an occlusion of the ureter and a distended bladder. Although we had for years seen moth-ers licking their young about the genitals I thought that this was a matter largely of cleanliness. On closer observation, however, it appeared that during such stimulation the young defecated and urinated. Consequently, about twelve years ago, we started to stroke the genitals of the young after each hourly feeding with a wisp of cotton and we were able to elicit elimina-tion. From this point on we have had no trouble with this problem.[29].

It is interesting that it is a civilized concern for cleanliness that prompts us to overlook this far more profound benefit of licking. If it weren't for our own squeamishness, we would probably recall more readily that no part of the body is as germ-laden as the mouth, and that the tongue can scarcely be regarded as a proper cleansing agent since its touch innoculates the skin with a thousand bacterial horrors. The stimulation of internal functions is far more important to the infant's health than the removal of a little surface dirt, because it empowers the organism itself to eliminate toxins and wastes on its own. Let not this fact be an invitation to allow our children to wallow in filth, but let it protect us from the fastidious delusion that if baby is clean and fed, then all is well.

Touch and Maturation

The importance of cutaneous stimulation is not confined to the critical period of early infancy. Animals who receive this early contact develop superior func-tions and immunological resistance that last them their whole lives, and these effects are strengthened if they are fondled throughout their maturation.

In the early 1920's Frederick S. Hammett, of the Wistar Institute of Anatomy in Philadelphia was interested in discovering the effects of total removal of the thyroid and parathyroid glands in a genetically homogeneous stock of rats. It was a foregone conclusion that these rats would die from the removals; presum-ably the specific causes of death would shed light upon the functions of the two glands.

Making inquiries, Hammett discovered that his rats had been raised in two different groups — one in a rich environment where they were customarily petted and gentled, and one in much more isolated circumstances where their human contact was incidental to routine feeding and cage-cleaning. The handled group were relaxed and yielding when picked up, and not at all easily frightened. The latter group, by contrast, were highly excitable, tense, and tended to bite their handlers in fear and rage.

In subsequent experiments, Hammett kept close track of the subjects from each colony. After the thyroid and parathyroid glands had been removed from the excitable rats, 79% of them died within forty-eight hours, while in the same time period after identical surgery only 13% of the gentled rats died.

> Hammett concluded that the stability of the nervous system induced in rats by gentling and petting produces in them a marked resistance to the loss of parathyroid secretion.[30]

Similarly, another group of investigators discovered in another series of experiments that gentled rats survive an injection of leukemia cells for a considerably longer time than do ones that have been raised in isolation.[31]

It would seem that these wide disparities in organ function, in resistance to trauma and disease, and in dominant personality traits suggest that the actual physical constitution of the gentled animals had been benefited in some material way by tactile stimulation. In the late 1950's Seymour Levine began looking for concrete physiological changes that could be attributed to stimulation or deprivation during infancy. His findings were astonishing.

In almost all respects, Levine's fondled rats developed more rapidly than his isolated ones. The fondled rats opened their eyes earlier, and achieved motor coordination earlier. Their body hair grew faster. They were heavier at weaning, and continued to gain weight at a faster rate even after their experimental stimulation ceased at three weeks of age. They had a stronger resistance to disease. These superior gains were not related to food consumption — which was identical for both groups — but to a brisker, more efficient metabolism, a "better utilization of the food consumed, and probably a higher output of somatotrophic (growth) hormone from the pituitary."[32]

Levine concluded that all of these strengths were the result of an accelerated maturation of the central nervous system in the fondled animals. In particular, he discovered that the cholesterol content of their brain tissues was distinctly higher.

> Since the cholesterol content of the brain is related to the brain's white matter, this is evidence that in these animals the maturation of structure parallels the maturation of function.[33]

"White matter" consists of the fatty myeline sheaths which surround some nerve axons, insulating their electrical activities, and making them more efficient transmitters; it is opposed to "grey matter," cell bodies and axons with no myelin sheaths, and is therefore regarded as a sign of more highly evolved neural development.

These increases in brain cholesterol and pituitary activity were clues that were rich in their implications, and in the late 1960's a research team at the University of California at Berkeley began to look for specific differences in the neural structures of gentled and ungentled rats. They found that greater tactile stimulation resulted in the following differences: These animals' brains were heavier, and in particular they had heavier and thicker cerebral cortexes. This heaviness was not due only to the presence of more cholesterol — that is, more myeline sheaths — but also to the fact that actual neural cell bodies and nuclei were larger. Associated with these larger cells were greater quantities of cholinesterase and acetylcholinesterase, two enzymes that support the chemical activities of nerve cells, and also a higher ratio of RNA to DNA within the cells. Increased amounts of these specific compounds indicates higher metabolic activity. Measurements of the synaptic junctions connecting nerve cells revealed that these junctions were 50% larger in cross-section in the gentled rats than in the isolated ones. The gentled rats' adrenal glands were also markedly heavier, evidence that the pituitary-adrenal axis — the most important monitor of the body's hormonal secretions — was indeed more active.[34]

Many other studies have confirmed and added to these findings. Laboratory animals who are given rich tactile experience in their infancy grow faster, have heavier brains, more highly developed myelin sheaths, bigger nerve cells, more advanced skeletal muscular growth, better coordination, better immunological resistance, more developed pituitary/adrenal activity, earlier puberties, and more active sex lives than their isolated genetic counterparts.

Associated with these physiological advantages are a host of emotional and behavioral responses which indicate a stronger and much more successfully adapted organism. The gentled rats are much calmer and less excitable, yet they tend to be more dominant in social and sexual situations. They are more lively, more curious, more active problem solvers. They are more willing to explore new environments (ungentled animals usually withdraw fearfully from novel situations), and advance more quickly in all forms of conditioned learning exercises.[35]

Moreover, these felicitous changes are not to be observed only in infancy and early maturation; an enriched environment will produce exactly the same increases in brain and adrenal weights and the same behavioral changes in adult animals as well, even though the adults require a longer period of stimulation to show the maximum effect.[36]

A Basic Need

How is it possible that mere tactile stimulation, in the form of fondling, cuddling, licking, and stroking, produces such dramatic changes in the neural anatomy and physiology, the organ functions, and the behavioral patterns in these animals?

The answer is quite simple: Tactile stimulation appears to be a fundamentally necessary experience for the healthy behavioral development of the individual.... The raw sensation of touch as stimulus is vitally necessary for the physical survival of the organism. In that sense it may be postulated that the need for tactile stimulation must be added to the repertoire of basic needs in all vertebrates, if not in all invertebrates as well. Basic needs, defined as tensions which must be satisfied if the organism is to survive, are the needs for oxygen, liquid, food, rest, activity, sleep, bowel and bladder elimination, escape from danger, and the avoidance of pain. It should be noted that sex is not a basic need since the survival of the organism is not dependent upon its satisfaction. Only a certain number of organisms need to satisfy sexual tensions if the species is to survive. However that may be, the evidence points unequivocally to the fact that no organism can survive very long without externally originating cutaneous stimulation.[37]

Touch is food. Vital food.

Harlow's Monkeys

In fact for the infant, this food is as vital as mother's milk. Just as the stimulatory significance of licking was obscured by the assumption that the animal mother is motivated by a desire for cleanliness, so the infant mammal's dependence upon its mother's breasts for nutrition has obscured some of the crucial nurturing qualities derived from nursing.

It was long held to be an obvious fact that the infant's affectionate attachment to its mother was generated by the satisfaction of feeding. But studies of rhesus monkeys in the late 1950's by Harry Harlow, head of the Primate Laboratory at the University of Wisconsin, demonstrate clearly that the close body contact involved in clinging, nuzzling, and nursing is the main factor in forming the affectionate bond, and is as important to the infant's physical and mental development as is the milk itself.

Harlow provided baby monkeys with two surrogate mothers — one made of cold wire, and the other made of soft terry cloth. For some monkeys, only the wire mother had a nipple and milk, while for others the milk came only from the terry cloth one. Regardless of which mother fed them, all of the monkeys displayed a dramatic preference for spending their time clinging to the soft, cloth one — about eighteen hours a day compared to an hour or less spent with the wire mother. Both sets of infants drank the same amount of milk, but this nutrition coming from the wire surrogate produced no affectionate bonding whatsoever.

Harlow found that this preference became even more pronounced and psychologically significant when the monkeys were subjected to stress. A wind-up noise-maker was put into a cage containing a monkey and its two "mothers." Again, despite which mother had been dispensing milk, the frightened infants overwhelmingly chose the soft one for comfort and protection. Even if they first blindly rushed to the wire one, they quickly abandoned her and clung desperately to the cloth one. Then, when their fears were assuaged by the bodily contact, they would turn around and look at the noise-maker with no evidence of fear.

Very similar results were obtained by placing the monkeys in a strange environment, such as a room much larger than their cages and provided with a variety of unfamiliar objects: If the cloth mother was present, the infant would rush wildly to her, clinging tightly and rubbing against her; its initial fear then subsided, and it began to explore the room, returning periodically to the cloth mother. But if the cloth mother was absent, "the infants would rush across the test room and throw themselves face down on the floor, clutching their heads and bodies and screaming their distress."[38]

These behaviors are, of course, very like those of human infants who find themselves in new circumstances either with or without their mothers. But when only the wire mother was provided in the strange room, starkly contrasting behavior resulted. The bare wire mother provided no more real assurance in this "open field" test than no mother at all. Control tests on monkeys that from birth had known only the wire mother revealed that even these infants showed no affection for her and obtained no comfort from her presence. Indeed, this group of animals exhibited the highest emotionality scores of all. Typically they would run to some wall or corner of the room, clasp their heads and bodies and rock themselves convulsively back and forth. Such activities closely resemble the autistic behavior seen frequently among neglected children in and out of institutions.[39]

Certainly no one would deny that being clean and well fed are desirable conditions for babies, but studies such as this one, along with those previously cited, make it clear that when these factors are stressed and adequate handling and comforting are short-changed, the infant suffers consequences just as grim as those of malnutrition or disease. Psychologically at least, cleanliness and food are clearly secondary to the infant, since supplying these things alone creates no affectionate bonding and offers no security in times of stress. "These data make it obvious," Harlow concluded,

> that contact comfort is a variable of overwhelming importance in the development of affectional responses, whereas lactation is a variable of negligible importance. With age and opportunity to learn, subjects with the lactating wire mother showed decreasing responsiveness to her and increasing responsiveness to the non-lactating cloth mother. . . . We were not surprised to discover that contact comfort was an important basic affectional or love variable, but we did not expect it to overshadow so completely the variable of nursing; indeed, the disparity is so great as to suggest that the primary function of nursing as an affectional variable is that of insuring frequent and intimate body contact of the infant with the mother. Certainly man cannot live by milk alone.[40]

Freudian Confusions

These findings have a frightening significance for our culture when we reflect upon the routine treatment of newborn infants in almost all of our hospitals. They are thoroughly cleaned, provided with a bottle, plopped into an isolated crib, and given just enough physical attention to avoid the most obvious symptoms of deprivation, with little to listen to their first days but the cries of their neighbors in the nursery. "Nursery" as the name for baby's room is revealing in itself on this head; where, we might ask, is the "touchery"?

The sort of contact that has been shown to stabilize the young rhesus monkey is often not even offered to the young human, a lamentable situation that is even further darkened by the fact that no animal is as dependent as the human being upon early experiences for the formation of personality traits. Nor is this shaky beginning compensated for by popular child-rearing theories which stress the importance of not "indulging" children when they clamor for attention.

Freud at least understood the biologically compulsive nature of these needs. Unfortunately, one of the principle features of his theory — the specifically erotic Oedipus complex — led him to term this hunger for contact "infant sexuality," a heavily loaded term which has ushered as much confusion as clarity into our understanding of basic human needs and motives.

Montagu suggests that "infant tactuality" would be a far more accurate phrase to describe the child's endless delight in exploring and experiencing its body, an activity that we can call "sexual"[41] only by broadening the term to the degree that its primary meaning dissolves. Self exploration and the response to external touch do not have nearly as much to do with preparing the child for adult genital expression as they do with organizing the nervous system as a whole, preparing it for all of its potential contacts — not merely genital ones.

There is no question of the enormous significance of sexuality as such to human behavior; but the failure to differentiate between specifically sexual development — along with the erotic qualities of the skin — and the more generalized neurological maturation induced by physical contact has served to keep the whole subject of touching under a threatening shadow in our culture. Such confusion confounds to a damaging degree the pleasures and implication of genital activity with pleasures and implications which are equally sensory but utterly different in their physiological and psychological significance to the individual. The lack of important distinctions in these matters has by and large reduced touching to foreplay for coitus, rather than recognizing it as preparation for living.

It is this confused point of view which associates every sort of contact with sexual innuendo, which regards touching as dangerously intimate, which views bodywork as a modified form of prostitution. The adult's sexuality may or may not be painfully crippled by these blurred lines; the sense of well-being that should come from all the rest of his contacts is almost certain to be.

This confusion of tactuality and sexuality prompted Freud to maintain militantly one of the weakest features of psychoanalysis as a therapy — the strict taboo against touching the patient. It was his fear that this would, at the very least, enormously complicate the problems of the patient's process of transference (the attaching of libidinous desires to the analyst). In addition, he felt that the analyst himself must guard stiffly against his own ambiguous sexual motives in dealing with the patient, and not present *himself* with the temptations that were thought to be implicit in physical contact.

Be these fears as they may (and it is certainly possible for therapeutic disasters of a sexual nature to be perpetrated), Freud unquestionably cut himself and his patients off from much that has proven to be helpful. Wilhelm Reich's separation from Freud and his pursuit of techniques to liberate the patient from what he called "body armor" was a recognition of this limitation by one of Freud's own pupils. It simply is true that by means of various kinds of touch therapies many breakthroughs have been made with mental patients who have proven to be completely refractory to classical analysis. Alexander Lowen has pointed out that the central crisis in all forms of schizophrenia is the loss of the identity of the *ego* with the *body.*

> The feeling of identity arises from a feeling of contact with the body. To know who one is, the person must be aware of what he feels. This is precisely what is wanting in the schizophrenic. There is a complete loss of body contact to such an extent that, broadly speaking, the schizophrenic doesn't know who he is.[42]

And the withdrawal of the schizophrenic from contact may, just as in the cases of malnutrition previously mentioned [see pp. 45-46], create a whole range of secondary developments which we confuse with his primary disturbance. Such experiences have been recorded by many therapists, and there are those who insist that the analyst's strict adherence to Freud's taboo can only

> confirm the patient's own convictions that words are good and touch is always erotic or destructive and bad. Both therapist and client need to learn tolerance for their own excitement and realize that fantasies need not lead to action. Thus the therapist's nonerotic touch may break through the client's defenses and help him separate and tolerate the two kinds of experiences.[43]

Touch and Disposition

Violence and rage have demonstrable relationships to tactile deprivation as well. There is scarcely a study that has been done on violent crime which does not solidly associate it with harsh or isolated childhoods. Dr. James Prescott, a developmental neurophysiologist at the National Institute of Child Health and Human Development, writes that "recent research supports the view that the deprivation of physical pleasure is a major ingredient in the expression of physical violence." His experiments have convinced Dr. Prescott that the presence of physical pleasure categorically inhibits violence, that rage is not possible in the presence of pleasure. This has been demonstrated by instantly calming down raging experimental animals with electrical stimulations of the pleasure centers in their brains. He states further that

> I believe that the deprivation of body touch, contact, and movement are the basic causes of a number of emotional disturbances which include depressive and autistic behaviors, hyperactivity, sexual aberrations, drug abuse, violence, and aggression.[44]

Certainly there are other contributing causes. But — no less certainly — lack of physical contact is a major one.

Cultural Norms

Since different degrees of tactile experience in early life play such a large role in neural maturation and the development of differences in personality traits and behavioral patterns, it would seem reasonable to assume that similarities in tactile experience among a given people go a long way towards conditioning their social and cultural norms. Observations of various national and tribal temperaments indicate that this is very much the case.

Widely recognized national contrasts in upbringing and social character leap immediately to mind, such as the spartan education of German youths in the nineteenth and twentieth centuries, or the mild and harmonious existence of the Tahitians and other South Sea Islanders prior to European colonization, the tightly controlled reserve spawned by the English public schools, or the jovial, rowdy fellowship common to those English lower classes who do not attend them.

Margaret Mead made a study of two New Guinea tribes that throws some light on this influence of touch in such markedly different cultural circumstances. The members of the Arapesh tribe take great delight in children, and fondle them regularly; an infant is rarely out of someone's arms. The mother carries it in a sling around her body all day long, regardless of her activities, and if she is absent for any length of time she is careful to devote enough attention to the child upon her return to make up for the lost hours. Nursing continues three to four years, and mealtime is a happy affair to both mother and baby, with nuzzling, tickling, rocking, sucking, playful pats, and laughter being usual parts of the ritual.

> The whole matter of nourishment is made into an occasion of high affectivity and becomes a means by which the child develops and maintains a sensitivity to caresses in every part of its body.[45]

Nor is the mother the only source of affection; virtually every adult treats every child in the same fashion.

The result is an easy, gentle, receptive unaggressive adult personality, and a society in which competitive or aggressive games are unknown, and in which warfare, in the sense of organized expeditions to plunder, conquer, kill, or attain glory, is absent.[46]

Living to the south of the Arapesh are the Mundugamors. To them, children are not a joy, and often before a child is born there is much discussion about whether or not to let it survive. If it is allowed to live, it is promptly placed in a hard, rough basket carried like a pack on the mother's back, or hung from the wall while she is working. Infants are suckled when their crying cannot be stopped by other means. The mother stands to nurse, and indulges in no fondling; as soon as suckling stops, the infant is put back in the basket on the wall. Thus the infant has to fight for its food, clamping the breast aggressively and frequently choking, which infuriates the mother. The nursing experience is "one of anger and frustration, struggle and hostility, rather than one of affection, reassurances, and contentment."[47]

It is time for weaning as soon as the child can walk, and this is done with abrupt harshness, as often as not by repeatedly slapping the child when it approaches the breast. The Mundugamors are "an aggressive, hostile people who live among themselves in a state of mutual distrust and uncomfortableness."[48] They are cannibals.

Conclusions

It is not true that every child, in any culture, who suffers tactile impoverishment or abuse grows up to be a catatonic or a violent misfit. Some are clever enough to devise alternative sources of stimulation, or lucky enough to stumble onto them. For many others, the tactile deprivations they suffer are not crushingly extreme, and so they are able to strike a compromise with the world based upon the sensory development they *do* receive.

It is simply impossible to estimate the numbers of these partially deprived individuals, or to calculate exactly the degree to which their physiologies, their attitudes, and their behaviors are conditioned by deficient tactile development. And yet, given what we know about the relative paucity of touch in our child-rearing practices and in our adult culture, and given the mass of evidence which points to the importance of touch in the chemical, structural, and psychological characteristics of the individual, it is equally impossible not to suspect that the number is probably vast, and that the amounts of more or less subtle retardation and compensation are probably considerable.

The overwhelming majority of complaints in the doctor's office are not congenital defects, traumas, or diseases. They are headaches, various somatic pains, digestive or eliminative dysfunctions, heart conditions, apathies, depressions, obesities, losses of appetite, emotional tensions, physical stresses—more often than not with no discernible specific causes. Unable to find microbial culprits for so many unhealthy conditions, researchers have begun increasingly to examine our life-styles for excesses and deficiencies that might trigger them.

Dietary deficiencies are unquestionably very significant factors. And bodywork — the form of "cutaneous stimulation" we are the most concerned with in this book — is *good food*. Or, more precisely, bodywork can have a marked positive effect upon the efficiency of the metabolism of whatever we eat. It does not matter how much of what we include in our intake if we do not have the internal capacity to extract and utilize its value; touch has been demonstrated to be helpful, even necessary, to the development of this metabolic efficiency.

Stress is presently receiving an enormous amount of attention as a powerfully contributing factor to these common complaints with vague etiologies. Soothing touch, whether it be applied to a ruffled cat, a crying infant, or a frightened child, has a universally recognized power to ameliorate the signs of distress. How can it be that we overlook its usefulness on the jangled adult as well? What is it that leads us to assume that the stressed child merely needs "comforting," while the stressed adult needs "medicine"? It has been the thrust of this chapter to suggest that there is nothing "mere" about tactile comforting, and that there is no gulf between "medicine" and simple contact.

Certainly we cannot rub away diphtheria, or leukemia, or botulism, or Hong Kong flu, or hundreds of other conditions, diseases, and mutilations for which allopathic medicine has made itself our lifesaver. Let us continue to increase the scientific training of the physician, the tools of the emergency room, the apparata of the intensive care unit, the range of effective pharmaceuticals. But if by "good health" we mean something more than the absence of a life-threatening emergency, if we mean rather the ongoing development of the individual, his awareness of his body, the optimum maturation of the nervous system, and the resilience of the tissues to resist toxins and repair damages, then let us not forget bodywork either.

There is a great deal of stress and discomfort and discord in our world that clearly do not yield to traditional medical procedures, but which require rather a coaxing into existence of new habits, new attitudes, new ways of relating to self and to others.

> Indeed, inside every failed individual there is a potentially warm, loving creature struggling to get out. The trick is so to interact with the individual who has been tactually failed as to release that potentiality for something resembling the kind of humanizing experiences he should have enjoyed in infancy and childhood.[49]

It is difficult to imagine a more direct way to rectify these failures than by supplying the touch that was missing in the first place.

It is the burden of the bodyworker to discover and to develop within himself or herself that quality of touch which will provide the emotional comforting, the tactile information, and the integrating experience so acutely needed by the distressed individual. This is not an easy task, but the developing therapist may take comfort in the fact that the surface he most directly stimulates, the human skin, has a marvelous intelligence of its own, and possesses the means of carrying his efforts to the very core of the person being touched.

3

Connective Tissue

When we command movement of an arm or a leg we establish all the conditions to effect the movement of several long levers in organized action. The wisdom lies not in man's "command," but in the various systems cooperating with the neuromuscular mechanisms to establish right conditions.

— Mabel Ellsworth Todd, *The Thinking Body*

Water Bags

"What is a human being?"

"A human being is a container invented by water so that it can walk around."

There is a good deal of truth couched in this little joke. According to one current theory, the first colloidal droplets and primitive living cells were probably formed in sea water, in which the elements and compounds necessary for the formation of organic compounds, nucleic acids, and finally amino acids — the primary proteins of living cells — were dissolved.

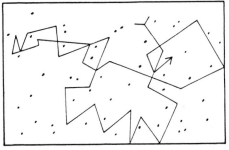

3-1: The random path of a molecule in a water solution. Single cell organisms living in the sea depend upon this diffusion of molecules in liquid, and the currents within the liquid itself, to deliver their nutrients to them.

We have already noted that in order for these complex compounds and their interactions to be sustained in a form which we can call "life," the development of a membrane was necessary so that they would not be scattered by every passing current or eddy. Then, as aggregates of these primitive cells became larger and larger, and as individual cells within the growing aggregates began to differentiate and establish more specialized functions and relationships with neighboring cells — that is, as groups of cells began to merge into primitive organisms — some kind of "metamembrane" became equally necessary in order to glue the whole mass together, again to resist the scattering forces of currents in the surrounding liquid medium.

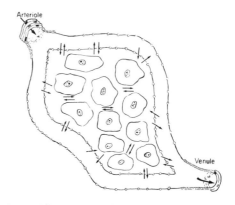

3-2: As long as an aggregate of cells is not too large, simple diffusion can deliver nutrients from the surrounding liquid and carry away wastes.

But the basic chemical building blocks that were dissolved in this supporting medium of water were still indispensable to the continued life of each and every cell, so when an aggregate became too many layers of cells thick for the simple seepage and diffusion of the water and its dissolved nutrients to reach the innermost members, the primitive organism's gluing agent had to help the mass of cells shape itself into invaginations, cavities, and circulatory tubes which kept the flow of sea water active and fresh in the interior, available to even the deepest cells. Complex digestive and circulatory systems are but further sophistications of these invaginations, cavities, and tubes. And before aquatic organisms could achieve those decisive moments when they took to the air and the ground, a means of permanently

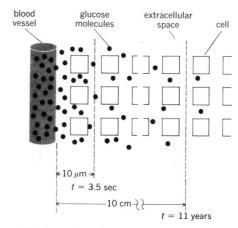

3-3: The larger the cell aggregate, the longer diffusion into the interior and out again takes. As the distance to be traveled by a randomly moving molecule becomes greater, the time required multiplies rapidly. It takes about 3.5 seconds to diffuse a distance of 10μm (the diameter of a typical cell). But it would take 11 years for simple diffusion to carry a molecule 10 cm (approximately 4.5 inches).

containing within the organism an adequate supply of the mothering and supporting medium of water had to be developed, so that each and every cell could continue to be bathed in the life-forming solutions of the sea.

The fluid bathing our own cells throughout every nook and cranny of our bodies can still be resolved into the basic proportions of elements, salts, and carbon compounds — the organic building blocks — that are found in the ocean. So we did not really leave the sea behind at all; we were, and are, obliged to carry part of it with us. We are, in fact, mostly water. As terrestrial organisms we may live on solid ground and breathe air, but as a collection of individual cells we still live within the same liquid medium from which we first emerged.

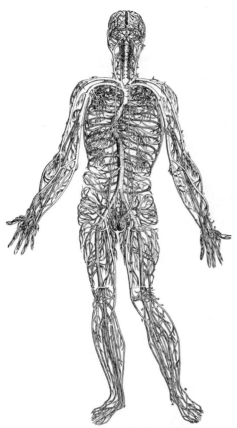

3-4: Large organisms must develop delivery systems so that fluid-borne molecules can reach their interiors and continually bathe their cells. The life of every cell in the human body depends upon the proper function of these circulatory conduits in its vicinity. Wherever this circulation is chronically blocked, areas of our internal sea stagnate and parts of us suffer or die. Connective tissue is one of the principal shapers and supporters of this circulatory network.

Every organ and system in the body supports in some way the containment, the renewal, and the circulation of this internal sea. The chemical, mechanical, and hydraulic problems involved in this packaging of sea water underlie the logic of all our physical structures and functions.

It is to this "metamembrane," this binding, containing, and shape-giving connective tissue of the body that we now turn our attention. We have had a brief glimpse of it as the fibrous mesh which gives the skin both its pliability and its tensile strength, and which fixes it firmly to the structures underneath. Now we shall follow this supportive network deeper, to see how it divides the body's different tissues and organs into separate compartments, how it provides the binding agent which unifies them into complex systems and finally into a single organism, and how it transforms the pull of gravity and muscular contractions into the controlled and harmonious movement of all our body parts.

The look and feel of connective tissue is familiar to any cook: It is the whitish, glossy sac which surrounds each individual muscle in a carcass, the smooth, slick covering over raw bones, the membranes that encase the internal organs and line the body cavities, the tough tendons, ligaments, and bursae which cook up into gristle.

The protein *collagen,* the thin white fibers which make up these familiar sheets and cords, is perhaps the most abundant protein in the animal kingdom.[1] It constitutes as much as forty per cent of all the proteins in the body.[2] Collagen is to animals what cellulose is to plants — the tough lattice in which all other kinds of tissue are developed and contained, the walls of the compartments which fill with fluid to give the plant its juiciness and its upright turgor.

In animal evolution, collagen appeared at a very early stage, during the development of the coelenterates (an ancient phylum including jelly fish and sea anemones) and the sponges.[3] In the development of the human embryo it appears

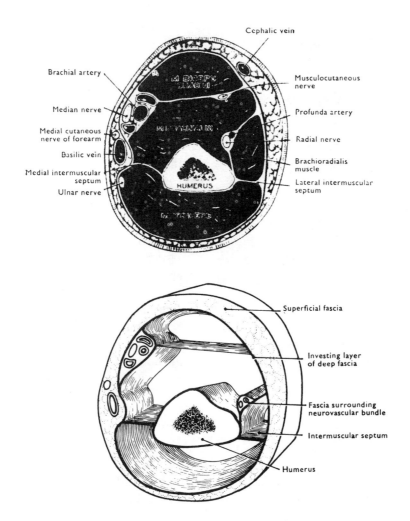

Cephalic vein

Brachial artery

Musculocutaneous nerve

Median nerve

Profunda artery

Medial cutaneous nerve of forearm

Radial nerve

Basilic vein

Brachioradialis muscle

Medial intermuscular septum

Lateral intermuscular septum

Ulnar nerve

M BICEPS BRACHII

M BRACHIALIS

HUMERUS

M TRICEPS

Superficial fascia

Investing layer of deep fascia

Fascia surrounding neurovascular bundle

Intermuscular septum

Humerus

3-5: A cross section of the upper arm, hollowed out in the figure at the right to reveal the connective tissue compartments.

as early as the twelfth day following fertilization.

It is derived from the mesenchyme — a subdivision of the primitive mesoderm layer of embryonic cells.[4] This same germinal layer subsequently produces tendon, ligament, cartilage, bone, marrow, muscle, blood, lymph, blood and lymph vessels, and the surface linings of the body cavities, the joint capsules, the kidneys, ureters, gonads, genital ducts, and the adrenal cortex.[5] Developmentally, all of these structures are as closely related to one another as are the skin and the nervous system.

There is no tissue in the body that is as ubiquitous as connective tissue, and as it migrates and develops in various forms in various locations, its "connective" qualities cannot be overstated. It binds specific cells into tissues, tissues into organs, organs into systems, cements muscles to bones, ties bones into joints,

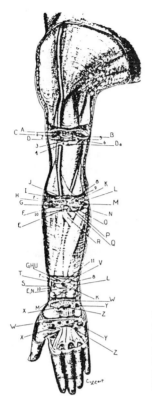

wraps every nerve and every vessel, laces all internal structures firmly into place, and envelops the body as a whole. In all of these linings, wrappings, cables, and moorings it is a continuous substance, and every single part of the body is connected to every other part by virtue of its network; every part of us is in its embrace. George E. Snyder, Professor of Anatomy at Kirksville College of Osteopathy and Surgery, remarks that "The connective tissues not only bind the various parts of the body, but, in a broader sense, connect the numerous branches of medicine."[6]

3-6: Connective tissue wraps virtually every other tissue formation in our bodies. If everything else were removed, the network of empty connective tissue compartments would preserve our physical forms in detail. All the arrows are pointing to different connective tissue structures in the arm—tendons, ligaments, nerve and blood vessel sheaths, sheets of fascia wrapping individual muscles, the coating of all the bones, and so on.

The Main Ingredients

Although this connective tissue is continuous throughout the body, it takes on many different shapes and properties, and has local qualities that differ widely from place to place. It can be quite diffuse and watery, or it can form a tough flexible meshwork. In the tendons and ligaments its tensile strength is superior to steel wire; in the cornea of the eye it is as transparent as glass; it accounts for the toughness of leather, the tenacity of glue, the viscosity of gelatin. Invest it to various degrees with hyaline, a nylon-like substance exuded by chondroblasts, and it becomes the various grades of cartilage; invest it with mineral salts, and it becomes bone.

All these forms of connective tissue are in a very active state of flux in the growing individual, with hardenings and dissolvings continually producing changes in shapes, sizes, and consistencies of all the structural members which it composes. This process slows down after we reach our full growth, but it continues uninterrupted throughout adult life. Such remarkable versatility and plasticity are possible because this complex tissue is made up of a number of fluid, cellular, fibrous, and crystalline constituents which can alter radically their relative concentrations at different times and in different places to produce very different kinds of building materials.

Ground Substance

A transparent fluid *ground substance* is found to one degree or another in all of the body's connective tissues, and it may be thought of as the basis for the production of all their other forms. This viscous liquid, much like raw egg whites in appearance and consistency, surrounds all the cells in the body, and is a part of the internal ocean.

However, it is quite distinct in its origins and its functions from the other intercellular fluids which seep from capillaries or from other tissue cells. These latter fluids are mostly plasma, nutrients, and hormones diffusing from the blood vessels to the cells, and metabolic wastes from cell activity diffusing back to the capillaries and the lymph vessels. The ground substance of connective tissue is different from all of these fluids; it is the liquid medium through which these other fluid exchanges take place. It is the retort in which all extracellular activities occur. It does not come from the capillaries or from other tissues, but is produced by cells which are among the earliest specialized cells to emerge from the embryonic mesoderm, the *fibroblasts.*[9]

Technically, we must speak of ground substances in the plural, because even the make-up of this basic fluid varies from location to location. Essentially all these varieties consist of a carbohydrate combined with a protein chain —a mucopolysaccharide—but the chemical variations are complex, and they account for many different properties found in the ground substance in different locations. Some of the mucopolysaccharides are large and dense, creating a ground substance with a more viscid gel-state, while others are smaller molecular units which create a more fluid sol-state. And these compounds themselves are not inert, but are constantly changing their arrangements in healthy, active tissue.

These fluid ground substances are the immediate environment of every cell in the body, and they undoubtedly have a wide range of effects upon every cellular membrane which they contact. Their chemical activities are legion, and so they also directly influence the passage of all sorts of gases, nutrients, wastes, hormones, antibodies, and white blood cells between the capillaries and the tissues they irrigate.

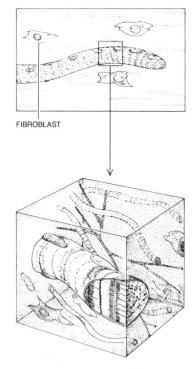

FIBROBLAST

3-7: A muscle cell surrounded by ground substance. This fluid constituent of connective tissue flows around every cell and every connective tissue structure in the body. The smaller banded strands next to the muscle cell in the magnified section are collagen fibrils.

We can regard these ground substances as both facilitators and as barriers between the blood and all the cellular surfaces, chemical filters which regulate many interactions. Damage to their elements through malnutrition, trauma, fatigue, stress, and the like results in the impairment of these supporting functions by the depletion of fluid volume, by the build-up of foreign particles and toxins, or by altering the chemical properties of the mucopolysaccharides. Such disruption strikes at the very basis of healthy metabolic activity. Healthy ground substance works constantly to help maintain a supportive chemical and physical equilibrium between all the body's tissues.

The Collagen Fibers

Although this ground substance forms the fluid medium for the ongoing activity of connective tissue, the most abundant constituents of this intercellular support system are the long white fibers of collagen. As we have noted, these fibers account for more than one third of all animal protein. They are the chief fibrous content of skin, ligament, tendon, cartilage, bone, vessels, and all organs, and their tough strands give to these tissues their shape, their tensile strength, their resiliency, and their structural integrity. These fibers can be arranged in any number of ways to produced a wide variety of properties; they may be criss-crossed randomly in

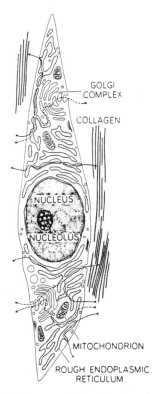

GOLGI COMPLEX

COLLAGEN

NUCLEUS

NUCLEOLUS

MITOCHONDRION

ROUGH ENDOPLASMIC RETICULUM

3-8: A fibroblast exuding molecules of collagen (called tropocollagen) which are forming collagen fibrils in the surrounding ground substance.

3-9: Criss-crossed collagen fibrils that have been carefully pulled away from human skin. The ground substance has been removed during preparation for this electron micrograph.

blocks or sheets, carefully stacked in alternating layers like plywood, spun into loose areolar webs, or packed into dense parallel formations.

These fibers are not living tissue, but are made up of protein chains that are produced by the same living cells which exude the fluid ground substance, the fibroblasts. In the embryo, these fibroblasts develop in the mesenchyme, from which they are dispatched throughout the growing organism. When they settle in a particular area, they begin manufacturing and secreting collagen chains, which then respond to local chemical properties and specific stresses in the area to form the appropriate kind of fibers and arrangements — sacs around the muscles, ligaments across the joints, the walls of blood vessels, the cornea of the eye, and so on.

Of all the cells in the body, these fibroblasts are the only ones which retain throughout our lives the unique property of being able to migrate to any point in the body, adjust their internal chemistry in response to local conditions, and begin manufacturing specific forms of structural tissue that are appropriate to that area.[10] No other cell exhibits this wide range of regenerative activity, and this makes the fibroblasts the key element in wound healing of all kinds; scar tissue is new collagen that has been secreted by fibroblasts which have migrated to the injury.[11]

Fluid Crystal

We have already seen that the ground substance can vary considerably from a watery sol-state to a viscous gel-state. Now this variable fluid base may be combined with varying proportions of collagen fibers to produce a truly remarkable array of different properties and structural functions that can be observed in different kinds of connective tissues. Where we find mostly fluid and few fibers, we have a watery intercellular medium that is ideal for metabolic activities; with less fluid and more fibers, we have a soft, flexible lattice that can hold skin cells or liver cells or nerve cells into place; with little fluid and many fibers, we have the tough, stringy material of muscle sacs, tendons, and ligaments. If chondroblasts (cartilage-producing cells) and their hyaline secretions are added to this matrix, we obtain even more solidity, and in the bones this cartilagenous secretion is replaced by mineral salts to achieve a rock-like hardness.

Taken as a whole, then, connective tissue in its various forms can be regarded as a *fluid crystal,* a largely non-living material that can be adjusted over a wide range from sol to gel — here watery, there gelatinous, here dense and elastic, there hard as a stone. In the growing animal, this liquid crystal is in its most active state, changing from sol to gel and gel back to sol with great facility, as rapid growth demands.

Organ shapes are outlined by genetic codes and then continually modified by the specific stresses and strains experienced by the developing organism. Flaccidity, flexibility, and rigidity emerge as the conditions of life demand. This sol-gel activity should continue, albeit at a slower rate, throughout the adult life of the animal. It accounts for a great deal of our metabolic efficiency, is responsible for the healing of all injuries, and provides the physical adjustments necessary for new muscular bulk, new habits, new skills.

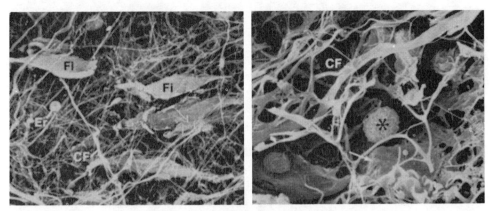

3-10: Once formed outside the fibroblasts, collagen fibrils are arranged in many possible ways. Pictured here are loose areolar networks such as are found in internal organs and glands.
Fi: fibroblast; CF: collagen fibrils.

3-11: Loose areolar networks of collagen fibrils compose frames which hold other cells and structures into place. Pictured is a cross-section through the inner surface of a bronculus in the lungs, but the connective tissue which binds the layers of skin together and fixes them to the underlying structures is very similar in appearance.
ES: epithelial sheet; LP: Lamina propria; Fi: collagen fibrils; BV: blood vessels.

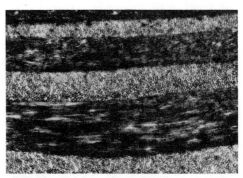

3-12: Dense connective tissue has more collagen fibrils and less ground substance or other kinds of cells within it. These irregular criss-crossings of fibrils make sheets of fascia, such as are found within muscles, or on the surface of the bones.

3-14: A high degree of specific arrangement. These alternating plywood-like layers of collagen fibrils compose the cornea in the eye of a fish.

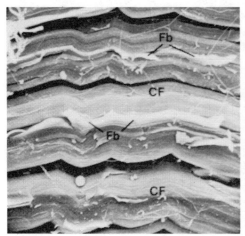

3-13: Dense connective tissue more regularly arranged. These parallel zig-zag relationships are found in ligaments and tendons.

However, there are some limiting factors in these fluctuations, factors which tend to become even more problematical as our tissues age and our varieties of physical activities decrease. First of all, connective tissue shares with many other gels a phenomenon called *thixotropy:* It becomes more fluid when it is stirred up, and more solid when it sits without being disturbed.[12] This is exactly the case with common gelatin, which solidifies while sitting in the refrigerator and melts when it is whipped vigorously, or when it is exposed to low heat.

In the human body the heat energy and movement required for appropriately solvent states of connective tissue can be provided by rapid and efficient metabolism (a chemical "burning"), by physical work, aerobic exercise, stretching, and the like. Unfortunately, these are the very things which we gradually begin to avoid as we get older, less spry, more sedentary. With disuse, the connective tissues become a little colder, less energized, and the thixotropic reaction makes them gel more, become sluggish, lose their full, juicy quality and their ability to soften, stretch, and flex.

There is no way that we know to prevent the eventual drying and stiffening of the connective tissues, a process which eventually produces the wrinkled skin and cranky joints of old age. Connective tissue does seem to have some final limits upon its ability to regenerate and maintain its resilient properties, limits which make us the mortal creatures that we are. But there can be no doubt that poor nutrition and sedentary habits weaken all the connective tissues of the body, stiffen them, and can significantly accelerate their biological aging, even in a young adult.

Thixotropy and Bodywork

This thixotropic effect provides one of the cornerstones for effective body-work. Since connective tissue is largely non-living, it is the mechanical motion and friction caused by muscular activity which provides much of the energy and warmth that maintains its fluid qualities. When a part of the body loses some degree of movement and vitality through trauma or disuse, it will not be as inviting, as comfortable, perhaps not even as possible to move that part with the vigor it requires to keep the connective tissue warm, moist, and resilient. At this point, manipulation of the tissues by skilled hands can provide a pleasant and extremely effective means of introducing freer movement and higher levels of energy into the connective framework. The hands of the therapist can literally supply the mechanical activity which a sluggish limb fails to produce, raising the metabolic rate and restoring some of the fluidity of its connective tissues.

R. B. Taylor, an osteopathic physician, has stated that, "Manipulative pressure and stretching are the most effective ways of modifying the energy potentials of abnormal tissues."[13] Note well the principle involved here: Nothing chemical or structural has been either added to or subtracted from the connective tissue. Rather, by means of pressure and stretching, and the friction they generate, the temperature and therefore the energy level of the tissue has merely been raised slightly. This added energy in turn promotes a more fluid ground substance which is more sol and ductile, and in which nutrients and cellular wastes can conduct their exchanges more efficiently. In addition to this mechanical stimulation of pressure and stretching, a powerful thermodynamic effect can be produced upon the bioenergetic field of the patient by the stronger and healthier bioenergetic field of the therapist. This comes partly in the form of literal body heat transferred by the therapist's penetrating touch, and partly from subtler forms of energy such as galvanic skin responses or vibratory rhythms.[14]

Thus, with regard to its effects upon the connective tissues, bodywork accomplishes its ends in an utterly different fashion than do the additive and subtractive means of pharmaceuticals and surgery. Skillful manipulation simply raises energy levels and creates a greater degree of sol (fluidity) in organic systems that are already there, but are behaving sluggishly. The effect can be analogous to that of turning up the temperature and humidity in a greenhouse that has been too dry and cold.

Hydrogen Bonding

Besides this thixotropic gelling effect of reduced energy potentials, there is another limitation to the flux of connective tissue in the mature and aging body. The collagen fibers derive their tensile strength and their structural shapes from their ability to bond tightly to one another, knitting firmly into a continuous web. As these fibers lie side by side over the years, they tend to pack more and more tightly and bond more firmly together, particularly in areas which receive the most compression and strain. As life goes by and habits develop, these areas of chronic stress in the connective tissue thicken and rigidify, bunch up, lose their range of motion, and impose their limitations on the movement of the body as a whole. To understand just how this happens, we will have to look more closely at the structure of collagen molecules and at the way in which they knit themselves together.

The Collagen Molecule

Collagen is the longest molecule that has ever been isolated; if one were as thick as a pencil, it would be a yard long.[15] If ever a molecule was designed to make netting and cable, this is it.

It begins its formation within the body of the fibroblast as long chains built up from free amino acids in the cell's protoplasm. These aminos are linked together in a particular order, so that the entire chain is composed of repeated units, each unit being a regular number of aminos.

As with most proteins, the tensions of atomic attraction and repulsion among the amino complexes twists this chain into a left-handed spiral, or helix, making it look like a long corkscrew.

At this point in their formation, the protein chains are very fragile, and

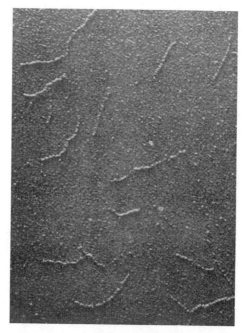

3-15: Collagen molecules.

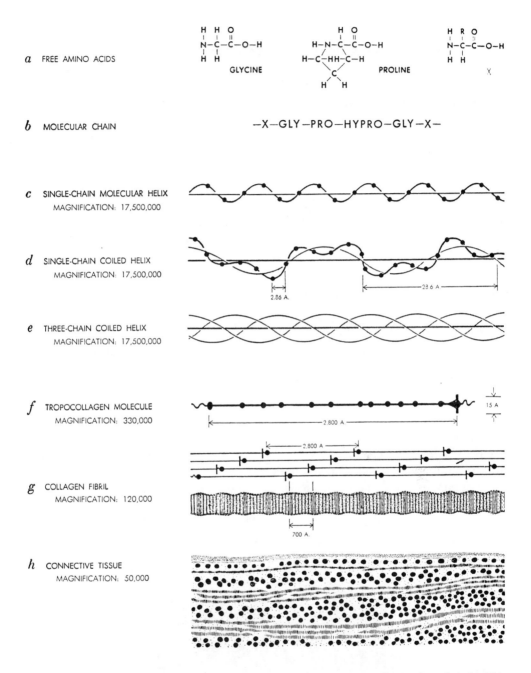

a FREE AMINO ACIDS

GLYCINE   PROLINE

b MOLECULAR CHAIN

—X—GLY—PRO—HYPRO—GLY—X—

c SINGLE-CHAIN MOLECULAR HELIX
MAGNIFICATION: 17,500,000

d SINGLE-CHAIN COILED HELIX
MAGNIFICATION: 17,500,000

e THREE-CHAIN COILED HELIX
MAGNIFICATION: 17,500,000

f TROPOCOLLAGEN MOLECULE
MAGNIFICATION: 330,000

g COLLAGEN FIBRIL
MAGNIFICATION: 120,000

h CONNECTIVE TISSUE
MAGNIFICATION: 50,000

3-16: Stages of connective tissue formation. Free amino acids (a) are strung together in a long chain (b). This chain is in the form of a left-handed spiral helix (c), which also coils its overall length (d). Three of these left-handed coils wrap together in a right-handed triple helix (e), which composes an individual molecule of tropocollagen (f). These individual molecules are then exuded from the cell into the surrounding ground substance, where they assemble into fibrils (g). The fibrils then assemble into structures (h).

float about the cell in random coils. However, as these single helixes come into contact with one another, they begin to spiral around one another in groups of three, twisting this time to the right, until they form a compact triple right-handed helix that is analogous to three-stranded rope. These triple chains are bonded together by hydrogen molecules which attach to oxygen radicals sticking out from the sides of the individual strands of protein, forming what are referred to as *hydrogen bonds*. This triple helix chain is a single collagen molecule.

Once these complex molecules are formed within the fibroblast, they are exuded out into the fluid ground substance as separate units. At this stage, they

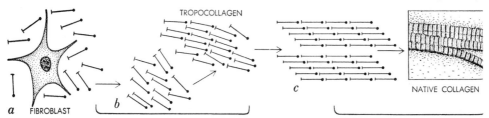

3-17: A fibroblast exuding tropocollagen (free collagen molecules). Outside the cell, they assemble into fibrils, and fibrils assemble into structures.

are free-floating collagen with no over-all structure, and ready to take on whatever form is appropriate in the local area. Specific local qualities of the ground substance and the surrounding tissues then dictate the specific manner in which these molecules join together to create the collagen structure with the qualities needed for that specific area. The process is in general the same for all varieties of connective tissue in all locations of the body. The individual molecules line up side by side, and attach to one another with exactly the same kind of hydrogen bonds which hold together the triple helix, overlapping themselves like bricks and forming a very tough and stable fabric.

The collagen *fibril* that is thus formed by the overlapping molecules is hollow, and is stronger in tensile strength than steel wire — it requires a load ten thousand times its own weight to stretch it.[16] This fine cable is then the unit which bands together with others to create the various fibrous structures of the connective tissues.

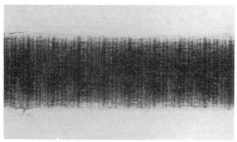

3-18: A section of a collagen fibril.

The hollowness of the collagen fibril is one of its fascinating aspects. Most tubes in the body circulate fluids of one kind or another, and this one seems to be no exception. These billions of fine tubules constitute one of the circulatory systems of the body, along with those for blood and for lymph. What one researcher reports finding inside the collagen fibrils is not blood, nor lymph, nor ordinary ground substance, but rather cerebrospinal fluid![17] This finding suggests that the connective tissue framework may play a role in the body's chemical messages and balances that is even more sophisticated than we can presently document.

Again, these fibrils glue themselves together into cables, nets, and sheets by means of hydrogen bonds uniting the oxygen molecules thrusting out from the collagen chains. Each of these individual bonds is very fragile, but as chains, molecules, and fibers stack together, millions of them are formed, creating an extremely stable building material.

Gluing

It is, unfortunately, this very hydrogen bonding which gives connective tissue its tremendous fibrous strength that also creates one of its most common degenerative problems. Through the years, the fibrils pack more and more closely together, forming tighter and tighter fits and more and more hydrogen bonds. When first assembled in a young animal or a fresh wound, they can be easily

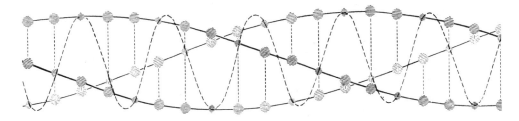

3-19: A section of a collagen molecule triple helix. All of the broken lines indicate hydrogen bonds which link the three single coils together into a strong cable.

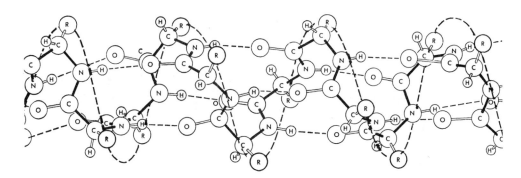

3-20: A detail of hydrogen bonding within a molecular chain. The broken lines linking hydrogen atoms (H) to oxygen atoms (O) are the hydrogen bonds.

dissolved by a neutral salt solution; with some aging, an acid solution is required to separate them; and with further aging and compression they cannot be divided even by acids.

Both chronic pressure and chronic immobility facilitate this bunching and gluing process, so that areas that are under constant stress and areas that have fallen into disuse both tend to fall prey to it. In this way structures that were originally designed to be functionally separate — such as two muscle bellies lying side by side, or a tendon within its connective tissue sheath — begin to form

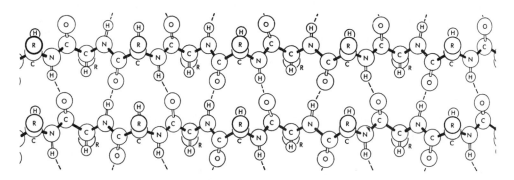

3-21: Separate molecular chains linked with hydrogen bonds. This is how single tropocollagen molecules tie together to form fibrils, and how fibrils tie together to form structures. Exposed surfaces of collagen structures also have hydrogen and oxygen atoms sticking out, and this is how structures meant to be separate may become "glued" together.
The molecules pictured in figs. 3-20 and 3-21 are not collagen, but the principle of hydrogen bonding is identical.

adhesions which impair their ability to glide freely over one another. Thus ranges of motion may be progressively curtailed, and any number of finely differentiated movements may be blurred or even lost.

This unwanted bonding is one of the major factors in the stiffness associated with old age, repeated strain, or poorly healed injuries. As a matter of fact in injuries of the knee, where a great deal of connective tissue supports the joint, the scarring and adhesions which develop during the time that the leg is immobilized in a cast may actually impair the function of the joint more than the original injury or surgery. For this reason, exercise and competent bodywork should be given to the injured knee just as soon as the healing tissues will tolerate them.

These excessive deposits of connective tissue can be felt as thick, lumpy bandaging around the joints, as fibrous masses throughout an entire area, or as tough fibrotic ropes and cysts in the muscle bellies. It is this thickening, shortening, and gluing which can eventually prevent erect posture and graceful motion.

Because the bonded tissues themselves limit the very activities which would help restore their flexibility, it is usually very difficult for the individual to work his or her way out of such a degenerated situation. In these instances, manipulation of the connective tissue can be of invaluable assistance in helping to restore some energy and some resilience to an area, so that more normal feel and function can begin to foster more normal activity.

The Supportive Network

Connective tissue, then, in its various shapes and consistencies, forms a continuous net throughout the entire body. It contains many specialized structures, but it is really all one piece, from scalp to soles and from skin to marrow. If all the other tissues were extracted, the connective framework alone would preserve the three-dimensional human form in all its details.

Just beneath the skin it forms an envelope that wraps the body as a whole. Beneath this outer layer it organizes the muscles into functional groups, wraps each individual muscle, and also honeycombs the interior of the muscle belly with supporting septa. It gathers the ends of the muscles into tendons, and tendons blend into the fibrous sheath that covers the bones, the periosteum; the periosteum is continuous with the ligaments, and even with the inner coating of the hollow bones, the endosteum.

Connective tissue binds other specific tissues into their organ shapes and supplies them with vessels and ducts. It ties these organs together, and suspends

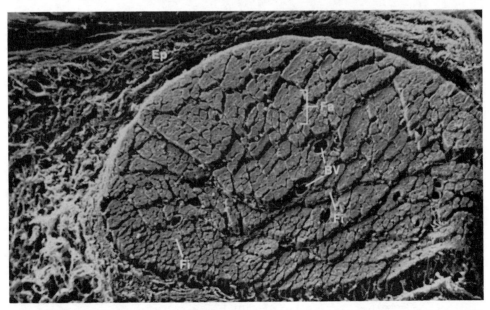

3-22: A cross-section of a muscle belly. The muscle is wrapped in a sheet of fascia, called the epimysium. This same sheath extends inward, where it is called the perimysium, and divides the belly of the muscle into many septa. These septa bundle the long muscle cells together, preserve their parallel arrangement, and determine their angles of pull upon the tendons and bones.

them in their proper relationships in the body cavities. It forms the walls of the blood and lymph vessels, it surrounds these vessels, and it anchors them into place among the muscles, bones, and organs. It anchors the nerves in the same fashion.

In short, connective tissue constitutes the immediate environment of every cell in the body, wrapping and uniting all structures with its moist, fibrous, cohering sheets and strands. Because of its complexity and continuity, this network can well be regarded as a full-fledged organ, one of the largest and most extensive organs in the body. And few organs have such an impressive variety of functions.

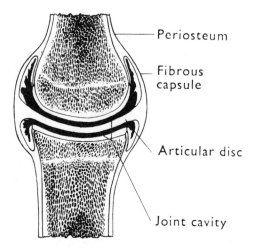

3-23: Connective tissue around the bones is called the periosteum. It is continuous from all bone surfaces to joint capsules to ligaments to tendons to muscle sacs, and thus wraps every bone, joint, and muscle in one seamless collection of compartments.

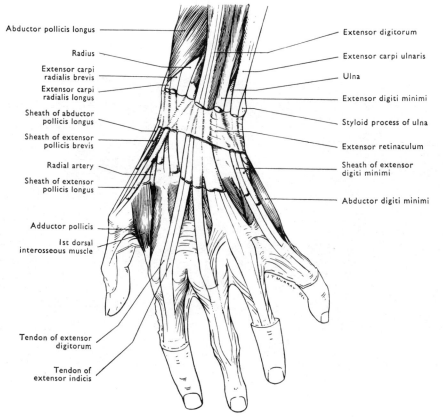

3-24: Tendons, ligaments, tendon sheaths, and the retinaculum holding the tendons into place are all continuous structures of connective tissue.

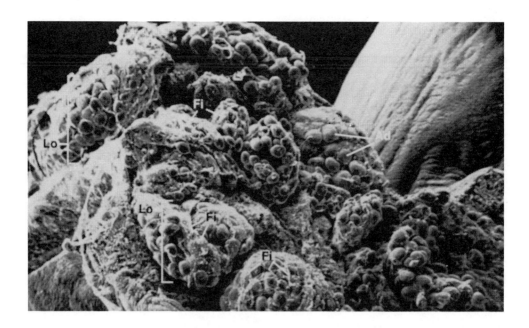

3-25: Fat storage cells are surrounded and held together into adipose tissue by collagen networks.

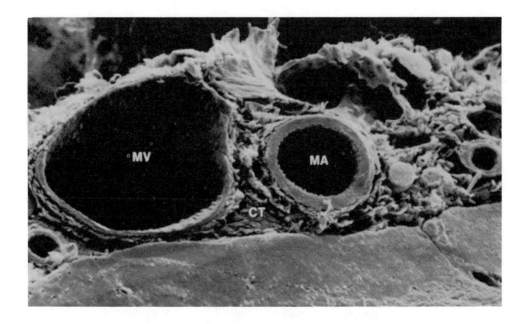

3-26: Arteries and veins are surrounded and held into place by connective tissue.
MV: muscle vein; MA: muscle artery.

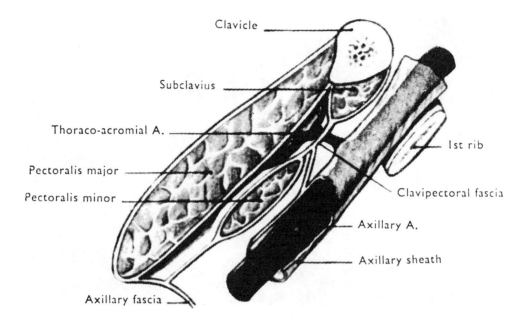

3-27: All separate parts are bound together into organs, systems, limbs, and so on by fascia sheaths continuous with one another.

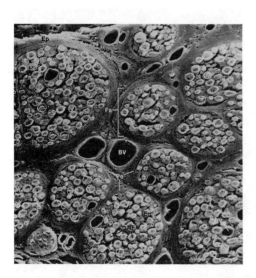

3-28: Nerve trunks are bundled together by connective tissue, and within the bundles fine connective tissue webs wrap each nerve fiber, much like the septa within muscle bellies.

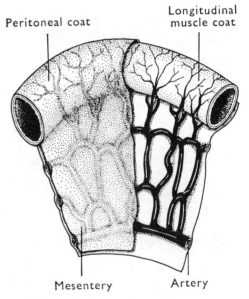

3-30: The peritoneal coat of connective tissue holds the blood vessels which serve the intestines into place, and defines the curves of the digestive tubing.

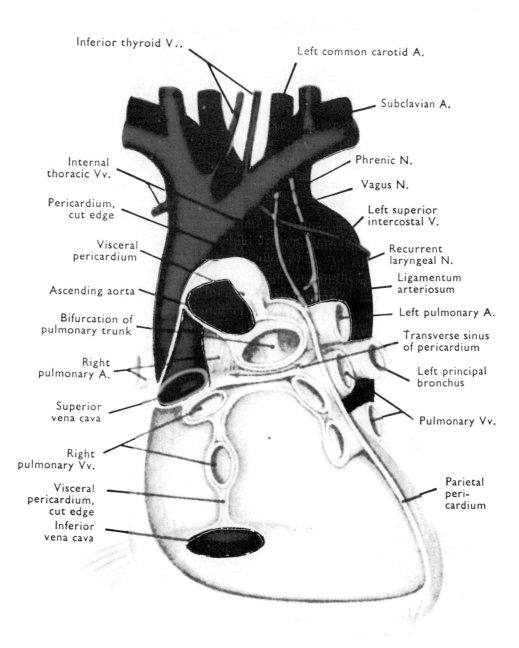

Inferior thyroid Vv.

Left common carotid A.

Subclavian A.

Internal
thoracic Vv.

Phrenic N.

Vagus N.

Pericardium,
cut edge

Left superior
intercostal V.

Visceral
pericardium

Recurrent
laryngeal N.

Ligamentum
arteriosum

Ascending aorta

Left pulmonary A.

Bifurcation of
pulmonary trunk

Transverse sinus
of pericardium

Right
pulmonary A.

Left principal
bronchus

Superior
vena cava

Pulmonary Vv.

Right
pulmonary Vv.

Visceral
pericardium,
cut edge

Parietal
peri-
cardium

Inferior
vena cava

3-29: Connective tissue forming a sheath around the heart is called the pericardium.

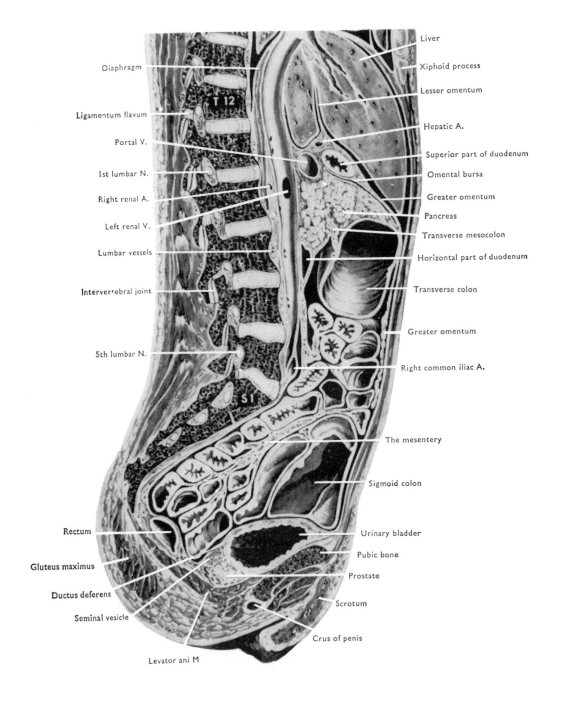

3-31: Each of the internal organs is encased in connective tissue, which then binds them to one another at appropriate junctions, and attaches them to the internal walls of the body cavity so that they stay in place.

Hydrostatic Pressure

In addition to supporting individual cells, tissues, and other organs, this connective organ serves an over-all structural purpose as well — it is woven together with the bones to create the movable frame which supports our posture and from which everything else is suspended. We normally think of the ligaments lacing around the joints, and perhaps the tendons which tie the muscles to the bones, as being the chief support that connective tissue offers the skeleton. But the situation is really much more complex than that; not only joint capsules and tendons, but literally *all* of the connective tissues — together with the fluids they contain — aid the weight-bearing capabilities of the skeleton.

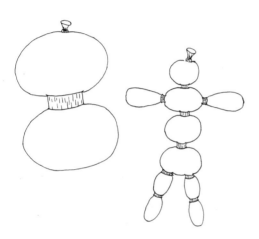

3-32: Binding water balloons to illustrate the forces of hydrostatic pressure in the connective tissue framework.

To see how this works, we can view the body as a large bag filled with water. If the surface and interior of this bag were perfectly uniform, like a filled balloon, then this bag would rest on the ground in the shape of a slightly flattened sphere. However, if we circle this sphere with cords and tighten them up, an interesting thing happens: The sphere is transformed into a cylinder, and can be made to stand erect. And if we continue adding cords, we can make the cylinder taller, thinner, and modify it into any number of shapes — all without adding a single rigid member to the interior. Given a tough enough bag (and remember that connective tissue is *very* tough), we can keep lacing and squeezing until we have created enough hydrostatic pressure to make the cylinder quite rigid.

This is exactly the same kind of water pressure that holds a flower stem up straight, exactly the same kind of forces that erect a penis when its *corpus cavernosum* is distended with blood. At this point, our cylinder does not really need an internal skeleton in order to remain upright; in fact, a skeleton could even be suspended inside the cylinder from the top, without its toes touching the bottom, supported solely by the tension of the pressurized walls of the bag.

This of course is what the various shape-giving cords and bands of connective tissue do to our own bags of liquid, trussing them up into cylindrical shapes and squeezing them tightly enough to give them rigidity.[18] When all the bands and cords are properly adjusted, and the hydrostatic pressure is strong and balanced, this tensional force alone goes a long way towards keeping us erect, and can give us that wonderfully light "skyhooked" sensation, as though our frames were suspended from the tops of our heads — as, to a degree, they literally are.

Tensegrity

Besides this hydrostatic pressure (which is exerted by *every* fascia compartment, not just the outer wrapping), the connective tissue framework — in conjunction with active muscles — provides another kind of tensional force that is crucial to the upright structure of the skeleton. We are not made up of stacks of building blocks resting securely upon one another, but rather of poles and guy-wires, whose stability relies not upon flat stacked surfaces but upon proper angles of the poles and balanced tensions on the wires. Buckminster Fuller coined the term "tensegrity" to describe this principle of structure, and his inventive experiments with it have clarified it as one of nature's favorite devices for achieving a maximum of stability with a minimum of materials.

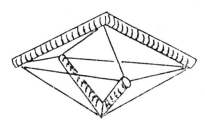

3-33: A simple tensegrity structure. The rigid tubes correspond to bones, and the cables to muscle/tendon tensional members. Note that most of the stress is borne by these tensional cables, and not by the rigid tubes. Connective tissue cable in the human body can support thousands of times its own weight without stretching or snapping.

A tensegrity structure, such as the one illustrated, is as rigid as a solid cube, and yet its solid members absorb much less than half of any compressional force applied to it. As weight bears down, the solid beams tend to spread; this they are prevented from doing by the tension of the wires, and it is this *tensional* force, not the compressional strength of the beams, which keeps the structure rigid. Fuller has demonstrated that this simple tensegrity cube can be expanded into large and complex cubes, spheres, cylinders, and masts, architecturally capable of supporting far heavier loads than solid supports of equal weight.

Note well here — it is the strength of the *connecting cables,* not the strength of the beams, upon which the superior function of the tensegrity unit principally relies. And of course, maladjustments, weaknesses, and imbalances in the lengths and tensions of the cables rob the structure of much of its stability.

This principle of tensegrity describes precisely the relationship between the connective tissues, the muscles, and the skeleton. There is not a single horizontal surface anywhere in the skeleton that provides a stable base for anything to be stacked upon it. Our design was not conceived by a stone-mason. Weight applied to any bone would cause it to slide right off its joints if it were not for the tensional balances that hold it in place and control its pivoting. Like the beams in a simple tensegrity structure, our bones act more as *spacers* than as compressional members; more weight is actually borne by the connective system of cables than by the bony beams.

This naturally makes the proper adjustments of length and tension in the connective tissues a matter of extreme significance in the distribution of gravitational forces throughout the body. A cable that is too tight in the lower back has great consequences for the structural integrity of the entire mast of the spine. And, conversely, an overbalanced superstructure demands hosts of tensional compensations throughout the lower members. It is the network of connective tissue — the pressurized water bags and the tension cables — and not the bones, that bears most of the structural responsibility for stable, upright posture and graceful carriage.

Connective Tissue as Retort

We have already remarked that insofar as the ground substance makes up a large part of the intercellular fluids, it is the site of a large number of metabolic exchanges as nutrients cross from capillaries to cells and wastes cross back. Its temperature, its relative states of sol or gel, its specific chemical variations, the presence of foreign particles or toxins, and its mechanical activity all have a direct bearing upon the efficiency of these exchanges. In turn, this ground substance is directly affected by the chemistries of diet, hormones, disease, and stress occurring in the body as a whole, and the action of these influences condition it to a great degree, for better or for worse.

Disease Containment

The many compartments of fascia throughout the body are of great assistance in the prevention of the spreading of infections, diseases, and tumors. Each separate compartment tends to contain the destructive agent, and prevent its spilling into adjacent compartments. This is accomplished partly by the fibrous walls of the compartments, and partly by specific chemical barriers in the fluid ground substance.[19] Of course this means that weaknesses in the fibers or disturbances of the protective chemicals in the ground substance contribute directly to the spread of infections and diseases from compartment to compartment.

The connective tissues surrounding some types of tumors—including cancerous ones—are particularly illustrative of this point. When these tumors are in remission or are growing very slowly, the surrounding connective tissue is dense and fibrotic, with a viscous ground substance. In contrast, the connective tissue around rapidly growing tumors is much looser and softer, and the ground substance is much more watery. Some of these tumors even secrete their own solvent — the enzyme hyaluronidase — which breaks up the complex mucopolysaccharides and creates this more fluid ground substance, allowing the growth to spread more freely. Thus the chemistry of connective tissue itself — above and beyond whatever antibodies or white blood cells may be swimming around in its ground substance — is an important part of the body's immune system.

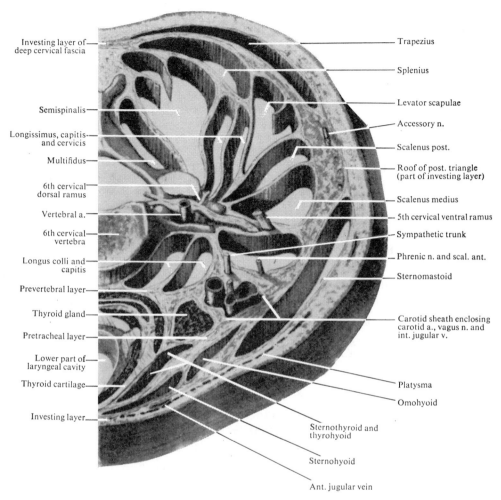

Investing layer of deep cervical fascia — Trapezius

— Splenius

Semispinalis — — Levator scapulae

Longissimus, capitis and cervicis — — Accessory n.

Multifidus — — Scalenus post.

6th cervical dorsal ramus — — Roof of post. triangle (part of investing layer)

Vertebral a. — — Scalenus medius

6th cervical vertebra — — 5th cervical ventral ramus

Longus colli and capitis — — Sympathetic trunk

Prevertebral layer — — Phrenic n. and scal. ant.

— Sternomastoid

Thyroid gland — — Carotid sheath enclosing carotid a., vagus n. and int. jugular v.

Pretracheal layer —

Lower part of laryngeal cavity —

Thyroid cartilage — — Platysma

— Omohyoid

Investing layer —

Sternothyroid and thyrohyoid

Sternohyoid

Ant. jugular vein

3-34: Planes of fascia—shown here in a cross-section of the neck—form various compartments which tend to contain the spread of infection and disease.

Growth Hormone

The hormone *somatotrophin,* the "growth hormone" secreted by the pituitary gland, has a powerful effect upon the connective tissues. It directly stimulates the fibroblasts and the mast cells, stepping up their production of both ground substance and collagen fibers. This is not surprising, since somatotrophin stimulates the reproduction, growth, and secretions of all our cells as we develop from embryo into adult. The inhibition of this pituitary secretion has to be very carefully timed and controlled, or else either the whole body or particular parts of it will continue to grow and grow, sometimes to grotesque proportions.

The adult consequently secretes a much smaller amount of somatotrophin than does the child, but this smaller amount remains very necessary. Anytime the adult's body calls for tissue expansion, such as an increase in muscular development, the growth hormone must be dispatched to stimulate the fibroblasts to produce the extra connective tissue necessary to structure and contain the new muscle bulk, and to build up the fascia and the tendons to withstand the greater tensions which the larger muscles can now deliver. And the secretion of somatotrophin is of special significance in wound healing, where rapid production of collagen fibers by a large number of fibroblasts is necessary to repair the damage.

We have seen in Chapter Two how bodywork is capable of playing a particularly fascinating role in this activity of the pituitary gland. Young children who receive inadequate tactile stimulation, we will recall, suffer from "deprivation dwarfism," the failure of the body to grow normally. This failure is correlated exactly with abnormally low levels of growth hormone in the bloodstreams of these children. Stimulation of the skin has been clearly demonstrated to be a crucial factor in healthy pituitary function. Lytton Gardner, discussing deprivation dwarfism, remarks that

> Patton and I have postulated a physiological pathway whereby environmental deprivation and emotional disturbance might affect the endocrine apparatus and thereby have an impact on a child's growth. Impulses from the higher brain centers, in our view, travel along neural pathways to the hypothalamus and thence, by neurohumoral mechanisms, exert influence upon the pituitary.[21]

Tactile stimulation (or the lack of it) produces certain emotional and conceptual responses in the higher brain; these feelings and concepts then exert an effect upon central nervous system activities as a whole, and upon the production levels of the pituitary in particular. These pituitary secretions in turn affect the health of the connective tissues and their ability to respond to trauma. These effects are all very different from mere "relaxation," the result most commonly associated with bodywork.

Cortisone

The adrenal glands are very closely coordinated with the pituitary, each of them sending out a variety of hormonal messengers which alternately stimulate and inhibit the activities of the other. Together, they form what is called the *adrenal pituitary axis,* the body's most important mechanism of hormonal control.

From the adrenal cortex comes an important hormone, cortisone, which has an effect upon connective tissue quite the reverse of that of somatotrophin: cortisone inhibits the activity of fibroblasts, the very sources of connective tissue. Prolonged application of cortisone to the skin markedly reduces the number of fibroblasts, and the ones remaining are smaller than normal.[22]

This effect of cortisone has been used pharmaceutically in the treatment of fibrosis — dense connective tissue deposits created in the muscle bellies by repeated strain. This hormone also acts as an anti-inflammatory substance, and as such has been applied to inflammations of all kinds to reduce the swelling and discomforts they create. However, the negative effects of continued exposure to cortisone have revealed themselves to be substantially greater than the positive effects in the long run.

Anything that depresses fibroblast activity obviously interferes with the normal healing of wounds, bruises, fractures, and the like, no matter how effectively it reduces the swelling associated with these injuries. Nor has cortisone proved to be as useful in the treatment of infections and inflammatory allergic reactions as was once hoped; it simply removes discomforting symptoms, without affecting either the basic mechanism or the course of the infection. In fact, since it weakens the connective tissue, it has been shown to actually facilitate the spread of infection from previously localized areas. Animals who have been given large amounts of cortisone develop spontaneous and rapidly fatal infections.[23]

These effects have great significance, because cortisone — along with adrenaline — is one of the hormones automatically released into the bloodstream by the adrenal glands during times of physical and emotional stress. Individuals who are chronically stressed or disturbed in any way weaken their entire connective tissue network over a period of time, exposing it to pathological invasions of all kinds. And for most of us, our environments provide us with many situations which stimulate this adrenaline/cortisone release; we usually cannot drive across town without experiencing it at least once or twice.

We will have occasion later to look more closely at the entire alarm response of an individual threatened with real or imagined threat, in which the adrenal secretions play a major, and a potentially devastating role. But even now we can begin to see that bodywork could have positive effects upon our connective tissue framework far beyond those of merely stretching and warming and rendering more supple. This is because in addition to the mechanical effects of pressure and elongation, effective bodywork also utilizes the *sensations* which it is creating to stimulate and organize the nervous system as a whole. This stimulation and organization can positively affect the activities of the pituitary/adrenal axis, the most important monitor of the majority of the body's hormones, including cortisone.

And the prevention of the chronic flooding of the bloodstream with excess cortisone — which seriously inhibits the fibroblasts and the mast cells — bears directly upon all of the structural, the metabolic, and the immunological functions of the connective tissue framework. Again, we are looking at results that extend far beyond those we normally associate with temporary "relaxation."

Connective Tissue as Organ

The entire connective tissue system is large enough, complex enough, sophisticated enough in its varieties of forms, and important enough to our survival to be regarded as one of the vital organs of the body. Indeed, few vital organs fulfill as many necessary functions.

In the embryo, it is largely the proliferation of connective tissue structures which serves as a blueprint for the growth of the rest of the body's bulk. By and large, the other developing tissues fill in the appropriate compartments preformed by the collagen network to create bones, organs, and systems.

From jellyfish to human beings, connective tissue is the primary organ of structure, gluing cells into discrete colonies, defining their shapes, forming them into functional units, and suspending them together in the correct relationships within the organism. It supports and contains every cell and every drop of fluid that is in us. And in addition to simply *containing* all our fluids, the toughness of this tissue is responsible for creating the hydrostatic pressure which helps to hold the entire body erect and support its three-dimensional volume.

It is connective tissue that binds bone to bone, attaches bone to tendon, and ties tendon to muscle. The contractive power of muscle is absolutely useless without this extremely complicated system of sheets, pulleys, hinges, and cables which transfers muscular efforts to the levers of the skeleton. The relative lengths and resilience of these connections is often the decisive factor in the possibilities and limitations of our movements. And besides transferring contractile movements to the appropriate points on the skeleton, translating their pulls into infinite varieties of gesture and style, these same cables and sheets provide the tensional elements which support the bony spacers to form the rigid, weight-bearing tensegrity structures of the body.

The connective tissue also forms the fibrous bed for all the lymph vessels, blood vessels, and nerves, supporting them and keeping them in their appropriate channels. Nor does it merely suspend them statically; in addition, since all these supporting webs are continuous in their structure, it is the connective network which synchronizes the motion between muscles, vessels, nerves, and viscera, as well as transferring muscle action to the bones. In this way the movements of my muscles and my skeleton are continually stretching and massaging my internal organs.

As a collection of fibrous compartments, connective tissue serves as a physical and chemical barrier to the spreading of all sorts of toxic agents throughout the body. This it does both by means of its own structural and chemical nature, and by providing the fluid medium for the intercellular circulation of white blood cells, antibodies, hormones and other elements of our immune system.

And last but far from least, connective tissue constitutes the major repair mechanism of the body. Its chemistry monitors the inflammatory response, its fluids deliver the antibodies and white blood cells to fight infection, and its fibroblasts produce the fresh collagen to close the wound or fracture with new scar tissue. All in all, this stringy, gooey stuff provides us with a remarkable array of necessities and advantages; it is the container that allows us to take our precious sea water with us wherever we go, and that keeps our organisms intact in the midst of the many dispersing forces around and within us.

Therapeutic Manipulation of Connective Tissue

The majority of connective tissue is not made up of living cells, but rather of the fluids and fibers exuded by the living fibroblasts and mast cells. Direct pressure and stretching are among the most effective means at our disposal of increasing the overall energy levels of this substance in all its forms.

This increase happens in several ways. First of all, the warmth from the practitioner's hands contributes some thermal energy. But far more significantly, the pressure, motion, and friction created by deep manipulation raises thermal and thixotropic levels far beneath the surface. In addition, the squeezing, stretching, and contorting of the connective tissues creates a cleansing, flushing effect, similar to that of rinsing out a sponge or stirring the water in a ditch that is clogged with algae. It is not, in fact, uncommon for bodywork to make a person feel nauseous or headachy from the large amounts of toxins and wastes that can be thus moved out of the intercellular fluids and into the bloodstream, from which they can then be eliminated.

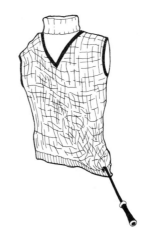

3-35: As with any continuous fabric, a pull on one corner of the connective tissue framework exerts a pull throughout the structure, affecting muscles, bones, nerve, blood vessels, glands and organs as their wrappings and septa are distorted and then released. We can see how healthy, active muscles alternatively stretch and compress all the body's compartments and the cells they contain, aiding the circulation of the internal sea considerably. In the same way, pressure and manipulations on the body's surfaces can be effective deep within the interior.

Over the centuries, many "deep tissue" techniques for the manipulation of the connective tissue framework have been developed and used with success. One of the most recent and popular of these is Rolfing. Based upon a realization of the profound systemic effects inherent in connective tissue's sol/gel properties and in its structural roles, these techniques all seek to energize and reshape this ubiquitous and fundamental material. They utilize the fingers, knuckles, forearms, elbows, and the feet of the practitioner, and sometimes even tools, to exert pressure or stretching forces on the connective tissues.

This pressure and stretching, carefully applied at specific points and in specific directions, results in a softening and lengthening of the connective tissues; they become more fluid and less gelled. The release of thickened and strained areas proceeds systematically session by session, area by area, until a new, more energetic overall equilibrium is achieved. In this way, the downward spiral of thixotropy is arrested and even reversed, improving the tissue consistency and the rate of metabolism of the entire system.

This type of therapy also addresses the structural properties of connective tissue. Lack of use, misuse, poor postural habits, chronic strains, and so on tend to create structures which reflect them. Some connective tissues thicken and shorten; others weaken and lengthen. Muscle sacs become fibrotic, or ligaments become too stretched to support their joints. Asymmetries and imbalances evolve which put the individual at a marked disadvantage for coping with gravity's constant pull. Sheets of fascia bunch up, and become glued to themselves or to their neighbors with hydrogen bonds. When these various related problems become widespread, it is extremely difficult for the individual to overcome their limitations on his or her own. The pressures and movements induced by the deep tissue practitioner are designed to rearrange and actually "sculpt" these deformed and glued structures, so that a more efficient distribution of weight and a broader range of motion are possible. The practitioner uses strong pressure in such a way as to physically separate unwanted hydrogen bonds and to introduce more effectively organized *lines of stress* through the body, which directly affect the future patterns of collagen deposits.

The exertion necessary on the part of the deep tissue practitioner and the discomfort that is not uncommon for the client during this kind of therapy are testimony to the tenacity of the connective tissue's formations. This tenacity must be sensitively and intelligently dealt with in order to avoid pain that is counterproductive, and even to avoid actual tissue damage. After all, many of the tissues that the practitioner must work *through* (skin nerves, organs) are not as strong as the tissue he is working *on* (fascia). And even fascia itself can be torn instead of stretched. The deep tissue manipulator's skill in warming, energizing, and preparing both the fascia and the reflex muscular responses of the client will have a tremendous effect upon the tissue's ability to change and upon the potential discomfort of the client. It is, in fact, this ability to ready the tissue before new stresses and demands are exerted that makes the difference between assimilable and long-lasting effects and those that are painful, rejected, and thus short-lived.

There is a tendency among some practitioners to regard brief periods of intense pain as being a worth-while price to pay for the promised long-term improvements. This justification is sometimes carried to the extent that excessive pressures are used which bruise or even tear various tissues. Such an attitude is ill-advised, and I would strongly urge any client to refuse to tolerate extreme pain during a bodywork session. For one thing, acute pain announces imminent tissue damage. For another, even in the absence of actual damage, acute pain creates a reflex neuromuscular contractile response which reverses the desired process of softening and lengthening.

The ability to defuse these reflex responses has at least as much, and probably more, to do with the positive structural and metabolic changes induced by deep tissue therapy as does anything done to the connective tissues themselves. We will be able to appreciate this more fully in later chapters when we examine the nature of the neuromuscular reflexes and their responses to pain.

We are touching here upon a most important feature of the interrelationships of the body's various tissues. There is no single system — sensory, connective, muscular, circulatory, neural, and so on — that is not intimately affected by the relative states of health and activity of all the other systems. Since there are no cells in the body that are not supported by collagen structures and bathed in the fluid ground substance, there is no place that we can apply warmth or pressure without affecting the connective network. And the converse is equally true: The manner in which we manipulate the connective framework will have distinctly positive or negative effects on the other tissues.

Our sense of the intimacy and complexity of these relationships can only increase as we add more to our knowledge of the body's tissues and systems. It is this intertwining of functions and relationships which makes it possible for the various kinds of manipulative therapy to be so potentially effective. If we can introduce a simple but positive change into any one element, that change will tend to support positive shifts in other elements as well, and in the balance of all elements, which after all is what constitutes good health.

But this is not to say that we can pursue our good intentions in one kind of tissue and ignore all of the others, confident that whatever improvements we make will naturally spread. The presence of other tissues, other parts, other functions, demands that we do not work on any one element in a manner which ignores the rest. If we do, we only reduce our effectiveness, and even court the danger of doing more harm than good. We must always resist the tendency to focus our attention upon localized and objectively predictable *effects*, and must always strive to include ever broader and more complexly interrelated *processes* in our ways of thinking and working.

There is no single key in the skin, in the connective tissue, or in any other system we will examine which will provide us with an infallible means of improving health. Whatever we do with our hands must be done with the knowledge clearly in mind that all of the physical and mental elements within the human being are inextricably related. On one hand, it is the interconnectedness of these relationships which gives bodywork its power; and on the other hand, it is this same interconnectedness which dictates the problems and limitations of each individual technique.

4

Bone

She is composed of those fleeting attitudes, of those Ninth-Symphony-like facial expressions, which, reflecting the architectonic contours of a perfect soul, become crystallized...and which, having been classified, clarified by the most delicate breezes of the sentiments, harden, are organized, and become architecture in flesh and bone.
—Salvador Dali, *The Secret Life of Salvador Dali*

Bedrock

Because the earliest colloidal droplets and primitive cells took their chemical constituents and their complex self-perpetuating processes from the conditions present in sea-water, life forms have from that time onward been obliged to carry their sea-water within them. And as these miniature, self-contained bodies of water emerged from the ocean and began to adapt to dry land, they began to internalize the elements and conditions of their new environment which forced them to adapt and evolve.

Indeed, since we have left the water and have become terrestrial creatures, subject now to the special forces of gravity, we have become miniature earths in the same way that we first became miniature seas; we have added more and more solid features to support our containers of fluid upon the ground. We have river channels and reservoirs in our circulatory systems, meadows and forests in our hair; we are mountains of flesh riddled with the caves and fissures of our pores and orifices; like enclosed valleys, we shelter our ancestral cultures, and like the open hillsides and plains, we teem with a microbial bustle of new citizens and migrants. We have become an ecology of earth and air, as well as one of water.

And before we could proceed very far in this development of terrestrial elements and structures, subject to the inexorable pull of gravity, we had to devise the same sort of foundations and internal supports that prevent any earthy structure from flattening, any upright from toppling. We had to develop our own geology, and learn to deposit our own bedrock of solid bone.

Due to their mineral hardness, bones are not often considered as subjects for most kinds of bodywork. There is not, after all, much there to soften, or lengthen, or relax, nor even many sensory possibilities. Most "bone" therapies deal primarily with seating the bones together properly at their joints.

Osteopathy and chiropractic have produced ingenious methods of dealing with dislocations and subluxations (minor dislocations), and they have demonstrated beyond a doubt that the correction of these disturbed bony placements yields positive benefits far beyond the simple mechanical advantage of properly fitting joint facets. The areas around joints tend to be densely laced with blood vessels and nerves, which can be seriously squeezed by the swelling and thickening of the surrounding connective tissues that often accompany the pinch and strain of poor joint alignment. This congestion can impair circulation and neural transmission through the area, affecting not only the joint itself but also organs and areas that are fed by the same vessels and nerves. This is especially true of the spine, where the spinal nerve trunks exit the spinal column through openings between the vertebral facets with narrow margins of clearance, and then radiate out to every organ and muscle of the body. Pressure on these trunks and their supporting blood vessels can be particularly debilitating.

However, even osteopathic and chiropractic manipulations do not usually seek to change bones themselves, but rather to adjust their relationships with their neighbors. Yet we will examine the possibilities of the skeleton with regard to bodywork, because the bones—seemingly the hardest and the most permanent formations in our bodies—have much to teach us about the wide ranges of plasticity that *all* of our tissues are capable of exhibiting. In spite of its apparent solidity, bone is always being formed and modified, added to and subtracted from, by a continual process—a process which normally lasts the entire lifetime of the organism. And like all biological processes, this one has wide margins of variability with any number of possible end results.

It is true that the rate of growth and the final size and shape of every bone is genetically coded to a very high degree of specificity, but once we leave the womb the inherited features of the skeleton become subject to the same array of additional influences which confront the genetic formations of all our other, softer tissues. Tactile stimulation, diet, exercise, trauma, the relative strengths and balances of various muscle sets, postural habits, proper use or abuse, or disuse, and all kinds of psychological factors enter into the conditions under which the bones are formed throughout infancy, youth, and adulthood. These things are knit into our bony frames as surely as is the genetic code. And many of these patterning influences are very susceptible to the positive effects of bodywork.

Particularly in children, adolescents, and young adults, and in older individuals during the healing phase following injury, bodywork can constructively intervene in the process of bone formation, and can help the individual avoid many skeletal aberrations. In this respect, bodywork may be just as beneficial to bones as it is to the softer tissues, because with all of them it is a *process* we are trying to shift, and long term consequences we are seeking to encourage or avoid.

Surgery and drugs are the readiest means to cause immediate local changes; bodywork, on the other hand, courts the body's abilities to change its own internal relationships over long periods of time. As the bones are an integral part of these relationships, their formation will certainly benefit from any overall positive shifts in the body's processes.

The Ingredients of Bone

Connective tissue is the mother of bone—its precursor as a structural material, and literally the womb in which it is formed.

One of the distinctive things about the different combinations of collagen and ground substance is their changeable sol/gel properties, which are responsible for connective tissue's wide varieties of fluidity and solidity as a liquid crystal. Bone is the most gelled form of this continuum. As vertebrates became larger and more complex, an even more rigid system of protection and support became necessary. To achieve this increased rigidity, the connective tissue matrix formed specific pockets, or molds; these molds were then packed with crystallized mineral—principally calcium and phosphorous,[1] but including vital traces of magnesium, sodium, carbonate, citrate, and flouride.[2] This mineral combination is not unlike that of many sedimentary rock formations, and it adds the compressive strength of its firmly interlocking crystals to the elastic strength of the collagen fibers. Thus bone is born, the hardest form of connective tissue.

The word "skeleton" is derived from the Greek *skeletos*, which means "dried." That is, when everything else has been removed by drying and brushing, it is the "skeleton" which remains. But this remainder is really only the skeleton of a skeleton; it is a brittle, white substance that has little in common with the remarkable properties of living bone. What are left in the dried bone are the sedimentary deposits of mineral salts that have added their rigidity to the plasticity of our connective tissues, rigidity which allows our bodies to maintain their shape and stature in spite of the gravitational compression of their own weight.

The mineral content and architectural properties of these dried remains are similar to those of marble. Their solid resistance to compressive forces is very impressive indeed, given their relatively light weight: The ends of the thigh bone will withstand between eighteen hundred and twenty-five hundred pounds of pressure.[3]

But this dried sedimentary deposit comprises only seventy-five per cent of a living bone; the remaining twenty-five per cent consists of connective tissue— collagen fibrils and their ground substance.[4] And a very tough variety of connective tissue it is, made up of merely three per cent watery ground substance and ninety-seven per cent collagen fibrils.[5]

These fibrils criss-cross in all directions, but in the main they are arranged along the stress lines of compression and tension within the bones created by the weight and activity of the body. The flat mineral crystals are deposited in place by being attached to the same hydrogen radicals on the surface of the fibrils that are responsible for the hydrogen bonding between collagen bundles. The presence of this collagen matrix in the living bone adds a great deal of *tensile* strength and *flexibility* to the *compression* strength of the mineral salts alone. Furthermore, the dense interweaving of the flat mineral crystals by the long fibers of collagen gives the bone a *shear* strength it would not otherwise possess.

The blending of these ingredients give to bone a unique combination of structural advantages that even most metals do not share—a simultaneously high degree of compression, tensile, and shear strengths. If the connective components are dried away, the remaining mineral salts are rigid, but relatively brittle to bending and shearing forces. If the mineral salts are dissolved away by acid, then the remaining collagen matrix is tough and flexible, but bends easily and provides no solid structure. Living bone requires all of its ingredients for its necessary strength and resilience.

Bone Formation

Living bone requires not only its connective tissue and mineral contents, but also a dense distribution of blood vessels and nerves. It is the removal of these that gives dried bone its porous surface. Since it would be impossible for these softer tissues to penetrate the solid parts of the bone, it is the bone which must grow around these nerves, vessels, and fibrils, accommodating them into its rigid structure. This process of surrounding the softer parts with the solid deposits of bone tissue is a complex one, and although it is of course most active during growth periods, it is continual to one degree or another throughout life.

Cartilage Molds

First of all, in the developing foetus, cells called *chondroblasts* appear in specific areas within the connective tissues, determined by genetic programming. These chondroblasts do not themselves produce bone, but rather they secrete tough, nylon-like cartilage, similar to the way that fibroblasts secrete collagen and ground substance. This cartilage is exuded into the surrounding connective

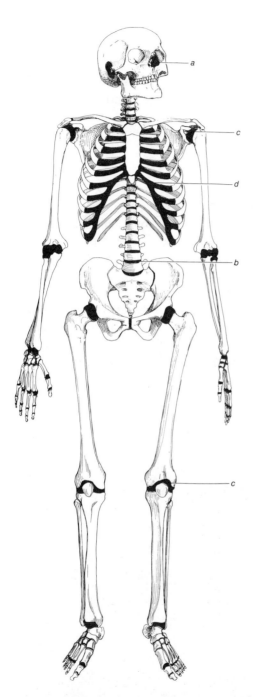

4-1: Most of the bones began as cartilage prototypes in the foetus. By adulthood, only the black areas on the pictured skeleton remain as cartilage—nose (a) and ears, portions of the ribs (d), joint surfaces (c), and the intravertebral discs (b).

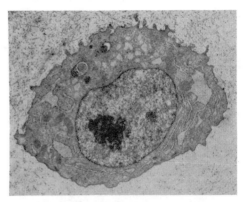

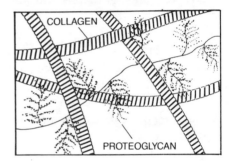

4-2: A chondroblast. This cell exudes mesh-like molecules into the surrounding connective tissue. These tough meshes, and the water they entrap, give cartilage its moist nylon-like consistency.

4-3: Cartilage. The chondroblasts are embedded in the cartilage matrix formed by the exuded proteoglycans.

tissue, which contains it in highly specific shapes that will serve as precise molds for the depositing of the mineral salts.

The connective tissue around the cartilage organizes itself into a collagen membrane which encapsulates the cartilage model, and the inside layer of this membrane becomes richly supplied with blood vessels and nerves. This membrane also continues to produce a high concentration of chondroblasts, which continue to add to the surface growth of the cartilage model; at this stage the membrane capsule is called the *perichondrium.*

Calcification

At a certain point in the cartilage's development (and this moment is different for different bones in the body), the chondroblast cells, which have been secreting cartilage, undergo a physical and functional change. They expand in size, stop producing cartilage, and begin to secrete the chemicals which precipitate into crystals the dissolved mineral salts delivered by the blood.

These cells are now called *osteoblasts,* or "bone producers." These precipitating crystals continue to aggregate along the hydrogen-bonding sites of the collagen fibers within the cartilage, gradually displacing the cartilage model, which is dissolved around their growing formations. Meanwhile, the chondroblasts in the encapsulating perichondrium membrane undergo a similar change, also becoming osteoblasts which begin layering the surface of the dissolving cartilage model with precipitated mineral crystals. This process is called *ossification,* or *calcification.*

It is worth noting that the mineral salts that are precipitated to make up bone are among the heaviest elements in the blood stream. It is this proclivity of the

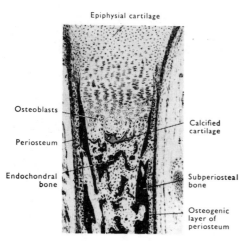

4-4: Ossifying cartilage. Chondroblasts in the upper cartilage are transformed into osteoblasts farther down. These osteoblasts then begin replacing the cartilage matrix with precipitated mineral salts, forming bone tissue.

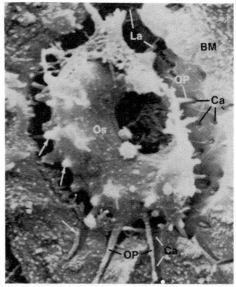

4-5: An osteoblast embedded in the bone matrix (BM) that it has precipitated around itself.

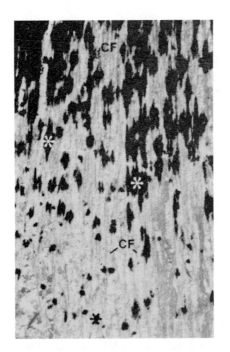

4-6: Calcification taking place. The long grey lines are collagen fibers, and the dark areas of calcification are occurring at points along these fibers. The bone crystals are attached to the collagen fibers by hydrogen bonds, forming an extremely tough complex. Calcification is more complete in the large dark areas.

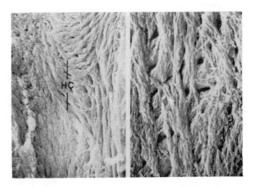

4-7: Bone crystals deposited along collagen fibers.

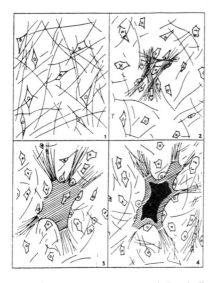

4-8: Some bones, including part of the skull, are formed in connective tissue pockets, without the benefit of cartilage molds.

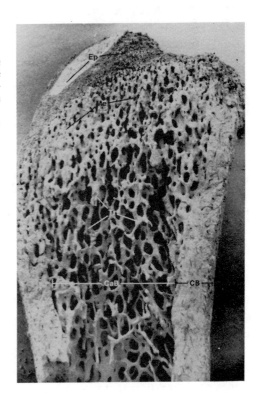

4-9: Trabeculae.

osteoblasts to fix into crystalline form the body's heaviest elements which allows them to build such a solid foundation. But this same proclivity also explains why some of the heaviest toxins find their way into the bones—lead, radioactive uranium and plutonium, and so on.

Calcification continues until a hard shell of bone has surrounded what was once the cartilage model, and the remains of the model have dissolved away, leaving the relatively hollow interior laced with thin filaments of bone, called the *trabeculae* (from the Latin *trabs*, a supporting beam). Although individually quite fragile, these trabeculae are highly organized along lines of stress, and collectively contribute a great deal to the compressive, tensile, and shear strengths of each bone as a whole. A small cube of trabeculae sawn out of the inside of the thigh bone will easily support a two-hundred pound man standing upon it. And the carefully placed struts of the trabeculae create this supportive strength with only a fraction of the weight of solid bone.

The Periosteum

Throughout these processes of calcification and of hollowing, the per-ichondrium remains as the tough, tightly clinging sheath which surrounds all surfaces of the bone, in the interior and on the trabeculae as well as on the original outer surface. It becomes more and more richly blooded and innervated, and for the life of the bone it carries the nourishment for the osteoblasts, the dissolved salts for the bone crystal precipitation, and the bone's nerve supply. From the time of actual bone formation onward, this membrane is called the *periosteum* on the outside surfaces, and the *endosteum* on the inside surfaces.

This connective tissue sheath, particularly the outer periosteum, is one of the most vital components of our bony frame. Lined as it is with blood vessels, nerves, and the bone-precipitating osteoblasts, it is analogous to the bark of a tree with its inner cambium layer, which carries the sap and which manufactures all the new cells which add to the girth of the trunk. The activity of the perio-steum has as much to do with the shaping of the bone within it as did the original cartilage model, and it continues to shape the growing bone long after the cartil-age has been dissolved away.[6]

If this periosteum is torn, growing bone slowly spills out of the tear in irregular globs, like pitch from a gash in a tree. If a bone is entirely denuded of its periosteum, the bony material decays. But on the other hand, if the bony sub-stance is hollowed out, leaving the periosteum, its blood supply, and its osteo-blasts intact, these will work together to produce a new correctly shaped bone.[7]

Some bones are in fact shaped *entirely* by the capabilities of this surrounding membrane, without the benefit of an original cartilage model at all. Most of the bones of the skull, the jawbone, and the greater portion of the collarbones are formed solely by the activities of the periosteum, which in these instances needs no previous cartilage model to follow.

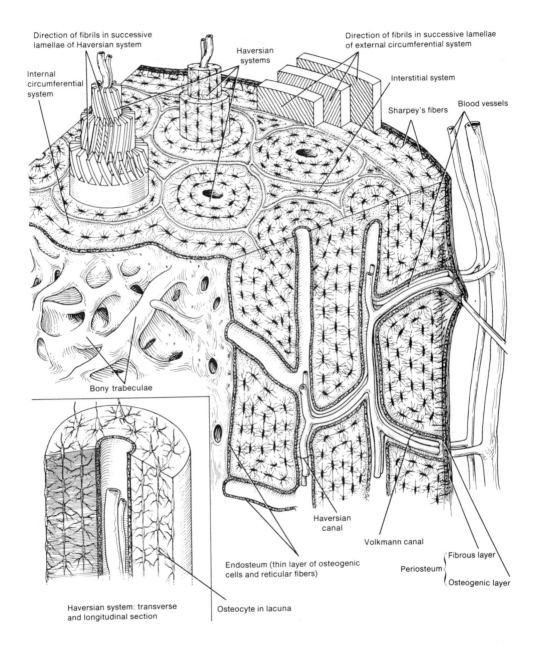

Direction of fibrils in successive lamellae of Haversian system

Internal circumferential system

Haversian systems

Direction of fibrils in successive lamellae of external circumferential system

Interstitial system

Sharpey's fibers

Blood vessels

Bony trabeculae

Haversian canal

Volkmann canal

Endosteum (thin layer of osteogenic cells and reticular fibers)

Periosteum { Fibrous layer / Osteogenic layer

Haversian system: transverse and longitudinal section

Osteocyte in lacuna

4-10: Schematic drawing of a section of bone. The periosteum is the outer wrapping that is pulled back by the dissecter's hook on the right. Note that the osteogenic layer of this wrapping—where new bone is formed—is continuous through the inner surfaces of the haversian canals, the interior of the hollow bone (where it is called the edosteum), and all the surfaces of the trabeculae.

The periosteum is also the means by which individual bones are joined together into a continuous structure. The matrix that holds it together and that contains the vessels, nerves, and osteoblasts is made up of collagen fibrils and ground substance, components that are no different than all the rest of the body's connective tissues. The collagen network which weaves together the periosteum of the bone blends into the fibers which compose a ligament; the collagen fibers of the ligament then span a joint and blend into the periosteum covering the next bone, and so on. Thus the sheath is actually continuous throughout the entire skeleton, each bone's membrane merging into ligaments and joint capsules, which in turn merge into the membranes of adjacent bones.

Muscle sacs and tendons, themselves made up of collagen fibers, attach to their appropriate anchors in the same way—not by contacting the bone itself directly, but by blending their fibers with those of the periosteum. A tendon is not, then, attached to any particular point on a bone, but rather to the membrane which surrounds the *entire* bone; and when we remember that this same membrane is continuous with the inner lining of the bone as well—the endosteum—we can appreciate the strength of the bonding between bone, ligament, muscle, and tendon. It is no wonder that the too-sudden contraction of a large muscle can sometimes dislocate or even break a bone before its tendon attachment pulls loose.

Bone Growth

These initial formations of cartilage and bone take place very early in an individual's life, many of them within the womb. All of the cartilage models are present by the fourth or fifth month of intrauterine life,[8] and by the time of birth many ossification centers are active as well.[9]

The date of the beginning of ossification is quite constant in different foetuses for any one bone, and the centers appear in a very orderly sequence that is the same for almost all members of the same species. The genetic intelligence displayed in these early developments is remarkable; the shape of a bone is evidently determined even in the initial limb bud of a tiny foetus. If a small piece of limb bud is removed from a foetal chick's femur and properly nurtured in the laboratory, a complete femur will develop from these few seed cells alone.[10]

This genetic program is contained partly in the connective and cartilaginous tissues themselves, and is partly the result of specific hormones (the major one being somatotrophin, the growth hormone), whose release into the developing system is closely timed by genetic factors unfolding in the central nervous system and the various glands.[11] Because the foetus is suspended in liquid in the womb, the bones are formed in their original size and shape in the absence of most of the gravitational and muscular stresses that will condition them in later life.

But this primary development of the skeleton is a small part of its lifetime of growth and modification. For the first four or five years, *secondary* centers of ossification appear as well within the cartilage molds, and from this time until the individual is about twenty-five years old they remain very active. It is true that this continuing development is directed in its general outlines by genetic factors, but as with all other growing tissue, the lengthening and thickening skeleton is also shaped to a large degree by sensory stimulation, diet, exercise, trauma, illnesses, emotional developments, and so on. As we have seen in cases of deprivation dwarfism, these factors can exert their most powerful influence in a child's younger years, but they continue to condition bone growth and repair all of our lives. And they can quite effectively overwhelm many genetic patterns.

Continual Building and Destroying

In order to maintain their highly specific shapes and yet increase their overall size by several times during the period of fifteen to twenty years, growing bones must continually exist in a delicate balance between cycles of depositing, destroying, and redepositing. If the precipitation of new mineral salts by the osteoblasts were the only mechanism of growth, it would merely add new layers of bone tissue onto the surfaces of the original models, making them thicker and thicker, and eventually obscuring their shapes like coral encrusting pebbles. Consequently, there is another type of cell that is necessary to orderly bone growth,

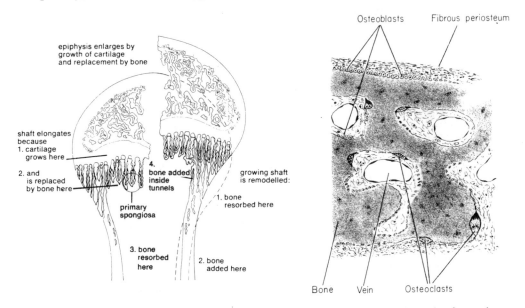

4-11: Both osteoblasts (which build bone) and osteoclasts (which dissolve it) are always in continual operation. The drawings indicate their finely coordinated efforts which allow the bones to grow in length more than they grow in bulk. The continual process also allows the bones to adjust the body's levels of free calcium, which is used by many tissues including muscles and nerves.

the *osteoclasts,* whose job is to dissolve calcified bone tissue that was formed earlier by the osteo*blasts.* It is the constant cooperation of these two counterparts that effects the unique growth pattern of the bones.

As osteo*blasts* in the outer periosteum deposit new layers of mineral salts on the outside surfaces of the growing bones, the osteo*clasts* in the inner endosteum dissolve the inner surfaces away, so that the hollow interior expands in size proportionately with the solid exterior. And this internal dissolving by the osteoclasts is extremely selective, leaving behind the proper thickness of bone wall and the lacy filaments of the trabeculae in exactly those patterns which most efficiently strengthen the bone to support the habitual stresses that are put upon it. In this way, the bones get larger without becoming disproportionately heavier and without altering their shapes. "The inherent skill of the tissue is uncanny. The gradual growth of a bone is a precisely balanced process of deposition and destruction."[12]

Epiphyseal Plates

At the ends of the long bones of the body, which must extend in length much more rapidly than they expand in girth, this dual process is particularly active. Here at the growth tips, new cartilage is continually added to create new length and larger articular surfaces. Beneath this new cartilage is a thin region called the *epiphyseal* plate, where osteoblast activity turns the cartilage into bone; as more new cartilage is added, the epiphyseal plate advances with it, leaving solid bone behind it which is eventually hollowed and trabecularized by the osteoclasts inside the bone shaft.

This lengthening and expansion in girth continues into the late teens for most of the bones in the body, and up until twenty-five for the bones of the spinal column. The height of the skeleton is then fixed.[13] But the dissolution and deposition activities of the osteoclasts and osteoblasts is far from over at this point.

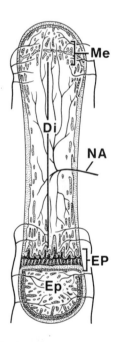

4-12: A growing long bone.
Ep: the epiphysis, the cartilage mold on the end of the bone; EP: the epiphyseal plate, where chondroblasts are converted to osteoblasts, and cartilage is displaced by bone mineral; Di: The diaphysis, the lengthening shaft.

They continue for our whole lives, keeping the bones responsive to dietary, mechanical, chemical, and psychological factors long after the original genetic growth program has retired from its work. And indeed, both the osteoblasts and the osteoclasts remain sensitive to a variety of influences. Various internal hormones continue to influence their behavior; in particular, two thyroid excretions are most important: parathyroid stimulates osteoclasts, and calcitonin stimulates the osteoblasts.

Improper diet will devastate a growing skeleton, and can even seriously erode the frame of a fully developed adult. Inadequate intake of calcium, phosphorous, and the various other trace minerals interferes with the strength, the renewal, and the healing of bone tissue. Without vitamin "D", the cartilage at the ends of the long bones becomes enlarged and painful (rickets). Without vitamin "A", the bones become abnormally thick and heavy. With no vitamin "C", the collagen matrix of the bones becomes weakened and watery, causing fragility and poor healing (scurvy).[14]

Bone Plasticity

The plasticity of seemingly fixed bones is dramatically illustrated by the ways in which mechanical stresses alter them over time. This resiliency and responsiveness to usage is lost only in old age, when the fibroblasts and osteoblasts begin to lose their vitality.[15]

Postural habits not only rearrange the relationship of the bones to one another, but also can alter the overall shapes and articulations of individual bones themselves. A lot of sitting in a squat, for instance, creates enlarged facets in the hip joints. A chronic slump can eventually alter the bony contours of the rib cage. Abnormal pressure on bone, such as that caused by tumors or by extreme postural aberrations, causes local erosion. Paralysis or long disuse of a muscle weakens bone where its tendon attaches, because even small local areas of bone are structured by the habitual demands that are made upon them.

In general, all sorts of habitual stresses, or the lack of them, alter the bones which are affected by them. "Bones are remodeled to meet increased exercise, or may atrophy from the lack of it."[16] In the healthy, full-grown adult, it is primarily the *load* placed upon the bone which determines the stimulation and the balancing of osteoblast and osteoclast activities, thus establishing the thickness of the bone walls. If a person habitually carries a heavy weight on the same side—say a heavy purse always carried on the same shoulder—the walls of his or her vertebrae and the supporting trabeculae will thicken on that side.

Athletes in general develop heavier bones. The bones in a leg that is kept in a cast become thinner, while those in the good leg thicken from the extra use they receive. Moderate stress that is consciously balanced, alternating with periods of rest, like dancing, athletic training, or healthy physical labor, produces strong, resilient, balanced bones; asymmetrical loads and unconscious abuse distort the internal symmetry of our frame. And stress that is con-

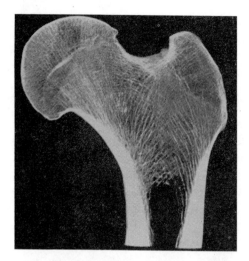

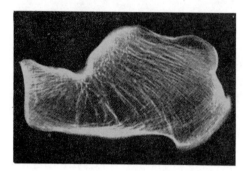

stant and long-term hampers growth and even erodes bone that has already been formed.

On the other hand, a bone that seldom or never receives any stress will atrophy just like an unused muscle. Clearly the absence of all stress is no healthier for the skeleton than is too much stress. Disuse harbors as many evils as does abuse. This is a theme we will encounter time and again in our examination of the body's plasticity, and of this plasticity's role in our physical health and well-being. It is not the complete absence of stress which creates optimal conditions for the development of our bodies and our minds.

Neither utmost endurance nor utter passivity is the condition which best serves our health; health is always a dynamic balance between the two. We suffer from too little, and we suffer from too much. We prosper when all elements are present in their proper proportions. The role of bodywork is not to eliminate stress, but rather to educate the individual to recognize

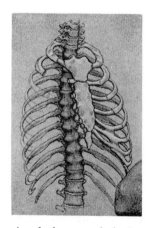

4-13: The thickness of the bone walls and the structural lines of the trabeculae are largely determined by the normal stress placed on the bones. These can change quite rapidly—in a matter of weeks—if stresses are increased or reduced markedly.

4-14: The spine of a slave unearthed at Pompeii. Most of his thoracic vertabrae have fused into a single bone due to the extreme stress which his labors placed upon his spine.

the right kinds and the right amounts. Just as with the productive rivalry between the osteoblasts and the osteoclasts, it is the intelligent balance between our many opposing tendencies that will reward us with natural forms of beauty and strength.

Most of the time it is the weight of our own bodies, our relative postural balance, and the tone of our muscles which create the loads and stresses which we place on our skeletons. Thus by altering poor posture and unbalanced muscle tone, body work can directly affect conditions which greatly influence bone growth, maintenance, and repair. Flaccid muscles result in weakly supported joints. Muscles sustaining a large amount of tension place constant strain upon bones and their joints, just as surely as does a heavy weight.

Growth at the end of long bones can be artificially retarded or even halted altogether by compressing with staples the two sides of the epiphyseal plate where the new bone is manufactured. Now this stapled bone is in exactly the same physical situation as a bone compressed against its joint by chronically tense muscles, and the growth pattern of the bone is disturbed by one as well as by the other. It seems reasonable to speculate that a large number of joint inflammations and erosions are initiated in such a fashion.

Scoliosis

One of the most common serious deformations of the spine is *scoliosis,* a gradual lateral bending and twisting of the vertebrae which creates a wide variety of abnormal curves in the spinal column. This condition usually develops slowly, over a period of several years, and eventually the vertebrae themselves can become distorted, along with the shape of the entire rib cage, producing a posture that is stooped, limited in movement, and often quite painful.

But "scoliosis" is not the name of a specific disease agent. It is the name of a characteristic corkscrew *shape* which is gradually imposed upon a previously straight and healthy spine. There are probably many different causes for many individual instances of scoliosis. Some of them may be genetic anomalies which surface at some point during development. Some of them may be particular diseases of the bones which strike like any other illness. But the fact of the matter is that no genetic factor has ever been isolated, nor has any disease agent been discovered, which can be said to be the cause of this twisting of the spine and distorting of the rib cage. It simply occurs. Nor has a successful standard course of therapy ever been devised; neither the radical spinal fusions nor the elaborate and excruciating body casts that are normally recommended can claim to offer anything like a satisfactory solution.

There are, however, some provocative clues that are associated with the onset and the development of many instances of scoliosis. To begin with, it almost always starts during adolescence, most commonly between the ages of twelve and sixteen. And it strikes significantly more girls than boys. In addition, about half of the children who are observed to be showing early signs of the characteristic bending and twisting of the spine seem to "grow out of it"; some go on to develop mild conditions, and only a relatively small percentage actually end up with severe skeletal distortions. And X-ray examinations most often indicate that the bones are pulled out of their normal placement *before* they actually begin to deform—or rather to *conform* to the stresses involved in this displacement.

This general profile of the "disease" is extremely interesting when we view it next to some of the the things we know about our physical and emotional development during adolescence. For one thing, these same years are the second most rapid period of growth in our lives, exceeded only by the years of infancy and early childhood. In particular, most adolescents are undergoing a considerable spurt in *height*. For another thing, this growth spurt coincides with the onset of puberty, and with all the physiological and emotional turmoil which characterizes our sexual maturation. This is the proverbial "awkward age." Adolescents have new bodies with which they scarcely know what to do. Those who are getting taller faster typically slouch their posture in an attempt to stay in line with their peers. Girls begin budding breasts, and many of them round their shoulders and shrink back their chests, trying to camouflage these confusing and embarrassing new appendages. And all adolescents are saturated with new hormones, new and acute kinds of self-consciousness, new kinds of desires, and confronted with the avalanche of new responsibilities that are associated with the threshold of adulthood.

All of this physical and mental turmoil creates a great deal of muscular tension in the adolescent. They squirm. They chew their fingernails. They tap their feet. They screw themselves up into the damndest kind of postures. They jump up and down and shout at the slightest provocation. They are like tightly wound springs. Might it not be the case that many instances of scoliosis are generated by the collision between this extra tension and the inexorable push of growth that is happening at the same time? If the growth processes are rapidly adding length to the spine while the emotional processes are building muscular tension, what can the spine do except to twist and turn, accommodating its new length to the space that the muscles will allow it?

This sort of cause and effect would explain why so many children who start to develop scoliosis "outgrow" it—these are the ones who somehow learn to control the violence of their emotions and to express their muscular activities in ways that are not so unconscious and self-destructive. Here is also a hint that suggests why girls are more frequently afflicted than boys: Girls are typically not given the same avenues of rough-housing, sports, and physical labor into which boys' energies are channeled. Girls are taught to be quieter and more prim, and often their hyperactive adolescent musculature is given little to pull against except their own internal parts.

Bones can only go where muscles pull them, and muscles can only respond to conditions which prevail in the nervous system. Scoliosis is a response to conflicting physical forces. It is not a bone disease until the bones have been pulled out of place and are then distorted by the architectural forces inherent in their new abnormal positions.

If these speculations are even partially correct, then here is the kind of situation in which we could expect bodywork to be very beneficial. Extra amounts of tactile stimulation would help the strained adolescent nervous system to organize itself and its monitoring of the body's growth patterns. Sensory input from all quarters of a rapidly changing body would help the teenager to clarify his or her own body image, and to be more consciously aware of the expansions and the shifts that are going on. The pleasurable side of this sensory input would make the emotional acceptance and integration of this new body easier and more complete. And the added neuromuscular education, the learning how to control tensions and how to discharge energy harmlessly, could go a long way toward making this explosive pubescence a time of enormous physical and intellectual progress, instead of a time of pressure, inhibition, and distortion.

Bones and the Body

The extensive fascia network and the bones provide the frame and the rigging for the support of the body and its parts. These two structural elements are woven into a single unit by the connective periosteum, ligaments, joint capsules, tendons, and muscle sacs. In fact, the strict distinction between all of these elements is somewhat arbitrary, since all of them—including bone—are merely slightly different forms of connective tissue which blend into one another with no sharp boundaries. Together they compose a complex and flexible tensegrity structure, made up of connective cables and bony spacers.

It is crucial to our understanding of the stability which this unit affords to realize that our skeletons do not support our posture in the same way that flat blocks stacked on top of each other support a building. There are no flat surfaces or securely stacked members anywhere in our frames. The whole collection of over six hundred bones would simply collapse into a pile if it were not held up by principles very different from those which support a stone wall or a marble column. Bones are not building blocks. They are a complicated and dynamic set of levers and spacers through which the entire musculature can act in order to constantly counterbalance the forces of gravity and of contraction, producing both stable erect posture and freedom of motion.

There is in the skeleton itself nothing inherently upright or even stable. And yet, when it is working in concert with the connective tissues and the muscles, it creates a rigidity without which we could not long survive. The skull, the rib cage, the spine, and the pelvis all protect soft organs vital to our existence. Without leg and arm bones, the muscles would be powerless to extend a limb or support a step. Boneless, the whole system would fall in on itself like a tent without poles.

In addition to these obvious structural roles, bones have indispensable functions in our body chemistry as a whole, which connect their healthy development to virtually every other tissue in the body. For example, the red blood cells which carry oxygen to every cell and organ, the various white blood cells which patrol our interiors for intruders, and the platelets which initiate blood clotting, all are formed in the marrow of the bones. Blood and bone therefore have an extremely intimate relationship, one upon which the health of the rest of the body depends.

The bones are also the major storehouse of the body's supply of calcium and phosphorous, two elements whose proper level in the blood, in the muscles, and in the nerves is absolutely crucial to the chemistry of those tissues. Phosphorous is continually necessary for the renewal of our supplies of ATP, adenosine triphosphate, the gasoline which powers almost all of our metabolic functions. The correct level of calcium in the blood is likewise a prerequisite for healthy cell function, particularly in the muscles and nerves. A low calcium concentration increases the excitability of muscle and nerve and muscle membranes, creating muscle spasms. These spasms may become widespread and severe enough to be fatal if calcium levels are sufficiently low. Conversely, too high a concentration of calcium depresses neuromuscular activity, causing flaccidity, lethargy, and even heart failure.[17]

In this light, the closely balanced activities of osteoblasts (which take calcium and phosphorous from the blood and fix them into bone crystals) and of osteo-clasts (which liberate the crystallized elements and discharge them into the bloodstream again) have as much to do with blood chemistry, muscular contraction, and nerve impulse propagation as they do with the growth of bones. And of course, the bones themselves, as storehouses of calcium and phosphorous, are affected by the needs of the nerves and the muscles; if these minerals are not adequately provided by diet, it is the bones that will be strip-mined to make good the deficiency.

Thus the skeleton provides more than one kind of support and stability. It has a highly plastic functional and chemical relationship with the body's connective tissues, as we have already seen. And it has equally plastic functional and chemical relationships with the musculature, to which we now turn our attention.

5
Muscle

For every thought supported by feeling, there is a muscle change.
Primary muscle patterns being the biological heritage of man, man's
whole body records his emotional thinking.
—Mabel Ellsworth Todd, *The Thinking Body*

The Largest Organ

When we think of bodywork, we usually think immediately of muscle. A healthily fed and exercised body is mostly muscle, and it is this bulk that we most obviously feel when we prod and stretch the torso and limbs with the various pressures, strokes, and movements that are used in body work. Muscle comprises the large majority of everything that is beneath the skin, and it is the localized discomforts and limitations that we normally associate with it which makes a person yearn for the soothing manipulations of a bodyworker. We have "stiff" muscles, "sore" or "achy" muscles; we suffer from muscular "blocks," from muscle "armoring," from "tense" muscles, "tired" muscles, "overworked" muscles, muscle "spasms," and so on. It is our muscles that are "in shape" or "out of shape," our muscles that make us "fit." And it is pliable muscle which gives us most directly the physical feel and appearance of "relaxation."

Depending upon one's physique, muscle makes up between seventy and eighty-five per cent of the body's weight. It is the presence of this remarkable tissue, just as much as it is the absence of chlorophyl, that defines us as animals instead of plants. Its unique ability to contract and lengthen is responsible for almost all of the movements of the body, from the most minute adjustment in the diameter of a tiny artery, to the broadest expressive gestures, to our locomotion on the surface of the planet. These activities consume the vast majority of all the food and oxygen that we take in. Our musculature is by far the largest and most metabolically active organ of the body. It, more than any other tissue (except for the excessive deposits of fat in obesity), fills the visible outlines of the body. Its abundance or scarcity and its flaccidity or tension define our size, our contours, our "feel," and the quality of all our physical actions.

As is the case with the skin, then, the conditions and activities of our musculature have enormous implications concerning our biological development, our mental states, and our competence in dealing with the world around us. It is no wonder that uncomfortable muscular conditions have been such a powerful impetus for the invention of bodywork treatments and such a formative influence upon bodywork techniques. Taken together, the skin, the connective tissues, and the muscles offer the skilled manipulator a wide and a direct path into an individual's psychological and physical responses to the world, and the relief of local problems in these tissues can have a most profound effect upon that individual's sense of well-being and effectiveness as an organism.

The Dynamic Structural Tissue

It is usual to think of the bones and the connective tissues as being the major elements of structure in the body, and to think of the muscles as primarily adding *motion* to this structure. The collagen network knits together the cells of all the organs and limbs, gives them their shape, and arranges them in the proper relationships to one another. The bones and ligaments give this network the rigidity necessary for erect posture, while the tendons connect the bones to the muscles so that muscular contraction produces specific motions.

But something very important is obviously missing from this notion of structure, if what we mean by "structure" is the fixing of parts in relation to one another and the maintaining of the overall shape of the whole. Even though it is true that if all other bodily substances were emptied out of the collagen matrix it would remain as a complete and detailed mold of the body and all its parts, it is also true that this matrix—and even the rigid skeleton within it—has no power to maintain the shape of the erect human being which once filled it. If it were hollowed out, this "structural" network would collapse and flatten itself on the floor like a handful of limp balloons piled on a jumble of sticks.

The structural element that is missing from the collagen/bone complex is muscle; it is primarily muscle that gives substance, shape, and stability to the body. Connective tissue and bone do not provide a solid framework from which muscles and organs are suspended, like the girders and beams of conventional architecture. Rather, the muscles actively *utilize* the cables and levers provided by the sheets of fascia, the ligaments, the tendons, and the bones in order to suspend the limbs and organs aloft in an erect and weight-bearing form.

Muscles supply the crucial *tension* in our tensegrity structure. The skeleton is

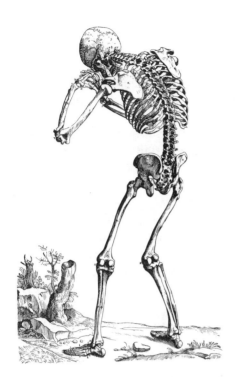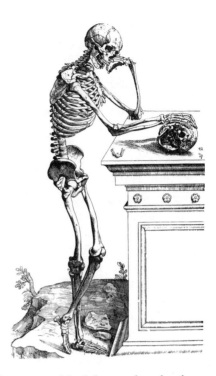

5-1: To a large degree it is the musculature which supports the postures of the skeleton, and not the other way around. A dynamic supporting medium, our musculature can change our stature, our shape, and our orientation in countless ways, and can fix us in any position that is anatomically possible. It provides us with an enormous range of utilitarian and expressive poses and gestures, none of which could be achieved by the "structural" bones or connective tissues.

The very arrangements of the bones in these two figures suggest distinct activities and states of mind. It is the musculature, using the levers and cables of the skeleton and the connective tissue, which creates these activities and these communicative attitudes. Only muscle can move bones and organs and fix them in space. The irony is that most aches and pains are routinely examined by means of X-ray, a picture which erases the causative forces involved and focuses both the diagnosis and the prognosis upon the bony and ligamentous *results* of chronic muscular activities.

held erect by the musculature, and not *vice versa.* The rigid girders of bone and the marvelous connective qualities of ligaments and tendons are useless without this sine qua non of structural support: "Stability of the joints is maintained above all by the activity of the surrounding muscles. Ligaments also play a part, but they will stretch under the constant strain when muscles are weak or paralyzed."[1]

General Anaesthesia

Consider what happens to any body under a deep general anaesthesia. A robust and athletic young man, who could normally carry a nurse under each arm, is prepared for an appendectomy. As the anaesthesia takes effect, all of his muscles slacken and lose their tone. His body can no longer support any upright posture whatever, and even further than this, great care must be taken whenever he is moved so that his spine or limbs will not dislocate. Far from providing a solid frame from which his muscles are suspended, the ligaments and bones are now in danger of being pulled apart by the passive and pendulous weights of the young man's large muscles.

Another example: An indigent man in his seventies is admitted for stomach surgery.[2] He is old, and he is stiff. He cannot turn his head, but must turn his whole body around to see behind him. Indeed, in every step and every gesture he displays the aging and hardening of his "structure" and the resulting limitation of his movements. As he is anaesthetized, this rigidity falls away from him in a matter of moments. Like the young athlete, he can then hold no part of himself erect, and is in danger of dislocating whenever he is moved. He is not in this danger because he is old, or stiff, or particularly fragile, but only because his loss of muscle tone has temporarily made him too "loose" to hold his joints together.

And what is even more astonishing is that after we have seen him in this passive and flaccid state, when the anaesthesia subsides, we can see his original stiffness creep back, degree by degree, into his whole body. As his consciousness returns, it exerts its old patterns through the nerves of the muscles, creating once again the stiffness of the spine and the limbs, the inflexibility of the joints.

Clearly neither the bothersome stiffness nor the dangerous flaccidity—the awake and sedated extremes of the old gentleman's structural qualities—has much to do with his bones or his connective tissues. It is the combination of sufficient tone and appropriate reflex responsiveness in his muscles, and no other thing, which create the blend of rigidity and mobility in his framework that is necessary for normal posture, gesture, and locomotion. And it is here, in these highly variable combinations of tone and resiliency, tension and release, that we can feel the palpable qualities of youth and strength, or of aging, stiffness, discomfort, and disability. Some of the most tangible and troublesome features of age itself are simply conditions of muscular activity, activities that bow the posture and rigidify the joints, that collapse the chest cavity, that put the squeeze on peripheral circulation, that create all kinds of limitations to movement and that waste precious vitality.

And so it is that when we think of body work we usually think of muscles, because it is in the exploration of tension and release and in the evocation of improvements in these conditioned responses that manipulative therapy has some of its most significant opportunities to restore posture, function, vitality, and normal activity to the individual.

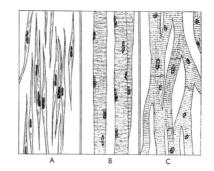

5-2: There are three general types of muscle tissue in the body—the "smooth" muscle of the internal organs and blood vessels (A), the "striated" muscle which powers the skeleton, also called "skeletal" muscle (B), and the cardiac muscle of the heart (C). It is the striated, or skeletal muscle with which we will primarily concern ourselves.

One Muscle, Many Compartments

The proper distribution of skeletal muscle tissue[3] into more than six hundred separate sacs or compartments is another one of the amazing feats that appears to be directed by the connective tissue framework in the developing foetus. Just as the foetal collagen matrix provides over two hundred molds of specific shapes and at specific locations for the deposition of cartilage and bone, so does it provide over six hundred pockets of appropriate size, shape, and location for the developing muscle tissue. And just as fibroblasts and chondroblasts migrated from their origins in the primitive embryonic mesenchyme to produce various structures of collagen, cartilage, and bone, so do the *myoblasts* migrate in an orderly sequence from the same source to their six hundred and more sites to fill their connective tissue molds with muscle cells.

Functional Unity

In examining the musculature and the effects of its contraction upon the various bones and limbs, there is some mental convenience in regarding each of these connective tissue compartments as an individual muscle with an individual function. However, this is a mental convenience which simplifies enormous complexities, and it has the danger of obscuring for us the real intricacies of the interplay of muscular tension and release throughout the entire system.

No matter how complicated they may be, the facts of the matter are that any given muscle compartment rarely works all of its fibers simultaneously, and that no one compartment ever acts independently of the many synergistic and antagonistic compartments which relate to its function. There is no one single muscle that controls one single motion; all the compartments must work in concert to produce graceful and efficient movement.

> The classic approach has been to study the effect of the individual muscles on joint action. But this is not physiologic. Normal muscle action is the patterned response of groups of muscles. Muscles may have anatomic individuality, but they do not have functional individuality.[4]

This is to say that if I pull on any part of a woven fabric, I create a pull over the entire warp and woof; a pull on one side of a tent affects the tension of the fabric clear across the roof and around all the sides. We will almost never find a single discrete muscle that is tense. Rather, we will find *areas* of tension, or body-wide *patterns* of tension, whose boundaries do not necessarily follow the anatomical divisions of muscle compartments. And we will never release a single muscle, but rather we will increase a *range of motion* that involves several, or many, separate compartments.

So we will be closer to the complex truth in our conceptualization of muscular activity if we regard the body as having only one muscle, whose millions of fiber-like cells are distributed throughout the fascial network and are oriented in innumerable directions, creating innumerable lines of pull. This single muscle may then contract or lengthen any number of fibers distributed in any number of compartments in order to change its consistency and shape in an almost amoeba-like fashion. Any contraction in one part will necessitate lengthening in other parts, and an extension of any one part will necessitate contractile bracing in other parts, so that the entire musculature must always utilize many of the different directions of pull afforded by the arrangement of its fibers and many of the cables and levers provided by the tendons and the bones in order to execute any single change of shape.

Raising an Arm

For instance, let us imagine ourselves observing a person who is standing erect and executing the simple gesture of raising their straight right arm to the side until it is horizontal. The fibers in the deltoid, the supraspinatus, and the upper trapezius will contract to produce the primary motion, while the fibers of the pectoral major, the pectoral minor, and the latissimus dorsi must simultaneously extend to allow it. But the contraction of the right trapezius will not only raise the right arm, it will also tend to pull the neck towards the right; therefore the left trapezius, along with other muscles of the neck, will have to contract as well in order to stabilize it. Furthermore, the extended right arm will overbalance the torso to the right, so the erector spinae muscles on the left side of the spine must contract to brace the whole torso and keep it erect. And since this contraction of the left erector spinae set will tend to pull the left side of the pelvis up as well, the gluteus medius and minimus of the left side must also brace to hold the pelvis level. Since not only the torso, but the body as a whole is threatened with tipping by the overbalancing weight of the extended arm, the right leg must brace as well, using fibers in the hip, the thigh, the calf, the feet, the toes. And of course our subject continues to breathe, so all of the muscles which cooperate to fill and empty the lungs must now make the necessary asymmetrical adjustments to continue their rhythm without disturbing the pose. And to further complicate the picture, if we add a weight (say a book) to the outstretched right hand, even more fibers from even wider areas will have to be called into play and instantly coordinated in order to preserve the position.

All of these muscular events must concur for such a small, isolated gesture, and even this description has been simplified considerably. It is the enormous complexity of this cooperative effort which drains the usefulness out of such singularized functional descriptions as "the deltoid raises the arm laterally," "the gluteus maximus abducts the femur," and so on. It is clear that muscle fibers from the occiput to the toes, and from both sides of the body, all must cooperate to "raise the right arm" It is utterly arbitrary, and possible only in the rarefied laboratory of the imagination, to say that it is the "deltoid" which is solely—or even primarily—responsible for any part of this gesture.

The idea of a single large muscle that can contract or extend any number of its millions of fibers and utilize any or all of its bony levers and tendon attachments to achieve an infinite variation of shapes offers, at least to my mind, a much more understandable and manipulable concept of the manner in which the system coordinates a gesture. There is only one muscle, controlled by one mind.

Tension and Release: The Sol/Gel of Muscle

As we have previously remarked, one of the most impressive features of connective tissue, the quality that makes it so widely useful as a structuring agent, is its ability to behave like a liquid crystal. It can be watery, elastic, wiry, or solid, depending upon local functions and stresses. It accomplishes this wide range of sol/gel properties in part by varying the proportions of liquid ground substance to tough collagen strands, in part by varying the number of hydrogen bonds which lock the long molecules of collagen together, and in part by investing its matrix with additional substances, like cartilage and bone.

Muscle tissue also displays this liquid crystal quality. We experience its extreme sol state in the passive flaccidity of complete relaxation, and its extreme gel state when it is flexed and hardened. Muscle tissue evolved out of connective tissue,[5] and it accomplishes its variations in softness and hardness in ways that are reminiscent of its precursor. Like collagen strands, muscle cells are quite long and thin, making them like fine threads when they are pliable and like tiny girders when they are stiffened. And like collagen strands, the interiors of muscle cells are composed of extremely long molecules arranged in parallel; these long molecules may be either loosely associated, as in our anaesthetized old man, or locked into one another in a fashion that is similar to hydrogen bonding in collagen fibers, giving the old man's muscles their stiff and hardened feel when he is awake. In both collagen and muscle, it is the varying degrees of loose association and chemical bonding of long, thin molecules which give us the varying qualities of sol and gel.

The major difference is that connective tissue adjusts to its degrees of sol and gel over longer periods of time—the millenia of evolutionary genetic development, the periods of rapid growth, the decades of drying out in the normal aging process, or the months and years of a chronic local strain. Its processes of hardening and softening, shortening and lengthening are relatively slow and conservative.

Muscle tissue, on the other hand, has developed the ability to shift its sol/gel states almost instantaneously to conform to the needs or whims of the moment. It can lengthen, shorten, soften, or harden all at the snap of a finger, creating an ever-changing kaleidoscope of structural conditions and movements. The activities of maturing connective tissue unfold our structure as a species Man; the activities of maturing muscles unfold the innumerable momentary structures that the individual superimposes upon his species—sitting man, standing man, walking man, dancing man, graceful man, awkward man, slumped man, erect man. Muscle is connective tissue that has learned how to quickly lengthen and contract.

Now no matter how soft or hard a muscle may be, or how complicated a coordinated movement, each individual muscle cell has only three options. It can shorten, it can lengthen, or it can lock into place preventing either motion. Yet from these meager choices muscle tissue produces all of the postures, gestures, and qualities of flesh of which we are capable. We stretch, we contract, and we lock into place; this is, in a nutshell, the entire gamut of our motor behavior. What are the arrangements and interactions of the muscle cells and their contents that make these simple miracles possible?

Myofibrils

If we examine with our naked eye an exposed muscle compartment—in a leg of lamb, say—we notice initially that all the long thread-like cells lie parallel to one another, and that the main compartment is divided into many long and narrow subcompartments by thin septa of connective tissue which hold the muscle cells in their parallel arrangement. And if we look even more closely with the help of a microscope, we will see that within each of these long parallel cells are many extremely fine strands, also parallel in arrangement, and running the length of each cell. These fine strands within the cell are called the *myofibrils,* and they are the actual contracting and lengthening units of the muscle cell.

In skeletal muscle, these myofibrils appear to be banded with alternate light and dark rings, giving these muscle cells their striated appearance, visually similar to the dark and light bands we observed on single collagen fibers. These myofibrils account for about eighty per cent of the contents of each muscle cell. So given the facts that we

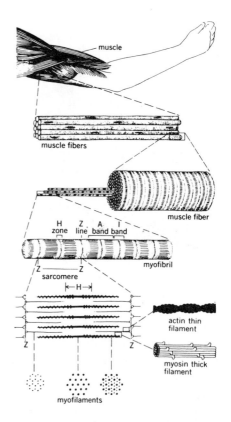

5-3: The composition of skeletal muscle fibers, or cells. The overlapping of the myosin and actin filaments gives the cells a striped appearance. A single unit of overlapping filaments is called a sarcomere; the arrangements within the sarcomere which create the striations are labeled—Z lines, I bands, A bands, and H zones.

are approximately eighty per cent muscle, and that each muscle cell is eighty per cent myofibrils, it follows that we are mostly made up of these fine strands; they comprise well over half of our weight and bulk.

Myosin and Actin

To find the actual mechanism of these fibrils' contraction, we must look even deeper into their molecular details with an electron microscope. This closer scrutiny reveals thousands of exceedingly fine strings of muscle proteins, also arranged in parallel, which fill each separate myofibril. These individual strings of molecules are called *myofilaments,* and they are of two distinct kinds.

One kind of myofilament is called *actin;* this is the thinner of the two. The actin filament is made up of two conjoined spirals of small, spherical molecules. These double spirals are then bonded tail to tail in clusters, with their lengths streaming in parallel and opposite directions.

The thicker kind of strand is called *myosin.* It is made up of much larger molecules, lollipop in shape. The shafts of the lollipops are bound together in parallel, creating a thick cable, while the heads protrude from this cable in a spiral arrangement along its length. These myosin filaments are not attached to each other in any way, but are stacked in between the free lengths of the actin filaments. This stacking arrangement is what gives the myofibril and the muscle cell their banded appearance, caused by alternately dense and sparse concentrations of molecules along the length of the fibrils.

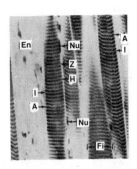

5-4: A photograph of muscle cells. The lines, zones, and bands are labeled.
Fi: muscle fiber; Nu: cell nucleus.

5-5: Electron micrograph of sarcomeres. The zones, lines, and bands are not labeled.

5-6: A highly magnified sarcomere. The heavy horizontal lines are myosin, and the lighter horizontal lines are actin. The dark vertical lines at the right and left edges are Z lines, and the light vertical stripe in the center is the H zone.

The protruding lollipop heads of the myosin strands are potential cross-bridges between the two kinds of filaments. When these heads make no chemical bond with their neighboring actin filaments, the myosin and the actin are free to slide back and forth along one another's length. This corresponds to the flaccid state of muscle in our anaesthetized patients. When the myosin heads do bond to their neighboring actin filaments, this sliding cannot occur; this is the condition of our old man when he is awake and cannot turn his neck because his neck muscles are locked into place. When the myosin heads move from bonding point to bonding point along the length of the actin filament, the muscle is actively contracting or lengthening at a controlled rate, and this is what happens whenever a particular motion is produced.

Obviously the activities of the myosin heads are central to the mechanism of muscular movement. When they are free, the muscle is flaccid; when they are attached to the actin chain, the muscle is locked into a fixed length; and when they are actively moving from place to place along the actin chain, muscle is contracting or lengthening. The exact chemical events which cause this behavior of the myosin head are quite complex and still being explored. There is currently no exhaustive and universally accepted theory.

But in the 1950's, H.E. Huxley advanced the idea that in the presence of certain catalysts available in the muscle cell these protruding heads act like ratchets against the actin chains, alternately locking, pulling forward, releasing,

5-7: A single actin filament. Note that it is two strands of pearl-shaped molecules that are twisted together into a double helix.

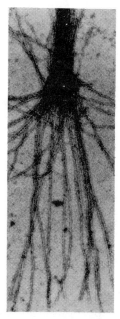

5-8: Actin filaments join end-to-end, creating the Z line—the junction between two sarcomeres. In this micrograph, the Z line appears as a dark region from which the actin filaments radiate.

re-locking, pulling forward, releasing, and so on, thus advancing their positions and creating contraction.

This theory has been repeatedly challenged and modified, and it still leaves many questions open. Nevertheless, it is our best available model, and it is worth looking at briefly both because it highlights several features of muscle chemistry and dynamics that are very relevant to bodywork, and because it increases our appreciation of the astronomical complexities involved in a single coordinated movement.

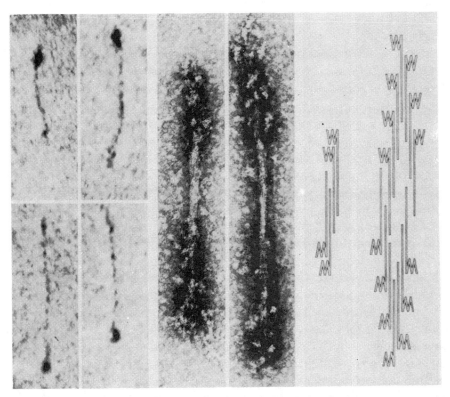

5-9: Myosin molecules and their aggregation into a filament. On the left, single molecules; center, a micrograph of a single myosin filament; right, a schematic drawing of the way in which the molecules stack together.

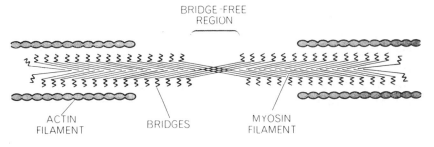

5-10: A schematic drawing showing the relationship between the myosin and actin filaments in a sarcomere.

Huxley's Ratchet Theory

When muscle tissue is in a completely flaccid state, the following conditions among its protein filaments exist: The myosin shaft (the lollipop sticks) are negatively charged, as are the actin chains; this means that they have no attraction for one another, and even experience a repulsion if they are brought too closely together. Attached to a bonding site on each myosin head is a molecule of ATP (adenosine triphosphate), also negatively charged. This ATP gives the myosin head a slight negative bias, so that it is both repelled straight out away from the negatively charged myosin shaft and pushed away from the the negative actin chain.

With the appearance of a crucial catalyst, positively charged calcium ions, this mutual stand-off is quickly altered. The strongly positive calcium ions bond to the actin chain (or, more precisely, to troponin molecules that are intimately associated with the structure of the chains), shifting the overall charge of the actin chain to a positive one. This positively charged chain then actively attracts the negative myosin head/ATP protrusions, bonding the myosin heads to sites along the actin chain. This constellation of molecules—myosin head/ATP molecule/actin bonding site—produces a strong positive charge, which is now attracted towards the negatively charged myosin shaft; and as they are pulled together the head of the myosin molecule is folded down flat between them.

This folding down of the myosin head advances the actin chain along the length of the myosin chain a small distance. Then, at a certain point (about a hundredth of a second after the myosin/actin bonding), the molecular tensions of the folding head become strong enough to snap apart the relatively unstable ATP molecule. When this snap occurs, the energy released is sufficient to break the tie between the myosin head and the actin chain. Almost immediately (about a thousandth of a second later) another ATP molecule bonds to the myosin head, charging it negatively once again and pushing it away from its negative shaft. As soon as it is erect, it bonds to yet another site on the actin chain, this one slightly ahead of the last one, and the ratchet process repeats itself, each repetition advancing the actin filament about one per cent of its length. Thus the myosin heads behave as "cross-bridges" to the actin chains, attaching, folding, and releasing in a continuous sequence much like the oars of a rowing team in a racing shell.

This process continues as long as the catalyzing positive calcium ions are present in large enough numbers. If their concentration is reduced, contraction weakens; if they are removed entirely, the bonding sites all revert to their mutually negative charges, connections are released, and the muscle becomes flaccid.[6]

This theoretical description has many holes in it, and begs many questions, so we will be well advised to regard it as an idea of how the thing *might* work, rather than as a precise description of reality. But whatever the actual molecular exchanges eventually prove to be, Huxley's general theory does suggest mechanisms for several incontrovertible features of muscular contraction. The

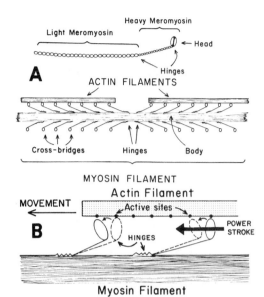

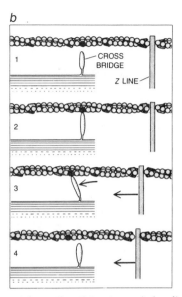

5-11: A ratchet mechanism. The myosin heads are hinged where they join the long tails which make up the body of the filament. These heads move back and forth, engaging active sites along the actin filaments, and sliding the two kinds of filaments past each other.

5-12: Myosin and actin filaments in a flaccid state, with no active bonds between them. Both the actin filament and the ATP/myosin head units have a slight negative charge, which holds them apart; the myosin filament is also negative, pushing the negative heads away from it.

5-13: An actin chain, showing the association of the troponin molecules and their partners, tropomyosin. When calcium ions (+) bond to the troponin molecules, the overall charge of the chain is changed to positive, attracting the negative ATP/myosin head units.

5-14: 1) Prior to the calcium/troponin bonding, the myosin heads with their negative ATP molecule are held away from the negative actin filament. 2) After calcium/troponin bonding, the negative ATP/myosin heads attach to the now positive actin. This positive bond is then attracted to the negative myosin shaft. 3) The myosin head folds downward, pulling the actin chain forward one notch. 4) The ATP/myosin bond breaks under the stress of the folding and pulling; this break releases the myosin head from the actin chain, and it returns to its upright position, immediately forming another myosin head/ATP unit, ready to make another bond with the actin chain.

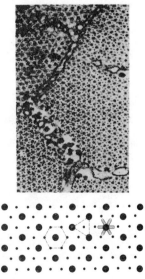

5-15: A cross-section of a muscle cell. Each thick myosin filament is surrounded by six thin actin filaments.

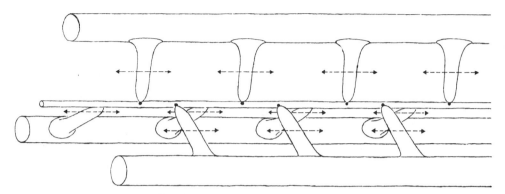

5-16: The myosin heads move from cross-bridge to cross-bridge, much like the oars on a racing shell contacting the water. The actin filaments are slid along between them.

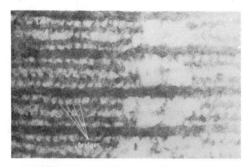

5-17: An electron micrograph showing the cross-bridges created by myosin heads bonding to the actin chains.

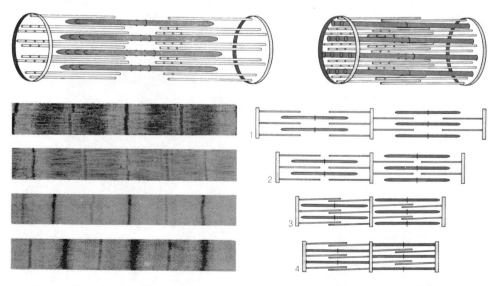

5-18: A three-dimensional model of contraction. As the actin filaments are pulled forward by the myosin heads, the sarcomere collapses as the Z lines are pulled closer and closer together. This shortens the overall length of the cell.

actin filaments *do* telescope back and forth along their parallel lines, and this sliding to and fro—and also the third alternative of locking into a fixed length—clearly *must* have something to do with an alternate bonding and releasing between the myosin and actin fibrils.

It is also clear that both of the two catalysts, calcium ions and ATP, play a crucial role in the process. Without these two humble ingredients, the fantastically complicated arrangement of muscle proteins is powerless, and each of these catalysts ties local muscular activity to larger processes that involve the whole organism. Calcium ions form the chemical bridge between the electrical events of the entire central nervous system and the response of every individual muscle cell, while the production of ATP is each and every cell's means of turning everything we eat and the oxygen we breathe into the fuel necessary to do its work.

The Calcium Ion

Why is muscle not always contracting, since it is the nature of myosin, actin, ATP, and calcium to continue to interact in one another's presence? Myosin, actin, and ATP are exposed to one another all the time in the cell's fluid, but what we discover about the calcium ions is that they are present only when a muscle is working; no calcium ions are found in the cellular fluid when the muscle is in a flaccid state. The appearance of the calcium ions, then, and their bonding to the troponin molecules on the actin chain—which changes the actin's overall negative charge to a positive one—is the event which triggers all of the rest of the process, while their retreat brings an end to contraction. Calcium is the immediate initiator of muscular activity.

But where is the muscle's supply of calcium stored? How does it disappear at the appropriate moments? When a muscle cell is not working, its supply of calcium is stored in a network of sacs within the cell, carefully segregated from the cell's main fluid contents. This network is called the *sarcoplasmic reticulum,* and it surrounds each myofibril. The whole reticulum is crossed in numerous places by fine little electrical conductors, called the *transverse tubules,* which are connected at both their ends to the outer membrane of the whole muscle cell. Now when a muscle cell is stimulated by its motor nerve—and *only* when it is stimulated by its nerve—the electrical charge from the membrane of the nerve cell is transmitted to the membrane of the muscle cell; from the membrane of the muscle cell, this charge is picked up by the transverse tubules which in turn transmit it to the interior of the muscle cell, where it contacts the membranes of the sarcoplasmic reticulum. When the reticulum is touched by the charge, pores in its delicate membrane open up, allowing the positive calcium ions to rush out into the cellular fluid, where they bond to the actin chains, initiating the cross-bridging cycles.

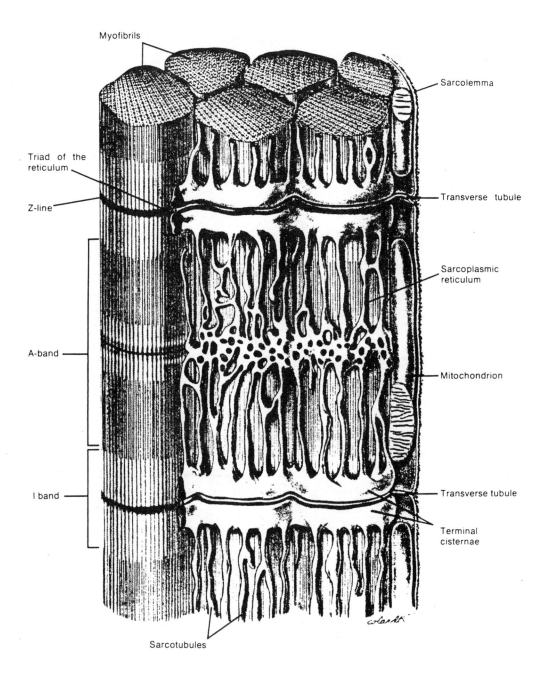

Myofibrils

Sarcolemma

Triad of the reticulum

Z-line

Transverse tubule

A-band

Sarcoplasmic reticulum

Mitochondrion

I band

Transverse tubule

Terminal cisternae

Sarcotubules

5-19: **A section of sarcoplasmic reticulum, with its transverse tubules. The tubules àre continuous with the muscle cell membrane, making pores in its surface which extend into the depths of the cell.**

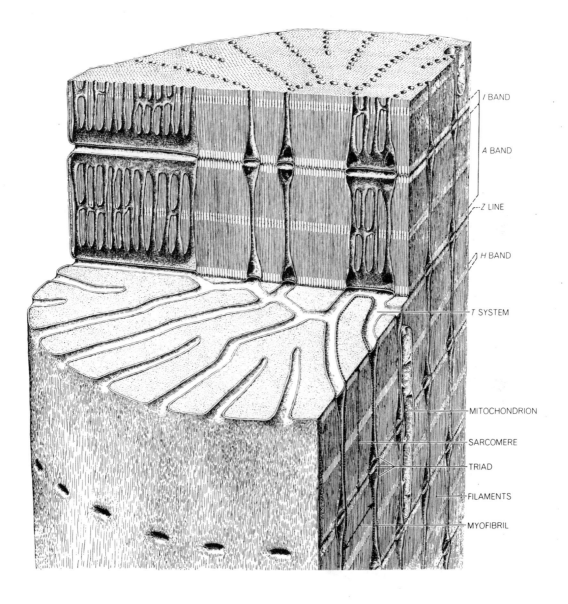

I BAND

A BAND

Z LINE

H BAND

T SYSTEM

MITOCHONDRION

SARCOMERE

TRIAD

FILAMENTS

MYOFIBRIL

5-20: The system of transverse tubules in a section of a muscle cell. We can see the intricate three-dimensional maze they create, surrounding every bundle of myosin and actin filaments.

Several drugs and diseases which affect muscular activity exert their effects upon this reticulum membrane, causing abnormalities of response. Some drugs can block the release of calcium from the reticulum, and keep the muscle in a relaxed state even when it is being stimulated by its motor nerve. High concentrations of caffeine do the reverse—they cause the spontaneous release of calcium from the reticulum, triggering contractions when no neural stimulation has occurred. Other drugs and toxins can interfere with the resorption of calcium by the reticulum, making it impossible to stop muscle spasms. The smooth and timely migration of these ions back and forth across this thin internal membrane is absolutely crucial to coordinated activity. In the healthy, undisturbed cell, this timing is monitored by the stimulation patterns of the motor nerve.

When the nerve ceases its electrical stimulation of the muscle cell's membrane, it is time for the calcium ions to be recalled back into the sacs of the reticulum so that contraction can cease. This is not accomplished as easily and instantly as was their release. The problem is something like gathering up a lot of helium balloons released in a room: It is one thing to let them go and watch them scatter, and quite another to fetch them all in again from their various contact points on the ceiling and stuff them back into their bag. Exactly how this so-called "pumping" action is done is far from clear, but we do know that one ingredient is necessary in order to do it. That ingredient is ATP, the same cellular gasoline that fueled the cross-bridging cycles of contraction.

ATP

Adenosine triphosphate, or ATP, is one of the most ubiquitous molecules in all of our cells. It is a relatively small molecule—only one adenine ring trailed by three phosphates—so it can circulate much more freely throughout the cells' fluids than can the larger, more cumbersome configurations of proteins. It is utilized in great numbers by these larger molecules and by membranes as a source of energy to maintain the very high degree of organization and rapid chemical exchanges that are typical of organic processes. It is the tension spring in the cell's clockworks, the fuel in the tanks, the steam in the turbine that keeps all the wheels moving and producing, and it provides this energy in a very simple, adaptable fashion.

ATP is a relatively unstable compound; the last phosphate on the tail can be popped off quite easily by a large variety of chemical interactions. When this third phosphate is separated, the bonds of energy that tied it to the rest of the molecule are suddenly released and radiated out into the immediate vicinity, like a tiny explosion. This released energy can then be snared by surrounding molecules to assist them in many of their active functions—forming bonds, breaking bonds, synthesizing proteins, maintaining the structure of membranes,

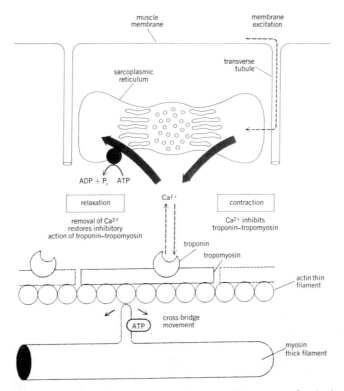

5-21: Calcium ion release and uptake by the sarcoplasmic reticulum. When a wave of excitation passes over the muscle cell membrane, it is conducted into the interior of the cell by the transverse tubules. When it reaches a sarcoplasmic reticulum, it opens gates in its fine membrane and calcium ions are released. These ions bond with troponin in the actin chains, and cross-bridge movement begins, causing contraction. For contraction to cease, the calcium ions must be pumped back into their sarcoplasmic reticulum storage bags, work which requires more ATP.

5-22: ATP (adenosine triphosphate.) When one of the three phosphates is broken off the unstable tail chain, energy is released. The resulting ADP (adenosine diphosphate) and free phosphate then must be rejoined to cock the trigger again for the next burst of energy. The nutrients we eat and oxygen we breathe supply the fuel for the continual reconstruction of ATP.

opening and closing their pores, "pumping" calcium and a host of other molecules back and forth through these pores, and so on. The number of ways in which ATP can be relieved of its third phosphate is enormous, and the uses to which the liberated energy may be put are equally legion.

Once this popping has occurred, the molecule is left in a disassociated state— ADP (adenosine diphosphate) plus a free phosphate. Now this ADP molecule is extremely stable, and offers no further energy release until its third phosphate is re-attached, cocking its "trigger" once again. Thus, the reconstituting of spent ADP back into potentiated ATP is one of the constant necessities of cell life.

The readiest way to do this is for the ADP molecule to contact another phosphorous compound found floating freely in the cell, *creatine phosphate*. When these two molecules interact, the ADP comes away from the skirmish with a new phosphate, once again become ATP, ready for new action. But creatine phosphate is present only in limited supply, and has to be reconstituted itself, so that a working muscle can only count on this source of renewed ATP for a few seconds.

Most of the cell's ATP reconstitution goes on in tiny but furiously active organelles within the cell, called *mitochondria*. They are the fuel-producing

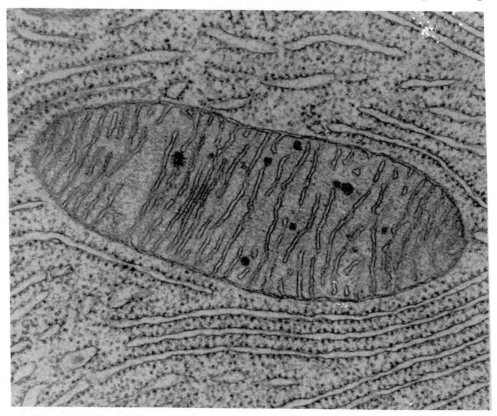

5-23: A mitochondrion, the site of ATP reconstruction. Tiny organelles within the cell (see figure 2-2), the mitochondria were probably once independent living units which migrated into primitive animal cells as parasites. They were able to strike a mutually productive bargain with the invaded cells, and have functioned as our fuel refineries ever since. One organism's waste product is another's life blood.

refineries of the cell, where everything that we have eaten is combined with the oxygen we have breathed in a combustion reaction that supplies the energy for the constant replenishing of ATP. The process is called *oxidative phosphorylation,* or, in plain language, the use of oxygen to maintain a slow burn that will provide the necessary energy for re-attaching loose phosphates.

The Krebs Cycle

By the time everything we have eaten has passed through its digestive and circulatory processes and is delivered to the individual cell, it has all been broken down and converted into glucose. And by the time it has migrated from the cell

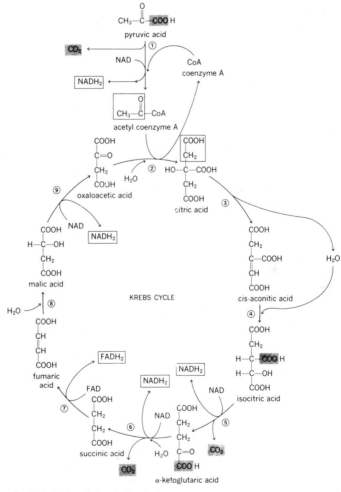

5-24: The Krebs cycle. This is the chain of chemical reactions which convert our food and oxygen into the energy required to reconstitute ATP. Oxygenation, or combustion, enters the picture once molecules of NADH are formed; the oxygen combines with the NADH, producing NAD + H₂O and liberating a burst of energy sufficient to reconstruct two ATP molecules. This is the slow burn which sustains the life of every cell in the organism.

membrane to the mitochondria, it has been further converted to pyruvic acid. This pyruvic acid then enters the mitochondria, where it encounters a series of enzymes, enzymes which initiate and maintain a long, circular metabolic reaction. This multi-phased reaction is called the *Krebs cycle*, first formulated in 1937 by the Nobel Prize winner Sir Hans Adolf Krebs.

Notice that at several points in this cycle, a particular molecule called $NADH_2$ is formed, an addition of two hydrogen atoms to NAD (nicotine adenine dinucleotide), one of the enzymes present in the mitochondria. It is at this point that oxidation, or combustion, takes place: One molecule of oxygen combines with the H_2 portion of the $NADH_2$; this reaction splits the H_2 complex away from the NAD, leaving a free NAD molecule plus water ($NADH_2 + O - NAD + H_2O$). This combustion and splitting releases a burst of liberated energy, and it is this energy which is utilized by ADP to attach its third unstable phosphate, creating two new ATP molecules for each $NADH_2$ molecule thus oxydized. Everything that we have taken into our stomachs and our lungs is refined and conjoined at this point, with the purpose of liberating pulses of bonding energy that are used to reconstitute ATP, which is then released from the mitochondria to do its work in the cell.[7]

ATP and Muscle Cell Functions

The working muscle's need for replenished ATP is enormous; it is crucial to at least three separate phases of the contractile process. First of all, in the relaxed state the myosin heads each require an ATP molecule to give them the slight negative charge that prevents them from bonding to the negative actin chain. If this ATP is absent, the solo myosin head has a positive charge, and it bonds very securely to the actin, and cannot be moved. This is what happens during the onset of *rigor mortis*—as soon as all of the ATP has been used up in a dead animal's muscle cells, the myosin heads bond tightly to the actin chains and cannot be moved from their positions.

Secondly, it is the popping of this same ATP molecule at the point of maximum stress on the "ratchet" which releases the myosin/actin bond and sets up the myosin head for its next contact in the cycle. Thirdly, more ATP is necessary for the "pumping" of the calcium ions back into the sarcoplasmic reticulum, so that contraction can cease when neural stimulation ceases. For these three reasons, maximum muscular effort consumes far more of our supplies of food and oxygen than does the activity of any other tissue.

It is the delivery rate of oxygen to the mitochondria that is the critical factor in sustaining a high level of work in the muscle cell. Most of the body's tissues can be broken down into glucose and fed to the cell, but the system's supply of oxygen must be constantly renewed with every breath. This is the key factor in all aerobic conditioning: The volume of oxygen that can be taken in, and the speed with which it can be circulated and delivered to the individual cells, dictates the absolute ceiling of muscular output by establishing the rate at which ATP can be reconstituted within the mitochondria.

The Fenn Effect

Muscle must be used in order to remain healthy and responsive; otherwise it will atrophy, decreasing its number of contractile fibrils until it is too weakened to be of much use. And the more it is used, the more important becomes the reliable delivery of oxygen and glucose via the circulatory system. The rule for the energy consumption of active muscle, called the Fenn Effect, is this: "The amounts of oxygen and other nutrients consumed by the muscle increases greatly when the muscle performs work rather than simply contracting without work."[8]

This seems obvious enough. But this simple fact is at the heart of the manifold difficulties that can be generated by chronic muscular tension. A muscle in a state of sustained tension is *working,* exerting a pull against a fixed resistance; and this very resistance is usually, of course, *another* muscle that is working in a static position. Therefore its nutritive needs are much higher than if it were at rest at a neutral length, or even contracting smoothly without resistance.

At the same time, the sustained contraction reduces circulation in the area by squeezing the small arteries and capillaries which service the working cells with glucose and oxygen. This is the first step in a circle that can become very vicious indeed—the more

The chart shows blood flow values (milliliters per second) by tissue across activity levels:

REST: BRAIN 750, HEART 250, MUSCLE 1,200, SKIN 500, KIDNEY 1,100, ABDOMEN 1,400, OTHER 600, TOTAL 5,800

LIGHT EXERCISE: 750, 350, 4,500, 1,500, 900, 1,100, 400, 9,500

STRENUOUS EXERCISE: 750, 750, 12,500, 1,900, 600, 600, 400, 17,500

MAXIMAL EXERCISE: 750, 1,000, 22,000, 600, 250, 300, 100, 25,000

5-25: Active muscles consume more ATP by far than any other tissue in the body (the numbers indicate milliliters of blood flow per second—that is, the delivery rate of nutrients and oxygen to the mitochondria). This is why chronically constricted, actively straining muscles can become such a significant energy drain upon the rest of the organism; they demand the lion's share of our blood flow, leaving less nutrients and oxygen for the activities of the other tissues.

work, the more need; the more constant tension, the less the fuel delivery; the less the delivery, the more difficult the work, and so on—until tissue exhaustion, with its discomforts, limitations, and toxic side-effects takes over in the area.

When an area of muscle is actively shortening and lengthening, shortening and lengthening, it actually *assists* its own circulation by pumping fluids through the capillaries and the intercellular spaces. But when it holds a contraction for an extended period of time, the pump becomes a squeeze, and fluid delivery is sharply decreased. This lack of circulation in a local area is called *ischemia*, a painful and potentially dangerous condition. It is ischemia that makes the feet blue and sensitive when they are cold, and that makes them decay when they are severely frostbitten. It is ischemia that makes the hands cold and painful when their circulation is poor. It is ischemia which creates bedsores when a person has lain so long in one position that the pressure of the bones against the surface of the bed pushes all the blood out of an area. All of these symptoms are the results of blocking the local blood supply, so that the cells do not receive adequate oxygen and nutrients to carry on their work.

> Muscular fatigue rapidly occurs with severe exercise, and, as an air cuff then applied to stop the circulation in a limb so exhausted prevents strength returning until the circulation is resumed, evidently the site of fatigue is in the muscle itself, not in the central nervous system. Even the light though skilled exercise of writing, if performed with the circulation occluded, becomes difficult in one and a half minutes, and almost impossible after another half minute.[9]

And in situations of chronic tension, it is the neediest cells that unfortunately get the least of what is needed.

Anaerobic Glycolysis

When there is not an adequate supply of oxygen for the work at hand, muscle cells have one more option in addition to creatine phosphate and the process of oxydative phosphorylation for the reconstituting of ATP, a process called *anaerobic glycolysis*. This is an enzyme-mediated reaction something like the Krebs cycle, but it functions without oxygen (an-aerobic), so that it is capable of extending a muscle's stamina somewhat beyond the limit established by the oxygen delivery rate.

However, it is a costly process, and builds up rapid debts. It is less efficient, so that it requires much more glucose to reassemble the same amount of ATP. And it produces far more lactic acid and other toxic metabolic wastes. It is probably the irritation of the tissues by this lactic acid and other waste products which produces the uncomfortable burning sensation in straining muscles, and which causes the later achiness in muscles that have been exerted beyond their aerobic capabilities. And these irritants undoubtedly contribute to the tenderness of muscles that have been under long-term stress.

So we can see that the health and efficiency of our muscle tissue—seventy to eighty-five per cent of ourselves—is dependent upon an adequate supply of ATP, and that this supply in turn depends upon the quantity and quality of everything we eat, the amount of oxygen we can inhale and circulate, and the kind of work we make our muscles do. Judicious exercise assists both oxygen intake and circulation, and hones muscle chemistry to a more and more efficient edge. Chronic tension, on the other hand, is worse than merely wasted effort; it initiates a vicious circle which plunges the area into deeper and deeper metabolic debts, draining energy from other parts of the body, producing ischemia and toxic wastes, creating discomfort, and eventual disuse.

Muscle and the Rest of Us

We can now begin to appreciate the extent of the interdependence of muscle with the rest of the tissues and systems of the body. Without the connective tissue framework, muscle fibers would not be organized into meaningful directions of pull, or be anchored to the bones, and their contraction could not produce the specific motions of the body. Without the digestive system, there would be no supply of glucose for the muscle cells, and without the lungs there would not be the oxygen required to burn this glucose to replenish the supplies of ATP. Without the circulatory system, this glucose and oxygen could not be delivered to the individual muscle cells that rely on them. And finally, without the appropriate electrical signals coming from the nervous system, no muscle cell would be able to begin or end its contractions.

Yet it is impossible to say finally whether these needs of muscle tissue constitute a slavery or a tyranny. Until all these things are supplied to muscles by the rest of the body, they can do no work; but until they are working, the organism can have no sustained existence. Every swallow of food, every breath of air, the distribution of every drop of blood, every exploration, every defense, even reproduction of the species—all are muscular activities. And the degree to which the muscular system is compromised, the entire organism is weakened, limited.

It is really not going too far to regard the musculature as the primary organ of the body, the dominant tissue of animal life.[10] It makes up the majority of our weight and bulk. It is by far the largest energy consumer. What we normally regard as our "vital organs" are, from another point of view, really only visceral support systems for the growth, function, and maintenance of the muscles. After all, it is these "vital organs"—lungs, heart, stomach, liver, intestines, various glands large and small, etc.—which must respond to the level of activity of the muscles, and not usually the other way around.

To be sure there are many factors contributing to the rate at which these digestive and circulatory "boiler works" take in, metabolize, and distribute: Age, sex, size, growth rate, menstrual cycle, infection or disease, body temperature, sleep or wakefulness, hormones, and emotional states all play significant roles in our energy needs. But, "all these influences on metabolic rate are small compared with the effects of muscular activity. Even minimal increases in muscle tone significantly increase metabolic rate, and severe exercise may raise heat production fifteen-fold."[11]

And when we reflect that muscle holding a chronic pattern of tension is working just as hard, and requires just as much metabolic support as does muscle that is exercising actively and getting actual work done, it becomes clear why and how muscular tension plays such a large and diffuse role in our physical and mental health. Nor is it difficult to understand why bodywork which effectively addresses chronic and wasteful muscular contractions can contribute so much towards changes for the better in our physical processes, our feelings, and our behavior.

Muscular Organization

We have observed within the muscle cell an astonishing degree of molecular organization. Billions upon billions of protein filaments are arranged end to end in precise sequences, and stacked together in precise geometrical arrangements. The resulting fibrils are laid side-by-side in parallel, and the whole collection is packaged together like a length of fine thread. Finally, within these fine threads are also contained the chemical catalysts which regulate a chain reaction of molecular exchanges so numerous that their number cannot even be meaningfully estimated, but which results in a smooth shortening and lengthening along one single axis.

We have also remarked upon the genetic wisdom of the formative connective tissues, which are responsible for our muscles' *spatial* organization. Fine walls of fascia hold the muscle fibers together, and maintain their parallel arrangements as they are collected into larger working units; thicker walls of fascia then organize these units into even larger packages and attach them to bones, giving them the leverage points and the directions of pull necessary for supporting and moving the whole frame. This spatial organization includes over six hundred compartments, large and small.

But these complex molecular and spatial arrangements are just the beginning of the organization of our muscle tissue. In spite of the precise alignment of its protein chains and the apt placement of its six hundred-plus compartments, muscle would be of no practical use if each of its individual cells could not turn "on" and "off," or if they were obliged to contract all at once, or if they contracted randomly. It is not merely contraction, but *patterns* of contraction that make muscle functional and fascinating. After all, it is not only engineering, steel, and gasoline which make an automobile a useful tool; it is the steering mechanism, and ultimately the driver, which bring purposefulness to its brute power. Since it is only the appropriate signal from a motor nerve that can make the contractile process start or stop, it is the pattern of these neuromuscular linkages to which we must look in order to begin to understand the *functional* organization of muscle.

The Motor Unit

Of the ten billion or so separate nerve cells in the nervous system, two or three million are designated as *motor nerves,* because they attach directly to muscle cells. The main body of each of these motor nerves is in the spinal cord, and this body is connected to particular muscle cells by means of a fine, long filament called an *axon*.

Both nerve cells and muscle cells reach a maximum number during foetal growth, and after that point no new nerve or muscle cells are produced for the entire life of the individual. Soon after their numbers are complete, the muscle cells and motor nerves are united by means of the nerve axons growing out from the cell bodies in the spinal cord, penetrating the musculature and attaching to their appropriate muscle cells. This connecting process is genetically controlled, and highly specific; presumably the blueprint of a normal human motor system is repeated in individual after individual by the subtle chemical attraction and repulsion between particular muscle cells and particular extending motor nerve axons, in much the same way that the sensory system of the skin is "mapped" by the orderly extension of sensory nerves.

Thus, a motor cell body which reaches its axon out to a portion of the "big toe" lies next to other cell bodies in the spinal column which also reach their axons out to muscles in the big toe; and these "big toe" cell bodies in turn send parallel connections *up* the spinal column to the brain stem and finally to the cortex, where once again the cortical cells whose channels connect to the "big toe" circuit lie side by side. And lying next to these cells in the cortex are the brain cells which belong to the "second toe" circuit, which runs adjacent and parallel to the "big toe" circuit all the way down from the cortex, through the brain stem,

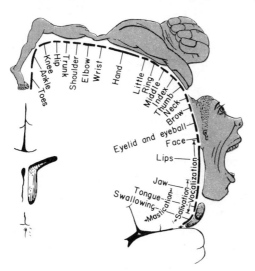

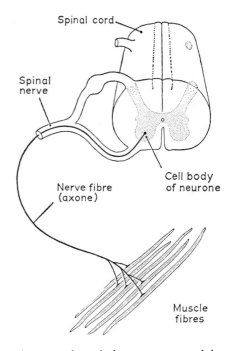

5-26: The map-like projection of the body's motor units upon the motor cortex, which creates a "homunculus" of associated cortical motor neurons. The parts of the body that have the greatest number of individual motor units (as opposed to the largest bulk of muscle tissue) have the largest representation in the cortex. The muscles of the hand and of vocalization have the most motor units, because they require the greatest precision in control.

5-27: A motor unit—a single motor neuron and the several muscle cells it contacts.

down the spinal column, and out the nerve trunk until they branch apart at their respective toes. In this way all of the peripheral neuromuscular connections are projected onto the highest surfaces of the brain, where they form a motor "map," or "homunculus" which faithfully represents their spatial relationships at the periphery.

At the peripheral end of these parallel circuits, the motor nerve axon attaches to a muscle cell by a motor *end plate,* creating a neuromuscular *synapse.* Each muscle cell receives one, and only one, end plate. Each motor axon, on the other hand, has a number of branches and attaches end plates to several different muscle cells in the same area. So each muscle cell receives commands from one nerve cell only, while each nerve cell stimulates several muscle cells. A single motor neuron with its group of attached muscle cells is called a *motor unit.*

Whenever a command is sent from the cortex, or a reflex is initiated in the spinal column, which stimulates a particular motor nerve, all of the muscle cells to which the nerve synapses are stimulated, and contract together. These motor units vary widely in the number of muscle cells involved, depending upon the strength or the delicacy that is necessary for the area in which they are found.

In the eye, for instance, where very small and precise movements are required, one motor nerve may control only ten to fifteen muscle cells. In the hand, where both precision and strength are needed, each motor nerve serves anywhere from ten to one hundred muscle cells. In the thigh, where added strength is more important than such fine control, a single nerve may synapse with one thousand to two thousand muscle cells. Each of these muscle cells can contract *only* when it is stimulated by its motor nerve; and any stimulation from a motor nerve obliges *all* of the muscle cells with which it is connected to contract simultaneously. These are the two basic functional axioms of the motor unit.

The Nerve-Muscle Partnership

These sets of motor units constitute the hard wiring of the neuromuscular system, and their connections are fixed in infancy. There will never be any additional muscle or nerve cells, and there will never be any additional synapses from motor nerve to muscle cell. From the time that our motor units are complete, we can make our muscles larger and stronger only by synthesizing more myosin and actin filaments within each muscle cell. And we can increase our motor skills only by learning to utilize the given wiring patterns with greater variety and subtlety.

Once this life-long bond is established, the muscle cell becomes extremely dependent upon its nerve, not only to initiate its contractions and lengthenings and to fix its resting lengths, but also to maintain its health as a living cell. Muscle apparently requires the constant play of electrical energy that it receives from its motor nerve just as much as it requires glucose or oxygen;

without its neural partner, a muscle cell cannot survive.

Even when there is no contractile stimulation, and we cannot discern any electrical activity, a constant exchange is evidently taking place between the neural and the muscular membranes, without which the muscle cells decline. If this exchange is arrested, by severing the connection or by somehow deadening the motor nerve, the muscle cell begins to atrophy, diminishing its myosin and actin contents. If the interruption lasts for three or four months, the muscle will be severely weakened, but it still may be returned to full function once stimulation is resumed. After four months the cells undergo further degen-

5-28: A muscle cell and its neural partner.

eration, and are only partially recoverable. If the situation persists for one and a half to two years, the cells break down altogether, disperse their myosin and actin, and are replaced by fat or connective tissue. Then nothing can regenerate them.

So the interdependence here at the cellular level is quite complete. Without its contractile partner, the motor nerve's electrical activity is without functional significance; and muscle relies upon its connection with the nervous system, not only for the stimulation of its practical functions, but also for the sustenance of muscle cell life itself. Without this constant contact, muscle suffers from a kind of "deprivation dwarfism," and eventually succumbs to a cellular "miasma."

Tone

In healthy, active muscle tissue, relaxation is a relative thing. The closest we can approach to being "completely relaxed" is under the influence of a general anaesthetic, and we have observed in our two surgical patients that this is a dangerous state. The primary job of the muscles, even in a body at rest, is to stabilize our structure and keep our joints from slipping apart. And in addition, muscle cells themselves will not thrive without a certain level of constant stimulation from their nerves.

Both this basic job and this basic need imply a degree of stimulation and tension-set in healthy muscle tissue, even when we sit motionless, or lie asleep. This basic level of tension required to brace our joints and maintain cell vitality is

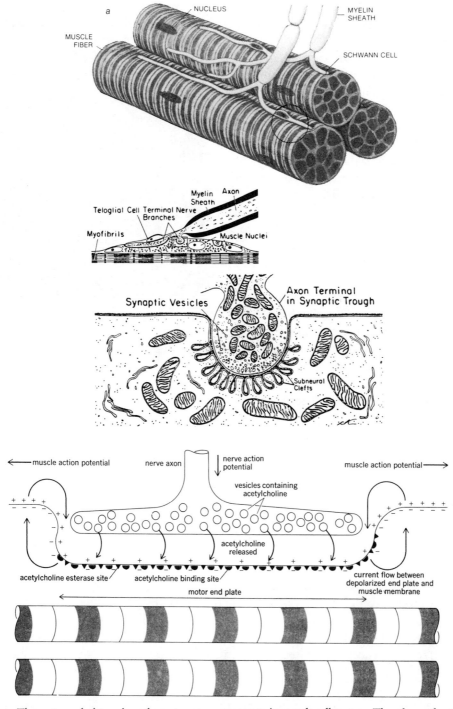

5-29: The motor end-plate, where the motor nerve synapses to its muscle cell partner. The release of acetyl-choline by the motor nerve stimulates the electrical wave of excitement which passes along the muscle membrane, into the transverse tubules and sarcoplasmic reticuli, and initiates muscle contraction by releasing the calcium ions.

called *tone,* or *tonus.* When this tone is absent, muscle has the helpless, flaccid feel characteristic of deep anaesthesia, paralysis, or of a corpse before *rigor mortis* sets in.

This tone's presence in healthy tissue, even in the "relaxed" state, is dramatically illustrated when a tendon is cut or torn: The muscle that the tendon fixed to the bone recoils forcefully and painful, and no amount of "relaxation" will passively lengthen it. It must be pulled back to its original length with equal force, and firmly held while it is re-attached. If the muscle belly is a large one, like a hamstring, this pulling and holding can require considerable effort.

Asynchronous Stimulation

Since any muscle cell will eventually reach a state of exhaustion—a depletion of available ATP—if it is kept constantly firing and working, one of the primary tasks of neuromuscular coordination is to maintain this overall tension set without exhausting any one cell. This is accomplished by a firing pattern called *asynchronous stimulation,* which alternates the working tonus contraction from motor unit to motor unit, so that some are always engaged while others are resting. The drawing illustrates a cross-section of muscle, say in a standing thigh. At this instant, the white-coded motor unit is firing, exerting enough base tone to maintain erect posture. Then the white motor unit switches off, and the

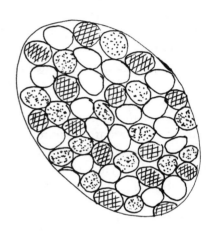

5-30: Asynchronous stimulation. Pictured is a cross-section of three motor units—white, dotted, and cross-hatched. When the cells of one unit cease firing, the cells of another unit take over, thus sustaining a steady tension without exhausting any one motor unit.

dotted one simultaneously takes up the tension load. As soon as this dotted motor unit tires—or rather, in anticipation of it tiring—the cross-hatched one takes over, and after the cross-hatched one has exerted itself, the white unit, which has rested in the meantime, is ready to work once again. Thus the postural load is transferred from motor unit to motor unit in a smooth and continuous trade-off.

If more weight is added, more units are engaged in each phase of the round-robin, each unit works a little harder, and each one rests a little more briefly. If one exaggerates the load on the thigh by holding a deep crouch, this flickering interchange can be seen and felt as a rapidly vibrating tremor in the thighs as the heavy load is passed from unit to unit. This flickering is happening constantly, and in all muscle compartments, in a pattern that is random insofar as there is no fixed firing order, yet not random at all in that the tonus load remains evenly distributed, with no jerks normally occurring.

It is the habitual level of this underlying tone which gives each individual's flesh its characteristic feel—the firm but pliable "juicy" quality of well-developed and maintained muscles, the hard, stringy feel of high-strung tension, the soft, flaccid feel of the phlegmatic.

What we know both about joint stability and about the demands of working muscle on the rest of the system, of course, suggests to us that these various levels of overall tone are fraught with significance for the entire organism. The level can be too low, giving a person a sense of weakness and lassitude, and making sudden exertions a danger to the joints. The level can be erratic and undependable, making precision and grace impossible. Or it can be too high, creating the hard inflexibility of our old man who could not turn his head without anaesthesia to release the tonal holding pattern being stimulated in the muscles of his neck. And with habitually high levels of tone, we must not forget that a chronically taut or fixed muscle is a constantly working muscle, one that exhausts itself, that hampers circulation, and that drains the reserves from the rest of the system.

All of these tone levels exist quite apart from any actual gesture, or external work being done. They vary widely from individual to individual, and even vary widely within the same individual, from a loosened state during sleep, to a calm waking state, to a state of anxious expectancy, all with only small visible changes in muscular appearance. Only when the muscles and limbs are *felt* and *manipulated* is the quality of this underlying tension-set clearly announced, both to the one touching to the one being touched.

Superimposed Movements

Now when any particular motion is to be executed, a separate series of neural signals must be sent which overlays this foundation of tone with additional stimulations. As we saw when we previously examined the raising of an arm straight out to the side, many muscles must contract and many other muscles must lengthen, while they *all* maintain a steady level of tone so that the motion does not cause dislocations from movement at the joints. And many muscles must neither shorten nor lengthen, but rather must *increase their tone* to create the bracing necessary to keep the structure from falling toward the over-balanced side. With the arm held out to the side, the whole musculature then readjusts to a different selection of tone settings, necessary for maintaining the new posture.

And if we now begin to walk forward while still holding out our arm to the side, a whole *new* set of motor commands imposes a *third* layer of stimulus pattern, which must adjust to the conditions of the fixed shoulder girdle, the over-balanced torso, and the compensations in the neck, the back, the hips, and the legs, in addition to adjusting to the original habitual pattern and level of resting tone.

We begin to glimpse the fabulous complexities involved in our simplest habitual motor patterns and movements. We are born with an initial level of overall tone, which keeps our bodies from collapsing or dislocating; as we grow, this tone level must gradually increase to compensate for the greater pulls from heavier and longer limbs. Ideally, this tension-set hovers around a compromise between stability of structure and freedom of movement, keeping things together while at the same time allowing for new weight, new strength, broader ranges of motion, quicker responses, more complicated combinations.

Then as we mature and develop our own ways of doing things, our habitual repetitions begin to build up a new sort of "norm," a more highly individualized "neutral," which overlays the primary tone levels. The unique ways in which each of us learns to walk, to hunch over a book, to stand with most of our weight on one leg or the other, to tilt our heads, to slump our shoulders, all require muscular changes to maintain these new poses and new patterns. So the local tone values all over the body must adjust to new base levels—tighter here to jack a shoulder up, more slack there to sag a chest, altered in another place to turn a foot outward, and yet in others to balance the spine on a twisted pelvis. Thus the musculature establishes new tone settings, new shapes and tensions, which eventually become our individual postures, as minutely unique to each of us as are our fingerprints.

Tonal Habits

Needless to say, these new local tone settings then condition the range and quality of all our motions. All future motor development builds upon this new base, incorporating its distortions into all our activities. As the seductive reinforcement of habit fixes these new local values more and more strongly, they begin to dictate what is easy, what is difficult, what is comfortable, what is painful or fatiguing, what is possible, what is not. And of course as we grow older, our habitual *emotions* as well as our habitual actions begin to make their stamp more and more visible.

Chronic anxiety, for instance, wreaks havoc with muscle tone, driving values everywhere up to higher levels without accomplishing a bit of practical work. Here the reassuring feeling of our muscles holding us together is seized upon and elevated to an exaggerated psychological significance: The tightening down of all the muscles in an attempt to combat the sense that things are about to "fly into pieces" is a normal neuromuscular response to the disoriented feelings associated with extreme emotional distress.

This is the way in which our entire histories become recorded in our flesh. Each stage of our motor development sets the overall tone and the idiosyncrasies which condition the next one, and we continually carry it all forward with us, becoming what we have created stage by stage. Nor is it possible to completely avoid this process; the accretion of habits is simply the way that motor development works, and it has as many positive results as it has negative ones.

But what we *can* do is to learn to recognize and to avoid the progressive development of *degenerative* habits, the ones that bow us, bind us, and trap us within ever narrower ranges of comfortable activity. We can exert our will, and use our conscious efforts to throw off destructive patterns and to develop more flexible, open-ended ones. What is needed first to begin this sort of change is awareness, awareness of the tendencies toward which our muscular habits are leading us, awareness of the local areas in our bodies that are conditioning those tendencies, and awareness of what it would feel like if our patterns were different.

It is exactly this kind of awareness that competent bodywork can impart to the individual, more quickly and more directly than any other means. Periodic treatments in the life-long course of development gives the individual the chances to pause, to actually *feel* what has come about through habitual usage, to reflect upon the present situation's progressive tendencies, to awaken insight into problem areas, to experience the relief of release, and—most importantly—to inspire the conscious will to take a more active hand in future developments.

"Optimal" Tone and Posture

I am not referring to merely the "fixing up" of a poor posture, the kneading and stretching of an individual until he can stand erect without any visible anomalies. This can indeed impose a new, more balanced "norm," and be helpful in its way, but it also runs the risk of establishing a false ideal. There really is no "proper" or universally "optimal" posture or activity. It is inevitable that we develop anomalies, because we will never use our bodies symmetrically. Resilience and vitality come not from preserving "ideal" postures, but from constantly renewed challenge and change. Only by continually varied use can general alertness and response—as opposed to a single dominant ingrained skill or habit—be brought about.

Any single repetitive pattern tends to distort the open-ended flexibility of the entire system. Let us be on our guard against adopting any particular posture, mode of exercise, or repetitive discipline as being perfect, or ideal, or best. Only *constant variation* calls the full alertness of the system into being. It is, after all, constant variation that we are called upon to cope with throughout our lives, a condition from which we can only partially insulate ourselves no matter how hard we may try to cling to models, and no matter how "right" those models appear to be from a particular theoretical point of view.

Any set pattern, no matter how good it looks on paper, is a fixation that threatens to be crushed by its own resistance to the onrush of things and events. The goal of bodywork should not be to impose universalized standards of posture and movement upon an individual, but rather to help the individual to cultivate the mental awareness and the physical flexibility to continually adapt to the changing needs of the moment.

"Voluntary" Muscle

The striated skeletal muscles are often referred to as the *voluntary* muscles, because they respond to our conscious commands (as distinct from the smooth muscles of the viscera and the cardiac muscles of the heart which, according to the traditional view, respond to unconscious, autonomic rhythms of stimulation). But our examination of even the most rudimentary levels of skeletal neuromuscular organization should indicate to us that it is only in a limited sense that we can call this system "voluntary."

It is true that my arm will respond appropriately to my mental command "Raise right arm out to the side," but the thousands of muscular events which actually accomplish that response are quite outside my normal conscious awareness, and quite beyond my normal ability to precisely control them. Perhaps a gesture is the most "voluntary" that it can be only when it is new and clumsy, and requires our full attention to every detail in order to initiate it and gradually perfect it. After that, training and habit begin to direct its course more and more firmly, until usage makes it as automatic as swallowing, and it itself becomes a factor which will condition the range and quality of all future gestures just as much as will their conscious commands.

I am aware that my arm conforms to my desire for it to raise; I may even be acutely aware of the contractions of the deltoid and trapezius which help to accomplish this; but I am not usually aware of the large number of muscle fibers in my neck, my back, my legs, my feet, and so on, which must also respond appropriately in order for my wish to be properly executed. The neuromuscular system is organized in terms of broad areas and general intents, not in terms of isolated one-to-one circuits connecting single cerebral to single peripheral cells.

True, it has been demonstrated that if I use a sensitive biofeedback device, I can learn to exercise exquisite control over a single motor unit. But this is not the way I normally control my movements, or else I would be continually inundated by the thousands and millions of informational bits that are involved in even the simplest gesture. These bits of information are not processed, and these motor decisions are not made, in the normally conscious levels of my mind. They are established over time by practice, usage, habit, development of style. And the more securely they settle into habits, the further they recede from my conscious awareness and my immediate voluntary control, until it can even become a considerable problem for my conscious commands to alter these familiar motor patterns, change their style.

The entire musculature *is* faithfully mapped in detail in the cortex, and each muscle cell *is* activated by one and only one motor nerve. But this one-to-one correspondence is very quickly lost in the millions of synaptic combinations which connect all the ascending and descending tracts of the spinal cord, the brain stem, and the higher brain, combinations which blend signals from all parts of the body and the nervous system with the activity of any given muscle cell.

We are not to imagine the idea of a motion conceived in the cortex, sent along a single circuit to an isolated muscle which then contracts dutifully to perform the task exactly as commanded by the conscious will. Rather this command, this will to move, leaves the conceptual precincts of the brain to filter out through a vast intermediate network of neurons, where it is weighed against all manner of feedback—limb positions, visual orientation, gravitational perceptions, available energies, and so on—as it proceeds from synapse to synapse. When the impulse for contraction finally reaches a muscle, it has escalated to an astonishing complexity from the simple mental command "Raise right arm straight out to the side," to a system-wide series of messages with literally thousands of components directing muscle fibers from head to toe in order to effect smoothly and accurately the motion desired. And our entire physical and psychological histories, our attitudes, memories, and present states of mind have just as much to do with the particulars of the final outcome as do any other, more "voluntary" factors.

Hence the term "voluntary muscle" is in many ways a figure of speech. I can consciously command a movement, but I cannot consciously command the recruitment of every muscle fiber which must be used, nor the precise order of their contractions and lengthenings which actually produce the desired effect. This is to say that every consciously willed movement is always conditioned by two things: genetically established organization and habitual usage. Our genetic organization is quite plastic, open-ended, filled with potential variations in behavior; on the other hand, habitual usage can become just as limiting as it is convenient, and can become a tyrant to exactly the degree that it becomes practiced, automatic, unconscious. We are free to train ourselves to act differently, but it is very difficult to suddenly act differently than we have been trained.

The tendencies in our motor behavior created by genetically determined patterns and by habitual usage do not lie within the muscle cells, nor even in the motor neurons that unite them into motor units. The search for the organizational factors of purposeful muscular control—whether it be action or relaxation—takes us deeper and deeper into the central nervous system, where we find that every muscular response is built up, selected, and colored by the totality of our neural activity, both conscious and unconscious.

It is to the functions of this neural network that we now must look, because just as no bone can move without shifts in the tensions and the lengths of muscles, so no muscle can make those shifts without neural stimulation. And the patterns of stimulation that are passed on by the nerves contacting the muscle cells are the ongoing summations of all of our sensory and mental events.

6

Nerve

You may not be what you think you are,
but what you think, you are.

— Jim Clark, Big Sur poet

The Connective Tissue Framework

Like the skeletal system, and like the muscular system, the nervous system is surrounded, organized, and supported by an elastic framework of connective tissue. The cells which form and maintain this frame are specialized connective tissue cells, called *glial cells*. They account for a considerable part of the bulk of "nervous tissue," making up the tough, elastic netting which cradles and anchors the delicate neurons.

The glial cells support every neural fiber, collect these fibers into bundles, and separate these bundles from the surrounding tissues and fluids. They give the nerve fibers the tensile strength and elasticity to stretch where stretching is needed, and they fix the nerve bundles securely to other structures where stability is needed.

This surrounding glial tissue also comprises the ground substance for the intercellular metabolic activities attendant to the needs of the neurons, mediating the interchange of nutrition, gases, hormones, and waste products between nerves and capillaries, and carrying the white blood cells, antibodies, and other immune factors which guard the lives of the irreplaceable neurons.

Other specialized glial cells—oligodendrocytes and Schwann cells—insulate the long axons with a tough, fatty coating called *myelin*. This insulation prevents signals from one axon inadvertently "leaking" into adjacent axons, and it also speeds up the passage of a neural impulse considerably. Myelin is whitish in color, giving the so-called "white matter" of the nervous system its name. "White matter" contrasts to "grey matter." the color of cell bodies and axons that are not coated with myelin sheaths.

The glial connective tissue, although it is specialized for its relationship to the nerves, is continuous with the ground substance and collagen net which make up the intercellular fluids, the fascia, ligaments, tendons, and the periosteum membranes throughout the body. It is a part of the connective structure of the organism as a whole, and as such is affected by the same kinds of factors—the movement of fluids, the diffusion of various chemicals, the rates of local metabolic activity, relative moistness or dryness, its own mechanical activity, age, and so on.

There is no reason to assume that the purely physical benefits resulting from

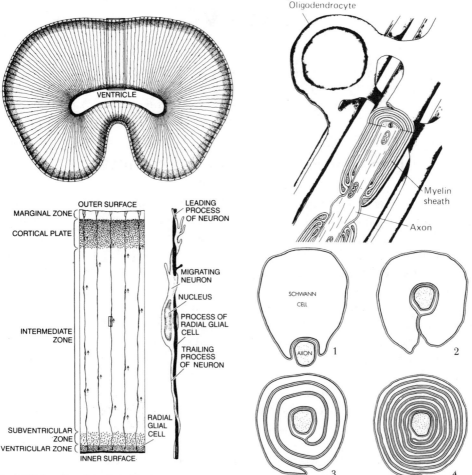

6-1: Glial cells of various kinds support nerve cells—structurally, nutritionally, and developmentally. Here a growing axon is seen following the lead of its glial cell. Various forms of these specialized connective tissue cells guide the growth of axons, bundle them together, support neural structures, and secure them to surrounding tissues.

6-2: Specialized glial cells, oligodendrocytes and Schwann cells sheath nerve axons with myelin, an insulating fatty substance which improves the transmission of neural impulses.

The bottom figure illustrates four stages in the growth of a Schwann cell around an axon.

effective manipulation would not apply to glial tissue as well as to any other connective tissue in the body. That is, its fluids can be helped to circulate more freely; its temperature can be raised and therefore its metabolic processes speeded up; pressure and stretching can stimulate its thixotropic effects, limbering up and lubricating tissue stiff from disuse; its wastes can be flushed more efficiently, and its overall resilience improved. All of these effects upon the glial cells could not help but improve the function of the neurons which they support.

Peripheral and Central

The nervous system is generally regarded as having, at least for the purposes of its topographical study, two major divisions—the central nervous system and the peripheral nervous system. The central nervous system is the deepest core structure of the body, being completely surrounded by the bones of the spine and the skull. This division contains the vast majority of our neural cell bodies and interconnections, making up the spinal cord and the brain. It is complex in the extreme, far too much so to completely unravel with current dissection technology, and its fragility is attested to by its need for the heavy, bony sheaths of the spine and the skull.

The peripheral system is made up of all those neurons that are not contained in the spine or the skull—the cell bodies, trunks, branches, and nerve-ends which reach out to all quarters of the body's interior and all areas of the skin. There are in this system two broad groupings of nerves—those which carry sensory information to the central system, and those which carry responses back out to the skeletal muscles, the organs, and the glands.

These last two divisions, together with the central portion, outline in the simplest possible terms the principal functional features of the nervous system, namely, a stream of information coming in from the sensory receptors of the body, a stream of commands going back out to the muscles and glands of the body, and an intermediate processing plant where responses between these two streams are coordinated.

> It is important to remember that in its simplest form the nervous system is merely a mechanism by which a muscular movement can be initiated by some change in the peripheral sensation, say, an object touching the skin.[1]

Data enters, responses are selected, and specific behavior results.

Such divisions are helpful to a degree, insofar as they serve to give us some basic sense of orientation within vast complexities and unknowns. The world ocean, after all, is really just one large body of water, but our analyses of currents and our global navigation is made far simpler by adopting the standard divisions of Atlantic, Pacific, Indian, Gulf of Mexico, and so on. Yet these same sorts of helpful divisions and labels can put us in grave dangers with regard to our understanding of the nervous system as a whole. They create the seductive analytical illusion that there is a separate part of us that "feels," another part which "behaves," and a more remote, interior part that "thinks." Ultimately this use of language leads us into the Cartesian dichotomy of the "mind" *versus* the "body," and into fiercely argumentative theories about where the "line" between them should be drawn.

We dupe ourselves into believing on one hand that "matter" is one thing and that "thought" is quite another, even while we theorize on the other hand that the various qualities of consciousness have various distinct locations within our physical systems. Ludicrous confusion, simplification run amok. There is no tissue that is not "body," and no response that is not "mind," and any analytical terminology which tends to divert one's attention from these irreducible facts must be used with extreme care, lest the limitations of our words become the limitations of our understanding, and we proceed to mutilate our concepts in the name of logical convenience.

"The basic unity of the nervous system must never be forgotten."[2] In the vast network of our neural impulses, there is no local activity which does not affect or is not affected by the entirety of the activities of the organism.

This fundamental unity of function is attested to by the very design of the individual cells. They may vary widely in shape and size, to accommodate a greater or lesser number of connections, and to achieve various anatomical configurations, but they all have the same basic parts: The cell body, containing the nucleus; dendrites, which receive connections coming from other neurons; and axons, which connect to the dendrites or cell bodies of other neurons. And even more importantly, their specifically neural function, the propagation and transmission of electrical signals—the *action potentials*—is accomplished in the same manner in every cell in the system.

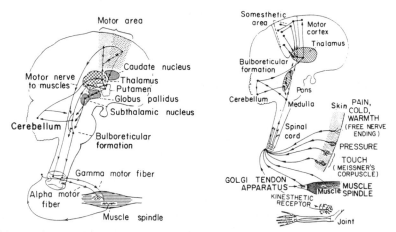

6-3: "It is important to remember that in its simplest form the nervous system is merely a mechanism by which a muscular movement can be initiated by some change in peripheral sensation, say, an object touching the skin." The human nervous system has sophisticated these simple elements tremendously, but the fundamental principle still holds true for us: Muscular activity is largely a response to the specific qualities of sensory input; as the quality of input changes, so does motor behavior.

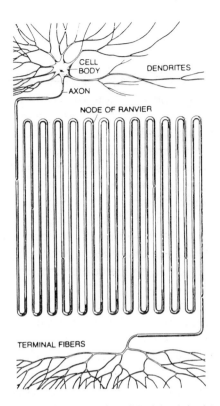

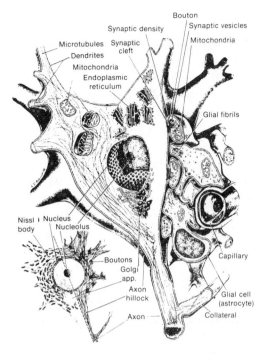

6-4: A typical neuron. The cell body and dendrites receive synapses and take in impulses from the axons of other neurons; the axon conducts these impulses away from the dendrites and cell body; the nodes of Ranvier are part of the myelin sheath which wraps the long axon; the terminal fibers synapse to the cell bodies and dendrites of other neurons, and pass the impulses on.

6-5: Detail of a neuron's cell body, with a capillary and supporting glial cells.

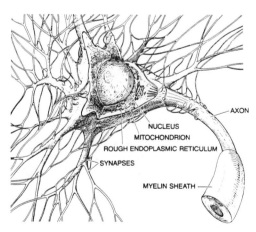

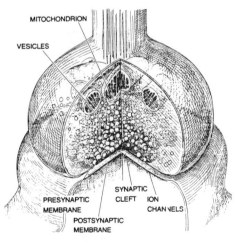

6-6: Terminal fibers from axons synapsing to a cell body and its dendrites.

6-7: A detail of a typical synapse. The impulse is passed on to the next cell when the seminal vesicles release a neurotransmitter substance into the synaptic cleft, which excites the membrane of the next cell (the postsynaptic membrane).

Action Potentials

The membrane of a living cell is one of the most remarkable barriers in nature. The decisive step towards the organization of life, this thin, fragile partition is like a veil separating two great worlds of matter, the living and the non- living. Outside, raw material, a hurly-burly of elements and compounds, swirls of currents, precipitations and dispersals, a prodigal indifference; inside, a slow measured stirring, compounds aggregating into proteins, proteins into proto-plasm, growth, reproduction, an intense discrimination and selection. The cell membrane is the threshold across which physics and chemistry become biology.

It is the membrane which constantly makes the decisions most fundamental to the life process—what is toxic, what is friendly, what must come in, what must stay out, what can remain, what must be eliminated, so that the interior may continue to live, grow, divide. It is not enough that a membrane be a wall; it must be doors and windows as well, doors and windows endowed with their own sense of discrimination as to who shall pass and who shall not.

In addition to these considerable skills, the membranes of nerves and muscles share another function which makes them unique among all the other cells of the body. Besides monitoring the passage of chemicals back and forth across their thresholds, these membranes conduct alternating positive and negative charges along their surfaces. These rapidly moving charges are called action potentials, and their ripplings along the neuromuscular network are the material essence of our experience and behavior. They are the only concrete, measurable bases of sensation, thought, and motor response.

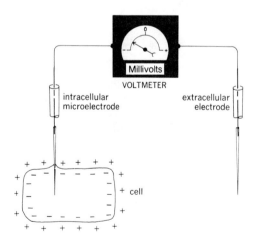

6-8: When measured with a voltmeter as illustrated, nerve cells—and most of the cells of the body—register a slight negative charge relative to the surrounding fluids.

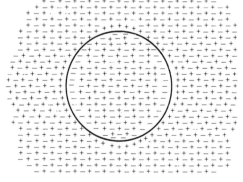

6-9: The interior negative charge of the cell is concentrated along the inner borders of the membrane.

Potassium and Sodium Ions

All of the body's cells, including all the nerve and muscle cells when they are completely at rest, retain a slight negative electrical charge. That is, the fluid *inside* the membranes contains a few more negative ions than does the intercellular fluid *outside* the membranes. This slight electrical gradient between the inside and the outside of the cell is maintained in the following manner.

The interior of the cell is richly saturated with positive potassium ions (K+), while the fluid outside the cell is equally saturated with positive sodium ions (Na+). Now in the absence of any blocking factor, both these concentrations of ions would tend to diffuse across the membrane until their solutions were equal on both sides, just like salt or sugar tend to diffuse themselves evenly throughout a liquid medium. But just such a blocking factor does exist: The cell's membrane is sixty to seventy times more permeable to the potassium ions (K+) than it is to the sodium ions (Na+). This is why there is more potassium *inside* the cell in the first place—it is sixty to seventy times easier for them to gain entrance.

This difference between the ease of passage of the two ions means that while they both follow their natural tendency to diffuse from a greater concentration towards a lesser concentration, there are more potassium ions which are able to exit than there are sodium ions able to enter at any given moment. The net imbalance is slight, but it is enough to create a deficit of positively charged ions inside that cell relative to the fluid outside, thus creating an overall negative charge within the cell. This charge varies from -5 to -100 mV. (microvolts), depending upon the specific type of cell and its chemical environment.

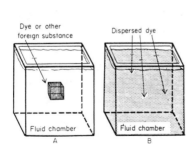

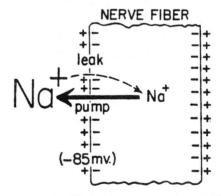

6-11: But the diffusion is not pure and equal. The membrane is 60 to 70 times more permeable to potassium (K+) than to sodium (Na+). Furthermore, the membrane actively "pumps" some sodium ions out of the cell, leaving a slight deficit of positive charges close to the inner surface of the membrane. The result is that the inside of the cell carries between -70 and -85 mV. of negative charge.

6-10: Pure diffusion. If the neural membrane passed sodium ions (Na+) and potassium ions (K+) with equal ease, then both of them would eventually equalize on both sides of the membrane.

Excitability

At rest, most nerve cells and muscle cells carry a negative charge rather towards the high end of this range, hovering at about -70 mV. And in marked contrast to all other cells, this slight electrical gradient does not remain stable. Whenever any nerve or muscle is stimulated, the polarities across the membrane reverse, the charge inside the cell jumping from -70m v. to -30mV. This reversal of charge, and the subsequent return to normal, happens in about one thousandth of a second. This rapid shifting of polarities is the action potential, the unique phenomenon of neuromuscular membranes. It is what we refer to as their *excitability*.

Once a nerve has been excited into this polarity reversal, a rapid and repetitive cycle of events is set into motion which carries the impulse on its way from nerve to nerve and finally to an appropriate motor unit.

At the synapse, where the axon of the first cell joins to the dendrite or the body of the next cell in line, this polarity reversal causes the quick release of a *neurotransmitter*. The most common neurotransmitter seems to be *acetylcholine*, although there are a number of other important ones. Once released into the synaptic cleft, this acetylcholine contacts the membrane of the next cell, with dramatic effect: The membrane, which at rest had been sixty to seventy times more permeable to potassium than to sodium, suddenly becomes six hundred times more permeable to sodium. The sodium gates are thrown open, and a volume of positive sodium ions rushes into the cell, making the interior region immediately beneath the synapse positive (+30 mV.) instead of negative (-70 mV.).

The acetylcholine is immediately reabsorbed by the first cell, to be reused for the next stimulation, but the disturbance of polarities it triggered continues along the membrane of the second cell in a self-perpetuating ripple. The positively charged sodium ions tend to be pulled out into a wider circle by the surrounding negative charge of the cell's interior; where this internal positive charge contacts the membrane, it opens more sodium gates, just like the acetylcholine did; this brings in another brief local flood of sodium ions, which in turn tend to diffuse their charge into even wider circles, which open more sodium gates, which expand the circles farther, and so on down the length of the nerve, until the depolarizing ripple hits another synapse, releases acetylcholine, and initiates the same process in the next nerve down the line. As the ripple passes along the membrane of the nerve, the sodium is pumped back out, restoring the original interior charge of -70mV., and setting the trigger for that nerve's next action potential.

When this impulse reaches a muscle cell via its motor nerve, it is conducted along the muscle cell's membrane in exactly the same way. But with the muscle cell, the action potential does not remain skimming along the outer membrane. The transverse tubules pick up the wave and carry it deep into the cell, where it disturbs the resting state of the sarcoplasmic reticulum, causing it to release its calcium ions and initiate the ratchet-like contraction of the myfobrils.

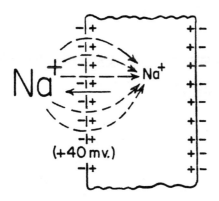

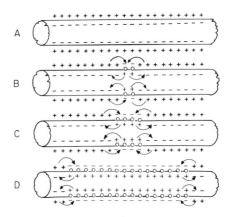

6-12: When a neurotransmitter is secreted across a synapse and contacts a neural membrane, the membrane suddenly shifts its diffusion priorities: it then becomes 600 times more permeable to sodium (Na+) than to potassium (K+). The result is an inrush of sodium, which tips the cell's interior charge to a positive one (but only in the immediate vicinity of the membrane).

6-13: Since the local influx of sodium ions (Na+) enters into the midst of a negatively charged environment just inside the membrane, they are naturally pulled to the sides, thus spreading the internal positive charge, opening more and more adjacent sodium gates, and spreading the depolarization across the whole membrane.

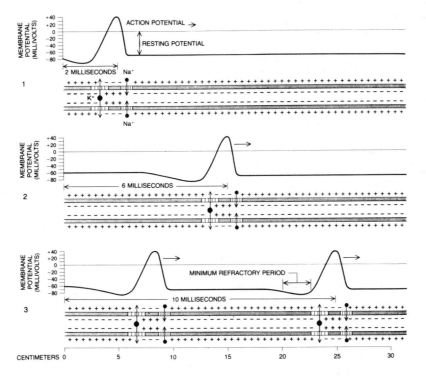

6-14: Each action potential ripples down the surface of the axon. The advancing internal positive charge continues to open more sodium (Na+) gates. Once a local area beneath the membrane has achieved a positive charge, this charge closes the Na+ gates and opens K+ gates, expelling K+ from the cell and restoring the net negative charge. This creates the rippling movement of the action potential along the axon. It then takes about four more milliseconds for the cell to re-establish the original ionic balances, setting the trigger for the next action potential. During this time, the membrane is refractory, and cannot generate another action potential.

Impulse Summation

Action potentials are propagated according to the law of "all or nothing." This means that there are not big action potentials and little ones; a stimulation is either sufficient to trigger a depolarizing wave, or it is not, and if a wave is triggered, its size and duration will be identical to all other action potentials for that cell.

Moreover, it is the rare case when a single action potential from a single nerve is sufficient to initiate an action potential in the next cell in line. Normally, a number of impulses must occur within a brief enough time for their effects to be accumulated before they can push the next cell into depolarization. This accumulating of impulses is called *summation.*

One way for this to happen is for a single cell to continue to fire rapid bursts of action potentials until all of the nerves contacted by its axons are excited to the level of an "all or nothing": response. This is *temporal summation,* and by this means the furious activity of a few neurons can cause the entire system to eventually respond.

Or alternatively, impulses from many cells may simultaneously stimulate many synapses connecting to a single cell and raise it immediately to the excitatory level. This is *spatial summation,* and in this way information from a broad range of sources can be funneled into a single circuit or motor unit. All nerve cells are capable of both types of summation, and in fact they are both occurring all of the time.

Excitation and Inhibition

These processes of summation are further complicated by the fact that not all synapses, and not all neurotransmitters, are excitatory. Some have the opposite effects; that is, they *hyper*-polarize the next cell's membrane (giving it an interior charge greater than -70mV.), so that the propagation of an action potential is even *more* difficult to achieve. Most nerves have several axons that contact several other cells. Some of these axons may end in excitatory synapses, while others end in inhibitory ones, so that the action potentials of one cell can have the opposite effect on various other cells which it contacts.

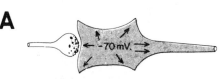

A   RESTING NEURON

-70 mV.

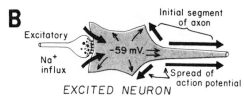

B   EXCITED NEURON

Excitatory
Na⁺ influx
-59 mV.
Initial segment of axon
Spread of action potential

C   INHIBITED NEURON

Inhibitory
K⁺ efflux
-75 mV.

6-15: An inhibitory neuron secretes a neurotransmitter which is different from that of an excitatory neuron. Instead of causing a sodium (Na+) influx, the secretion from an inhibitory synapse causes a potassium (K+) efflux, thus increasing the interior negative charge. This makes it more difficult for any excitatory neurons to overcome the negative charge within the cell and trigger an action potential.

And, of course, that single cell may itself be contacted by an array of stimulating or inhibiting synapses. It even appears that individual synapses can *change* their orientation, adapting themselves to release either excitatory or inhibitory neurotransmitters according to traumatic experience, training, conscious mental set, and the like. Summation, then, can get to be a rather complex affair, and the conflicting influences from many sources must be algebraically processed instant by instant in order to determine the net effect upon any given cell.

Most action potentials, then, do not serve simply to fire the next nerve or a motor unit. Rather they serve to facilitate or to inhibit, to a greater of lesser degree, the next cell's nearness to its threshold of excitation. Depending upon the frequency and chemical nature of the synaptic secretions contacting its membrane (which are, rémember, greatly influenced by experience, training, attitude), a specific nerve cell or motor unit may be very easy or quite difficult to spark into the transmission of an action potential.

> Muscular activity must therefore be viewed against a general background of neuronal activity, a balance of facilitation and inhibition resulting in muscular tone, upon which intermittent or phasic excitatory stimuli produce actual movements, the pathways concerned being at times facilitatory and at time excitatory.[3]

Neural "Electricity"

This conduction and transmission of action potentials is the only functional activity of our neurons. Their collective circuitry forms the fastest communication system within the body. Impulses ripple across their surfaces at an average of two hundred and fifty miles an hour, a speed which makes their traverses of the body very nearly instantaneous. Peripheral sensory signals, in the form of action potentials, encoded like telegraph taps, are forwarded to the central nervous system, where the new information is interpreted and evaluated against a computational background of old habits, present body positions, and mental attitudes. The transformed signals are then sent back out to the periphery along motor pathways, carrying the coded commands for specific responses to each specific stimuli.

There are features of this activity that suggest to us close analogies with what we know about electrical systems. The positive and negative charges, the polarization, depolarization, and hyperpolarization of the membrane, the propagation of wave forms, and the telegraphic coding of these waves, are all consonant with our understanding of the behavior of electricity. The high speed of the action potentials, and the fact that many axons are insulated with myelin in order to improve this conduction speed, deepen this analogy.

Add to these the fact that the observation and study of action potentials requires the same sort of voltage meter used to investigate all other kinds of electrical activities, and it indeed appears as though we are dealing with a familiar critter. The circuitry may be fantastically fine and complex, but the basic principle of operation is no more mysterious than the flipping on and off of a light switch. With this marvelously simple on/off option, the nervous system encodes, transmits, and stores its bits of information much like the diodes, circuits, and chips of a computer.

These electrical and computer models have been helpful in many ways. They have given researchers a direction to follow in the maze, and have allowed us to develop a "blueprint" of neural patterns, and to form some notion of their functional relationships. After all, many pathways do have a high degree of specificity, and often the cause of a reaction in a particular muscle or organ can be pinpointed to the activity of a particular group of neurons. And the "all or nothing" rule of action potential propagation is too strongly reminiscent of telegraph coding and of the dipolar switching mechanism of a computer's memory for us to resist their logical association in our minds.

But we risk forming serious misunderstandings of the principles and subtleties of our neural processes if we push this analogy of wires, switches, and chips too far. There are glaring differences as well as illuminating similarities. Electricity is produced by the transfer of outer electrons between the atoms on the surface of the

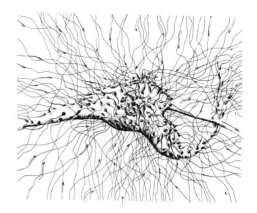

6-16: Terminal axon endings densely synapsing onto a cell body. It has been estimated that a single motor neuron may receive up to 15,000 synapses from other cells.

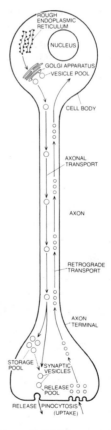

6-17: Not only must fluid be free to circulate around a neuron, fluid must flow all the way out its long axon and back again in order for the life of the cell and the conditions for action potential propagation to be maintained. Nerve cells are as hydraulic as they are electrical.

conducting medium. Action potentials are produced by the surging of whole ions across a selective membrane. Action potential impulses travel at about two hundred and fifty miles an hour. Electricity approaches the speed of light.

Furthermore, such anatomical "wiring diagrams" as we can establish often throw as much confusion as light upon the subjects of feeling, thought, and behavior. It is only muscle cells which receive one and only one axon, and which therefore have an easily predictable response to their connecting nerve's activity. As for both the peripheral and the central neurons, their cell bodies and dendrites may receive axons from tens, hundreds, even thousands of other nerve cells. It has been estimated that a single motor neuron leaving the spinal cord may receive as many as fifteen thousand synaptic endings!

Nor are many of these myriad connections anything like tidy and linear; there is a great deal of cross-connecting between many parallel nerve pathways. Indeed, in the higher regions of the brain this cross-connecting is even more dense than are the incoming and outgoing paths. If the system really were "electrical," such an arrangement could produce nothing but the static buzz of shorted circuits, and our consciousness would sound like a radio jam.

In addition to this bewildering schematic complexity, the synaptic events which transfer the action potential from one cell to another may be very different in nature from synapse to synapse, and even from moment to moment in the same synapse. Any action potential reaching the end of an axon may simply be transferred as a single impulse to the next cell and continue upon its way. Or it may be blocked at the synapse, to have no further effect. Or it may be changed from a single impulse into a rapid burst or a steady rhythm of impulses. Or it may be integrated with a large number of impulses acting upon the same cell, and contribute to effects far different than it could produce on its own.

Nor, as we have already noted, are the synaptic fluids, the neurotransmitters, uniform in a given pathway. Some are potent, some are weaker, some excite, some inhibit, and some merely accumulate, awaiting a moment of summation that is impossible to precisely predict. And there are undoubtedly others, as yet undiscovered, with effects we do not yet anticipate.

Neural Glands

The recognition of the crucial roles played by these synaptic secretions drives us even further away from our electrical model and its reassuring familiarity. It is not a "spark" which jumps the gaps from neuron to neuron, it is a fluid, a chemical agent, a kind of enzyme or hormone. And when the gap is jumped, it is not a stream of electrons that is initiated on the surface of the membrane, it is a liquid solution containing sodium ions which surges back and forth through the membrane.

Now these functional particulars are nothing at all like those of electrical wire; they remind us more of the secretion, circulation, and diffusion of all of the rest of our body's fluids. A nerve is not a wire; it is more accurate to think of it as a tiny gland, with the axon serving as the duct. From the tip of this duct, secretions are released in small quantities and circulate to contact the target tissue—the next nerve in line. So neural activity has really as much to do with the laws of hydraulics as it does with the laws of electricity. The action potential is the movement of fluids. It is only *like* an electrical signal in certain respects.

In this regard, the delicate cell bodies, dendrites, and axons are like the many other fluid-filled tubes within the body. The quality of their function is susceptible to changes in pressure, distortion, viscosity. Their need for constant irrigation is acute: If fresh oxygen is held back from a neuron for merely three to five seconds, it is rendered completely unexcitable.[4] And necessary substances must circulate inside the cell as well as around it. If a long dendrite or axon is pinched, closing its length off from the rest of the cell's fluid, the excitability of the isolated branch quickly decays and eventually the pinched axon or dendrite will atrophy.

You can park a truck on top of an electrical wire and it will continue to work nicely. It will work, in fact, until it is completely severed. In contrast there are many intermediate stages of malfunction in a nerve short of this final breakage—or lesion—most of them having to do with the relative effectiveness of the delivery and circulation of nutritional fluids and the adequate flushing of toxins and wastes. These intermediate malfunctions do not normally stop the system; they just make it less efficient. They confuse sensations, cloud thoughts, disturb the precision of our muscular efforts, make us numb in some spots, unaccountably sensitive in others, eliminate responses, force compensations. Insofar as effective bodywork can be of direct benefit to the circulation of bodily fluids, it can help to support the actual metabolic bases of nerve function, and this benefit is above and beyond the question of the value of any actual *sensations* it may produce.

No matter, then, how metaphorically useful electrical terminology may be in describing the nervous system, we must always remember that we have to do with fluids and membranes, not wires, switches, or chips. We will not be able to completely avoid the metaphors, but let us not allow them to obscure the realities, or seduce us into believing that by applying a well-worn analogy we have "explained" anything. Comparisons of the nervous system to a telegraph wire, or to a telephone switchboard, or even to a large main-frame computer, are gross oversimplifications, capable of severely distorting our view of the complexity and of the organic, developmental plasticity that is inherent in sensation, thought, and behavior.

Centripetal and Centrifugal Flow

The anatomical arrangement of the nervous system has suggested to us two major divisions of the whole, the central and the peripheral. In terms of physical organization, the logic of this division speaks for itself. The peripheral system is a diffuse network of trunks and branches reaching out to all the cavities and surfaces of the body; the central system is then made up of the spinal column and the brain, a far more dense concentration of neurons enclosed by the spine and the skull, and is the deepest, most well-armored core structure of the body. These spatial distinctions naturally lead our ideas towards functional distinctions, towards regarding the periphery as a relatively simple structure which "responds" to sensory and motor stimulation, and regarding the center as a relatively complex structure that "thinks."

But we have earlier noted that the skin—the outermost reaches of the periphery—and the brain are both generated by the same embryonic germ layer, the ectoderm. Furthermore, the connections of each circuit running out to the periphery is carefully arranged to reflect the organization within the core. The extreme coherence of this overall organization would seem to argue that *functional* divisions may not be nearly as tidy as are the visible *anatomical* divisions. We are equally justified in regarding the skin as the outer surface of the brain, and in maintaining that there is no fundamental difference between "responding" and "thinking." There is really no synapse to which we can point and say, "Here is where stimulation is turned into thought," or, "Here is where thought turns into behavior." The nervous system is *all* brain, and all action potentials are identical in form no matter where, at any given instant, they may be located, or in what direction they may be travelling. The terms "peripheral" and "central" are purely geographical, not functional or ecological.

There is another way of separating the whole system into major divisions, divisions which are not as visually obvious as peripheral and central, but which in some ways more accurately represent actual functional distinctions. Roughly half of our neurons have their dendrites reaching out towards the surfaces of our organisms, and their axons reaching in towards the core. This means that they propagate their action potentials in an *inward*—centripetal—direction. The other half are arranged with their dendrites reaching in towards the core and their axons reaching out towards the periphery; these neurons send their action potentials in an *outward*—centrifugal—direction.

The pathways created by these differently oriented neurons do not stop at the threshold of the central nervous system; both can be traced from the periphery to the spinal cord, throughout the cord's length, through the brainstem and the hypothalamus, and finally to the literal summit of the brain, the sensory and motor cortexes. This division is based, then, not on the separation of these different anatomical structures, but upon the direction of action potential flow *through* those structures and the specific pathways they take.

Afferent

The pathways with axons pointed towards the core, and which carry impulses inward, are called the *afferent* pathways. They originate in the various sensory endings of the body—the exteroceptors on the surface, the proprioceptors in the connective tissues (especially the joints), and the interoceptors in the the internal organs. Their final axons terminate in the sensory cortex. They are often referred to as the sensory pathways. Their job is to carry to all the levels of the nervous system information about everything that is *affecting* the organism—that is, all sensory stimulation.

Four of these afferent pathways are short and distinct, arising from the highly localized and specialized areas of the "special senses"—sight, hearing, taste, and smell. The fifth kind of sensory information, that wide array of sensations we refer to collectively as "touch," converges on the cortex from virtually every surface and cranny of the body. These are the "somatic senses," and they include all of the pathways and endings which inform us of our internal state of affairs and our relationship to the outside world.

The afferent, inflowing pathways of the nervous system constitute one of the principal tools of bodywork. It is by their means that surface contact and pressure enter into the deeper strata of the mind, where genetic potential and sensory experience are fused into behavior and character. Each successive afferent neuron is a finger reaching deeper and deeper into the interior, making its influence felt on *all* levels which influence behavior. It is sensory input which has conditioned our reflexes, postures, and habits into the patterns in which we find ourselves living. Nothing would seem to be more reasonable than the expectation that different sensory input can recondition these habits and patterns, alter them, improve them. This input can be different both in the sense of being more, giving additional nutritive contact to the various subtle degrees of "deprivation dwarfism," and in the sense of being more pleasurable, more caring, softening and dissolving compulsive patterns that have been created by pain and stress.

Efferent

The pathways with axons pointed towards the periphery, and which conduct impulses outward, are called *efferent* pathways. Their origins can ultimately be traced to the motor cortex, next to the sensory cortex at the top of the brain. Their axons descend the spinal cord and fan out into the body, terminating in synapses with the striated skeletal and cardiac muscles, and the smooth muscles of the vessels, internal organs, and glands. They are alternately termed the *motor* pathways. Their role is to carry out to the body, from every level in the nervous system, all of the impulses which control the appropriate contraction and lengthening of the entire musculature. Their signals are translated into the various behavioral *effects* which the mind chooses in response to input.

Sensorimotor

Because the efferent pathways lead directly to muscle cells, it is tempting to regard their activities as the cause of our motor behavior. But they are nothing of the kind until they are themselves stimulated by their numerous connections with the spinal cord and the brain (remember that an estimated fifteen thousand axons can converge upon a single terminal motor neuron). And these deeper, more central activities are in turn initiated and directed to a large degree by afferent, sensory stimulation.

In bodywork, it is often problematical aberrations of *motor* response that we want to change, but *sensory* affects are our only means of doing so. We know we are doing our job when our hands feel jumpy reflexes smoothing out, high levels of tone decreasing, pliability returning to stiffened areas, range of motion

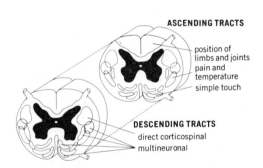

6-18: Ascending (sensory) and descending (motor) tracts in the spinal column.

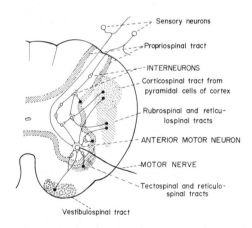

6-20: Impulses from all these sources — or any combination of them — converge upon individual motor nerves. Some synapses are excitatory, and some are inhibitory; their complex algebraic summation is what finally determines the motor response.

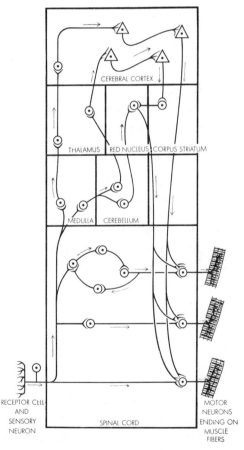

6-19: Various pathways through which a sensory signal can initiate a muscular response. Ultimately, every level of the nervous system may converge upon a single motor nerve, determining its excitation.

increasing. These are all quantitative and qualitative shifts in motor activity. But we also must know that it is only our skilled manipulation of sensory stimulation which can accomplish these things, because it is primarily sensory associations which have conditioned the muscular patterns in the first place. Until the body *feels* something different, it cannot *act* differently. Only when contact with the world is perceived as something other than jabs and buffets can the organism respond with something other than aggression and defense.

Hence it is that simple, pleasurable, soothing, and sustained touch can so profoundly affect motor patterns that have been established over a lifetime: These patterns are being influenced by exactly the same kind of forces which created them, only now influenced to reverse the contraction of aggressive or defensive reflexes. Muscular conditions can change only when feeling states change.

Fusion

We must resist immediately the logical tendency to view these afferent and efferent divisions as being in any way *opposed*. The flow of information which we are describing is not really back and forth, it is circular, simultaneous, mutually integrated on every level. Sensation evokes movement, movement produces new sensations, these sensations then evoke and modify further movements, and so on around the track. Each side of the circle has many synaptic connections with the other side, connections which weld them into a single unit, like the spokes of a bicycle wheel. Sensory and motor activities are everywhere and at all times interpenetrating one another to create the homogeneity of conscious experience.

It is difficult to imagine a stream that flows two directions at the same time, but this is just what the nervous system does. The failure to sufficiently appreciate this unity of seeming opposites leads us into separating too absolutely afferent from efferent, sensation from behavior, attitude from activity. And this leads us in turn into forgetting how powerfully touch, sensation, continually alters internal conditions and overt behavior.

The Internuncial Net

"In its simplest form," we have quoted earlier, "the nervous system is merely a mechanism by which a muscular movement can be initiated by some change in peripheral sensation, say an object touching the skin." Afferent messages enter and are interpreted; efferent commands are selected and sent back out to the appropriate muscles. In this regard, the sensory endings and the motor units are truly unified and symmetrical in their relationships, like the two ends of a pole, or two sides of one circle. It is possible to disassociate sensation from response only on paper.

But as neural activity moves from one end of the pole to the other, as it travels through the full circle, something very curious happens to it. In a creature as complex as man, the processes of *interpreting* information and *selecting* commands are at least as dominant, and often even more so, than are processes of pure sensation and response.

Once a stimulation passes beyond the sensory nerve ending that initially receives it, it bears no resemblance to its original quality. What was a physical distortion of tissue becomes a coded train of action potential impulses. As it continues through the afferent pathways, it can be inhibited, facilitated, rhythmically altered, or blocked at various stages. It has been *transduced,* and is no more like its origin than the current in a telephone wire is like the person talking at the other end.

Similarly, impulses in the efferent pathways cannot really be called "motor" until they leave the axon of a terminal motor nerve and excite the muscle cells in its unit, and are again transduced into tissue distortion—this time muscular contraction. Everywhere in between these two transductions, "messages" and "commands" are merely coded sequences of impulses, impulses that are individually identical. Somehow it is in the mingling and shuttling of these coded sequences that interpretation and selection occur.

The number of neurons in our nervous system has been estimated to be ten billion. There is no way to even meaningfully guess at the number of synaptic junctions which link them into a unified system, and even less of a possibility of following any one train of impulses through the maze. We know that all of our awareness and behavior consist of neural secretions and electrochemical waves coursing through our various pathways, but we can actually predict very few of these specific secretions and can follow any one wave only a very short distance.

> When the original impulse passes through more than one intermediary neuron, becoming related to other impulses, the final response is obviously more complicated and, when even a few of the millions of neurons in the cerebral cortex are called into play, the original simple muscular response to a sensory impulse may become modified out of all recognition. Consider what may happen in civilized man if his skin itches compared with the simple uninhibited scratch of an animal.[5]

Ninety Percent of Our Nervous System

It is within this network of intermediary neurons, arranged end-to-end and side-by-side between our sensory nerve endings and our motor units, that all of our tone levels, reflexes, gestures, habits, tendencies, feelings, attitudes, postures, styles have their genesis. It is called the *internuncial net,* and it has come into its fullest flower in the human being. (*Internuncios* were official messengers for the Pope, taking information and bringing back responses from the various courts of Europe.) This net composes roughly ninety percent of our nervous systems, including the entire spinal cord and the brain.

It is nothing less than the total activity of this internuncial net which influences the responses of the motor units. Let us recall our stiff old man who was anaesthetized for surgery. The anaesthesia had no direct effect on either sensory endings or motor units; rather, it interrupted the normal flow of signals in the internuncial network in the brain. The result was flaccid, unresponsive muscles, and a blanking out of all sensation produced by the scalpel and the probe.

It is only by influencing the flow of impulses through the vast internuncial net that we can have any effect upon tone, habit, and behavior. The conditions which direct that flow into specific patterns have been evolved through the handling of particular qualities and amounts of sensory experience and the repetitions of specific appropriate motor responses. One of the readiest means we have of actually influencing —rather than just temporarily interrupting—the conditions within the net is the introduction of more and more positive sensory experience, which elicits new kinds of motor responses, and can thus form the basis for the development of new habits, new conditions, new patterns of neural flow.

The complexities of the internuncial net are forbidding. The suggestion that body work might in some way make significant and lasting changes in its function may sound like the ravings of a necromancer turned amateur neurosurgeon. We know so very little, and would presume to do so much.

And yet we do know that very simple means can produce remarkable and demonstrably repeatable results in this fantastically complicated network. Infants who do not receive adequate physical stimulation die or are dwarfed and deformed. Laboratory rats who are handled on a daily basis develop markedly stronger resistance to fatal diseases, even to the loss of vital organs. Between these two extremes is a wide spectrum of quantity and quality of touching, all of which must certainly affect the health of the organism if touch in the orphanage and in the laboratory can be proven to be so crucial.

Common aspirin, along with many other useful reliefs and remedies, would have to be thrown out if we would insist upon knowing exactly how they work on the nervous system before we used them to good effect. If something works satisfactorily to improve our well-being, we should be—and medicine for centuries has been—content to use it and hope that we will understand later *how* it works. We make practical use of mysteries every day, and indeed become very adept at handling processes we understand dimly if at all. We would remain in the darkness for a long time if we refused go flip the light switch before we knew exactly what electricity is, or how it does what it does.

> The truth of the matter is that physiologists and psychiatrists have absolutely no idea of the mechanisms which give rise to conscious experience. Nor are there any scientifically meaningful hypotheses concerning the problem.[6]

And yet, given the sorts of things we *can* say about the neuromuscular system, perhaps we are able to form some insights about these demonstrable powers of touch in general, and of bodywork in particular. We must remember to be humble before the vastness of the problem, and we must realize that at this stage we probably will not find thorough explanations as to how the thing works. But we will have done no small thing if we can only unearth a few clues which give us some reliable indications about how to touch more effectively.

The Evolution of Neuromuscular Organization

The human nervous system, with its diffuse peripheral endings, its highly developed central cord and brain, and its afferent and efferent pathways which connect and organize the whole, comes to us as a product of millions of years of gradual additions and changes. Many features of its parts and their relationships are shared with all past and present animals, and the history of their collective development offers us some of our clues concerning the functions of our own system.

As an organism, a species, or a genus, faces the problems of adaptation to the surrounding world, how does its growing network of impulse-transmitting neurons continue to arrange itself so as to produce increasingly sensitive distinctions in perceptions, increasingly fine control of motor responses, increasingly complex feelings, memories, voluntary behaviors?

Hydra

In primitive coelenterates, such as the hydra, we find a cell that is both sensory nerve and contractile muscle all in one. Called *myoepithelial cells,* they are present in the hydra's surface layer, or epithelium. When the exposed portion of one of these cells is stimulated, its deeper, thread-like extensions (called myonemes) expand and contract. Since there are no synaptic connections between them, the stimulation and activity of each cell is purely local, conditions in the surrounding water affecting them independently. When the water is nutritious and inviting, their individual responses collectively create the slow waving of the feeding hydra's arms. When a toxic substance influences enough of them, they collectively constrict the feeding tubes. They are orchestrated only insofar as a general stimulus in the water strikes many of them at once; but even at this primitive stage of development, the total contractile response is directly related to both the *kind* and the *amount* of stimulation.

Anemone

In more highly developed coelenterates, like the sea anemone or the jellyfish, we discover that these myoepithelial cells are differentiated into two separate elements connected together—sensory cells in the skin, and muscle cells in the deeper layers. These branching, interconnecting cells form a diffuse neuromuscular network between the skin and the lining of the gut. Impulses from the skin can be transmitted through this network, and thus wider areas of response and more coordinated movements are possible. This diffuse network is well suited to waving the tentacles of a sea anemone, or to the spasm of constriction which protectively closes its orifice. It produces the rhythmic wave by which the jellyfish moves through the water, and it can even draw up a single tentacle.

But even though it is more complex and variable than that of the hydra, such a system is still purely reflex in nature, with any stimulation of a sensory neuron producing an automatic response in the connecting muscle cells.

Flatworms

The main features of this early diffuse system were not significantly altered until the appearance of flatworms. Here for the first time we find a second layer of nerve cells interpolated between the sensory neurons in the skin and the deeper muscle cells. The dendrites and the axons of this intermediate network of nerves are arranged so that the impulses stream towards the tail, originating in a group of neurons at the head end.

This is the primitive internuncial net, interposed in the midst of the pure reflex connections between single sensory cells and single muscle cells. It makes possible a much wider and a much more precise distribution of stimulation and response. Now for the first time, an impulse from the head end can initiate the contraction of a particular muscle in the tail. This is the beginning of the centralization of the nervous system, the first hint of spinal cord and brain.

It is natural enough that the densest distribution of these internuncial neurons developed in the head end, since this is the portion which receives the first and the most stimulations in the forward-swimming flatworm. Sensory cells have already begun to differentiate from touch into the special senses, and while taste buds are still scattered all over the skin, the eye spots and gravity-measuring organs are particularly associated with the head ganglion of nerves. Thus this head ganglion receives a greater number of sensory impressions, and is obliged to become increasingly more sophisticated in its integrative functions than is the rest of the internuncial net.

Earthworms

In still higher invertebrates, such as the earthworm, the diffuse nature of this intermediate net disappears, and it is now organized into a distinct nerve cord running the length of the worm. Sensory elements become even more differentiated, and the special senses—light sensitivity, smell, taste, gravitational orientation—become even more concentrated at the head end, swelling the size, capacity, and sophistication of the head ganglion even further.

The sensory elements still originate in the worm's skin, but now they send their axons directly to the central cord, without contacting any muscle cells; and for the first time distinct motor neurons appear, sending their axons out from the central cord to the muscle cells in the body. And in addition, we encounter another type of nerve cell for the first time, cells which are contained wholly within the head ganglion and the cord, cells which are neither really sensory nor motor, at last purely internuncial in their functions. They are arranged into afferent and efferent pathways, running the length of the cord towards and away from the head ganglion.

The overall structure of the earthworm is segmented, and each segment contains the sensory and motor neurons which directly control its sensations and functions. Where these segmental trunks of sensory and motor elements connect with the central cord, they form ganglia which orchestrate their local activities. All these segmental ganglia are connected to one another and finally to the head by the afferent and efferent pathways within the longitudinal cord, and in this way stimulation in any segment can influence movement in any other segment, while the coordination of the entire chain of segments can be orchestrated by the larger head ganglion.[7]

Encephalization

The further development of these sensory, motor, and internuncial elements has continued throughout millions of years and species. Higher forms of animal

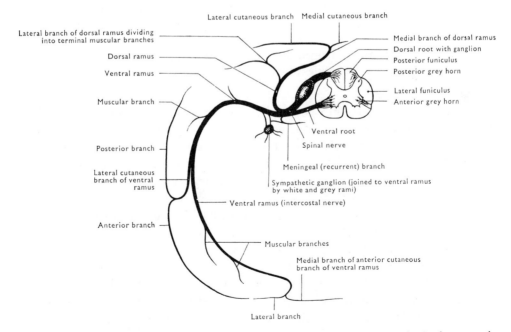

Lateral cutaneous branch   Medial cutaneous branch

Lateral branch of dorsal ramus dividing into terminal muscular branches

Dorsal ramus

Ventral ramus

Muscular branch

Posterior branch

Lateral cutaneous branch of ventral ramus

Anterior branch

Medial branch of dorsal ramus
Dorsal root with ganglion
Posterior funiculus
Posterior grey horn

Lateral funiculus
Anterior grey horn

Ventral root
Spinal nerve

Meningeal (recurrent) branch

Sympathetic ganglion (joined to ventral ramus by white and grey rami)

Ventral ramus (intercostal nerve)

Muscular branches

Medial branch of anterior cutaneous branch of ventral ramus

Lateral branch

6-21: In higher vertebrates—such as man—each spinal segment sends out a symmetrical pair of nerve trunks which serve all the sensory and motor functions for a segment of the body (see dermatomes, Fig. 2-19).

life produce more and more nerve cells, and in particular the number of internuncial cells in the cord and head ganglion have become more and more predominant. Increasing numbers of afferent and efferent pathways and increasing sophistication of their interrelationships has continued until a concentrated nervous system has evolved from the ancient diffuse net, which allows a more rapid and effective integration of sensory stimuli and a much finer coordination of all the parts of the body.

Moreover, these increases in number and sophistication have tended to be heavily weighted towards the head, which typically moves forward into the environment first, and which is therefore influenced by a mass of sensory information entering it directly from the various sources of the mouth, eye, ear, nose, and gravitational sensors.

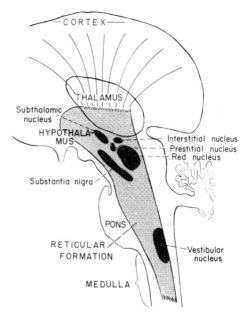

6-22: Major ascending levels of the central nervous system above the spinal cord. The brainstem is the area between the pons and the thalamus.

> The changes occurring in the fore-end of the neural tube are remarkable compared with those associated with the growth of the limbs. The head end and mouth region of an animal always make first contact with its environment, and here are developed the special senses, smell and taste, sight and hearing. Under their influence, the nervous system becomes profoundly altered.[8]

Thus what was once only a ganglion becomes the full-fledged brain. This process of enlargening the head—or *cephalic*—ganglion and increasing its influence over the rest of the system is called *encephalization*. More and more local reflexes become mediated by the cephalic ganglion's expanding means of integration and control, and older forms of organization are supplanted by newer, more sophisticated, and often more flexible ones.

This process will probably never be completed. The human cortex seems very much in the middle of its struggle to exert conscious control over instinctual responses and unconscious behavior, and we are just beginning to understand the awesome extent of influence—both for better and for worse—that the newer, higher portions of the brain can potentially exert over the older patterns of response. Encephalization is the slow creation of the means to develop greater awareness of, and greater conscious control over reflex response.

Throughout this evolutionary development, the kinds of structural and functional changes undergone by all the various elements in the nervous system have

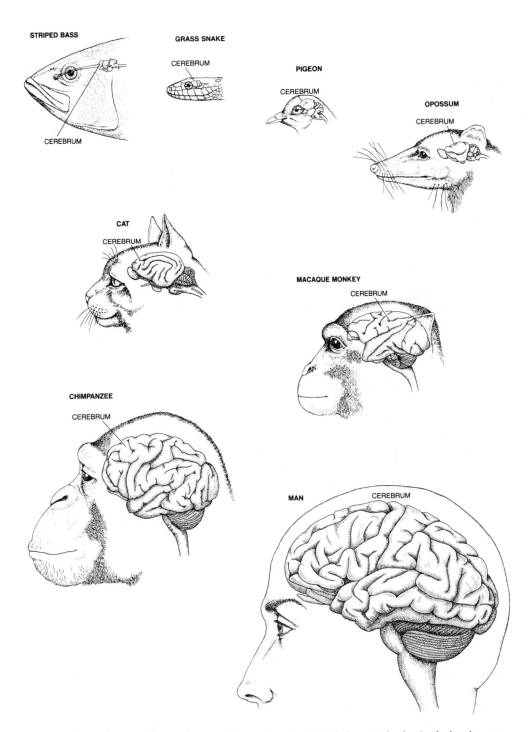

6-23: The expanding cerebrum and cortex. The cerebrum is relatively larger in land animals than in water animals, larger yet in those that walk or fly relative to those that crawl, and largest of all in the bipeds. In man it has erupted, dwarfing most of the other structures of the brain.

been inextricably interwoven with the kinds and amounts of stimulation these elements receive. While it is certainly true that no general coordination of movement was possible before the appearance of the internuncial network, it is also true that the subsequent patterns of coordinated movement greatly altered the nature of the entire neural organization.

It is really just as accurate to say that it has been the increasingly numerous and complex sensations associated with locomotion that has led to the development of the brain, as it is to state the case the other way around. As with all matter, usage dictates shape, specific function produces specific form. Thus as the animal kingdom has moved forward from jellyfish to worms to vertebrates, new behavior patterns have bodied forth new neural structures and circuits, just as surely as new neural mechanisms have opened up new modes of behavior.

> Once such a pattern has been established, the increasing sensory experience which arises from the assumption of a terrestrial existence is associated with further increases in the size and complexity of the brain, so that more complex behavior patterns are possible. Thus in man the assumption of the erect posture and the freeing of the upper limb from its supporting functions, with the consequent development of the hand as a potent new sensory organ, may well have played a significant part in the growth of the human nervous system to its present complexity.[9]

And note well: In this development, it has been *the kind and the amount of sensory input* which have been major factors in these profound changes in the structure and function of the central and peripheral nervous systems. Deep neuromuscular patterns most certainly are, and always have been, wide open to the influences of specific qualities and quantities of touch.

In this context, it does not seem at all impossible that effective bodywork could foster new levels of awareness, new patterns of response, and even new neural conditions that would influence all future responses in an individual. The entire developmental history of the centralization and encephalization of the human nervous system is evidence of the potent organizing powers of touch.

Ascending Levels of Organization

What we can say about the organization of the nervous system has largely been inferred from dissection, electronic or surgical probing, comparative anatomical studies of dead specimens, and various experimental techniques to isolate particular kinds of learning and behavior.

All these methods are disruptive of normal functioning. They were undoubtedly the only avenue open to us in which to begin our internal investigations, and they have in fact yielded a considerable amount of interesting information. But we must always approach this information with our means of getting it in mind. When we dissect and isolate in order to discern more clearly, we have altered our material so that we may simplify it, created anatomical distinctions where there may be no functional ones, separated elements which have no useful existence apart from one another, and we are therefore in constant danger of drawing conclusions which satisfy the limitations of our procedure rather than the realities of the situation we are investigating.

So while we are dissecting to examine, we must appreciate that even though we can discuss the functions of separate internal parts of the central nervous system, "these are developed from a single longitudinal tube and the line of demarcation between the parts is not sharp. Many structures pass longitudinally through the tube without interruption of any kind."[10]

The human nervous system has retained various characteristics that were typical of the various stages of its evolution. Four major ascending levels of the system reflect this developmental progression: 1)the peripheral sensory and motor neurons, 2)the spinal cord, 3)the lower brain centers, and 4)the cerebral cortex. Each higher level offers the organism wider options for behavior, and brings greater conscious awareness into the processes of the interpretation of information and the selection of responses.

The numbers of sensory endings and motor neurons in the periphery have multiplied considerably as fishes and birds and mammals developed, producing sharper and more complete sensory impressions, and making possible a finer and finer control of the musculature. But the essential features of the peripheral system have not greatly altered. In them the basic rudiments of the primitive nervous system are still preserved, "a mechanism by which a muscular movement can be initiated by some change in peripheral sensation, say, an object touching the skin."

What has changed enormously is the *relationship* between these ancient sensory and motor elements. In the hydra sensation and motor response were united in a single cell; in the jellyfish these two functions were divided into two specialized cells; and in the flatworm an intermediate network of cells was placed between them. The two more primitive elements have since become increasingly separated as the internuncial net has increased in bulk and sophistication, and their synaptic connections, once direct, have been distanced by more and more modifying circuits.

The Spinal Cord

The lowest level of this modifying intermediate network is the spinal cord. The cord still possesses many features that were first developed in the segmented earthworm. It is largely made up of neurons completely contained within it, which form bridges between the sensory and motor elements throughout the whole body. Each peripheral nerve trunk still innervates a specific segment of the body, and still joins the cord at a specific level, creating a ganglion.

Sensory signals entering into a single segment may be processed by its own ganglion, and cause localized motor response within the segment; or the signals may pass to adjacent segments, or be carried even further up or down the line, involving more ganglia in a more widely distributed response.

In this way, the cord can monitor a large number of sensorimotor reactions without having to send signals all the way up to the brain. Thus stereotyped responses can be made without our having to "think" about them on a conscious level. Most of these localized and segmentally patterned responses are not the result of experience or training, but of genetically consistent wiring patterns in the internuncial network of the cord itself. These basic wiring patterns unfold in the foetus during the "mapping" process of the nervous system, and they have been pre-established by millions of years of development and usage.

The spinal cord can be surgically sectioned from the higher regions of the internuncial net, and the experimental animal kept alive, so that we can isolate the range of responses that are primarily controlled by these cord reflexes. Almost all segmentally localized responses can be elicited, such as the knee jerk caused by tapping the tendon below the knee cap, or the elbow jerk caused by tapping the bicep tendon. These simple responses can also be spread into other segments, so that a painful prick on a limb causes the whole body to jerk away in a general withdrawal reflex. The bladder and rectum can be evacuated. A skin irritation elicits scratching, and the disturbance can be accurately located with a paw.

Some of the basic postural and locomotive reflex patterns seem to reside in the wiring of the cord as well. If an animal with only its cord intact is assisted in getting up, it can remain standing on its own. The sensory signals from the pressure on the bottoms of the feet are evidently enough to trigger postural contractions throughout the body and hold the animal in the stance typical of its species. And if the animal is suspended with its legs dangling down, they will spontaneously initiate walking or running movements, indicating that the fundamental sequential arrangements of the basic reflexes necessary for walking are in the cord also.

All of these localized and intersegmental responses are rapid and automatic, follow specific routes through the spinal circuitry, and elicit stereotyped patterns of muscular response. Most of them appear to consistently use the same neurons, synapses, and motor units every time they are initiated.

The Lower Brain

The lower brain—including the pons and the brain stem—is primarily responsible for our "subconscious" processes, those many activities which are more complex and integrated than cord reflexes, but of which we are seldom aware. To begin with, many more sequences of simple reflexes are possible if the pons and the stem are left intact with the cord. The lower brain clearly assists the cord in fine-tuning responses, and in arranging them in the appropriate order so that they produce more integrated behavior. The complicated sequences of muscular contraction necessary for sucking and swallowing, for example, are monitored at this level. These are skills with which a human infant is born; their underlying circuits—and even more importantly, the correct *sequence of operation* of these circuits—is a product of early genetic development, not individual experience and learning.

In general, the lower brain seems to share many of the "hard-wired" features of the spinal cord. Axons and synapses form organizational units that appear to be consistent for all individuals of the same species, and their activation produces identical, stereotyped contractions and motions. But the additional complexities of the lower brain appear to enable it to pick and choose more freely among various possible circuits, and to arrange the stereotyped responses with a lot more flexibility than is possible with the cord alone.

For instance, it is in the lower brain that information from the semi-circular canals in the inner ear—the sensory organ for gravitational perceptions and balance—is coordinated with the cord's postural reflexes. A stiff stance can be elicited from these postural reflexes by merely putting pressure on the bottoms of the feet; by adding information concerning gravity and balance to this stance, the same reflex cord circuits may be continually adjusted to compensate for shifts in equilibrium as we tilt the floor upon which the animal is standing, or as we push him this way or that. A rigid fixed posture is made more flexible and at the same time more stable, because compensating adjustments among the simple postural reflexes is now possible.

The lower brain coordinates the movements of the eyes, so that they track together. It directs digestive and metabolic processes and glandular secretions, and determines the patterns of circulation by controlling arterial blood pressure. And not only does it give new coordination to separate parts, it influences the system as a whole in ways that cannot be done by the segmental arrangement of the cord.

Relative levels of activity in the pons and the brain stem directly relate to the general level of arousal of the entire nervous system, and to the level of tonus of the musculature as a whole. The natural suppression of activity in this area corresponds to sleep; the chemical suppression of the same activity is anaesthesia. In addition to general arousal, some specific emotional states can be produced in an animal with only a lower brain and cord intact: general excitement, sexual arousal, anger, and typical responses to pain or pleasure. (This list of functions is by no means complete; it is intended only to convey an idea of the *kind* of organization and selection which the lower brain adds to that of the spinal cord.)

The Higher Brain

The cerebral cortex is the newest portion of the nervous system, and in man it has developed into the largest portion. In its anatomical connections, it is clearly an outgrowth of the areas of the brain directly beneath it—the hypothalamus—

and in its functional aspects it represents a great leap forward for the lower brain's abilities to integrate sensory information and to select and order new motor responses.

It is an immense addition to the internuncial net; fully three-quarters of the cell bodies in the human central nervous system are contained in the cortex. And the horizontal interconnections between these cells are far more dense than those in the cord or the lower brain. Schematically, it resembles a ball of cotton fuzz much more than it resembles even the most complex man-made electrical circuitry.

The cortex appears to be primarily a vast storage center of information and associations, preserved bits of experience which can be suppressed or recalled to modify behavior in truly innumerable ways. No single individual has ever come close to utilizing the enormous variety of responses made possible by the cortex, and we must look to the full range of collective human skills and experiences in order to form some notion of its capabilities of integration and selection.

As for explaining exactly how this information is stored, processed, and brought to bear upon future behavior, we are utterly at a loss. In the internuncial network of the spinal cord, we can isolate relatively simple neural structures that satisfactorily explain the paths which a stimulus must follow in order to create a reflex muscular response. But we are completely stymied when we regard the number of cells in the cortex and the density of their interconnections.

Nevertheless, it is clear that it is this highest and newest part of the brain which bestows upon us the cognitive faculties that are unique to our species, and makes possible the wide range of behavior patterns that is quintessentially human. Lots of animals can do a few things much better than we can do them, but no other animal can be taught to do as many things as man. Contained somehow in the cortical web is an enormous array of experiential memories and new, imaginative associations that lift our behavior out of the closed circles of reflex and instinct—lift us so high at times that it is entirely possible for liberated abstract thought to forget its own biological basis. It is only civilized man who can, in fits of inspiration or contemplation, forget that he is an animal, a body.

So in exchange for a relatively small but very stable variety of behavioral patterns, we have accepted the unknowns, the confusions, and the dangers of infinitely extended possibilities. This is the human experiment, and the cortex is its chief physiological element. Its mushrooming growth has been coincident with the rise towards erect posture, the development of the prehensile hand, the expanding use of tools, and the creation of verbal communication. Its increasing bulk and complexity have been the result of greatly increasing demands made upon the cephalic ganglion for the integration of these new postures and the perfections of these new skills. And as its relative size has come to dominate the rest of the central nervous system, it has made some demands of its own.

Further Encephalization

We are not to imagine that the older parts of the nervous system have continued to function in man exactly as they did before this prodigious mushrooming of the cortex, nor that the cortex simply adds new categories to consciousness without disturbing the functions which preceded it.

> Throughout phylogeny the control of the final motor neuron has become more and more complicated and refined as higher centers in the brain play a progressively greater part. The fate of the more primitive paths in this process must be clearly realized. Firstly, they continue to play a central and important role in muscle tone, secondly their function may be adapted in a modified form by pathways passing through the higher nuclei (the importance of which lies not only in their own intrinsic organization, but also in the fact that the afferent impulses reaching them are more highly correlated and the ultimate muscular response therefore more highly coordinated), and thirdly, a nuclear station, losing its original functions, may instead develop to serve a newer and higher nucleus. For example, the tectum in the frog is the visual receptor region and removal of the cerebral cortex produces little or no diminution of vision, but in man the occipital cortex has usurped this function and cortical destruction causes total blindness. As a result, the cell structure of the tectum in man has become simpler.[11]

The addition of new neurons to handle new operations is only a part of the process of encephalization. The other parts are the gradual modification of ancient reflex patterns, the diversion of neural flow from the older channels, and the creation of new chains of command in the ordering of specific sequences of motor activity. The net result has been that the higher cognitive centers have become increasingly influential, while the older time-worn patterns have become less authoritative, more variable. Conscious mental states have begun to condition the system just as much as the system conditions these higher states of consciousness.

But new powers and new subtleties do not appear without new complications, new conflicts. In bodywork we continually feel the muscular results of the intrusion of newer mental faculties into older, more stable response patterns. A good deal of the work is simply reminding minds that they are supported by bodies, bodies that suffer continual contortions under the pressure of compelling ideas and emotions as much as from weight and physical stresses, bodies that can and will in turn choke off consciousness if consciousness does not regard them with sufficient attention and respect.

It is possible—in fact it is common—for the mass of new possibilities to wreak havoc with older processes that are both simpler and more vital to our physical health. Thus with our newer powers we are free to nurture ulcers as well as new skills, free to inspire paranoia and schizophrenia as well as rapture, free to become lost in our own labyrinths as well as explore new pathways. We have unleashed the human imagination, to discover that there is no internal force as potent to do us either good or ill. With the addition of these new cortical faculties, the quality of our muscular responses—from digestion, to posture, to locomotion, to expressive gesture, to chronic constriction—is dependent not only upon stimulations from the environment, and not only upon patterns characteristic of the species, but also upon individual experiences, memories, unique associations, personal emotions, expectations, apprehensions, the entire legion of personal psychological states.

One of the main dangers inherent in the cortex is that its possessor can quite easily become persuaded that the physiological processes of the body have relatively little importance in comparison with the compelling presence of these emotional and cognitive states. More than personal health is at stake in this illusion. A very precious quality of contact, of social interaction, of collective experience is threatened by this turning away of abstract thought from its biological foundations.

The great significance of the facts that we all eat, grow, reproduce, react alike to pleasure and pain, and are linked together in the marvelous interdependencies of physical nature is all too often stifled and buried under the less fundamental observations that we do not all think or feel or worship in exactly the same ways. And, lost in our thoughts, we often omit to do those things that would turn our best ideas into physical realities, because we can easily forget that "the activities of the nervous system can have no external significance until they are expressed to our fellow men by muscular activity, be it action, writing, or speech"[12]. Helping to correct the solipsistic tendencies of abstract contemplation is one of the most important roles of bodywork.

Descending Motor Pathways

In spite of the fact that the cord, the lower brain and the cortex each make their distinct levels of contribution to the control of motor response, they are not really separate structurally or functionally. They were evolved from a single neural tube, and the longitudinal afferent and efferent pathways still pass through all three of them without interruption, making them a single unit in many ways. We can experimentally isolate the different qualities they add to motor control, but the truth is that the smooth and rapid coordination needed for normal motor behavior requires the constant correlation of information from all three of these levels. If their activities are not simultaneous and closely synchronized, muscular control is impaired in one way or another.

Direct

Efferent impulses may be conducted along one of two major pathways of motor neurons as they pass from the brain through the cord and out to the muscles, and together these longitudinal pathways provide for the convergence of the influences from all levels of the central nervous system upon the motor units.

The fastest of these descending routes is the *direct corticospinal pathway*. As the name suggests, the cell bodies of this path are in the cortex, and they send their long axons directly through the brain and down the spinal cord without any interruptions. These axons do not form any synapses until they reach their corresponding motor neurons in the cord, and thus they form direct connections between specific cells in the motor cortex and specific motor neurons at each level of the cord, making one-to-one relationships between cortical cells and peripheral motor units.

This pathway bypasses most of the intermediate circuitry of the lower brain and the spinal cord. This gives it the advantage of speedy transmission. The axons which are bundled together within it maintain a constant spatial relationship throughout their length, faithfully reflecting the spatial relationships of the cell bodies in the cortex. The longest axons, reaching all the way to the end of the cord, lie the closest to the center of the cord, and the progressively shorter axons which synapse to motor neurons in progressively higher segments, are carefully laid down in layers progressively far from the center of the cord, so that a "map" of skeletal muscle relationships is projected onto the motor cortex. This gives a high degree of specificity to this direct corticospinal tract.

This direct pathway is the mediator of fine, intricate movements, which

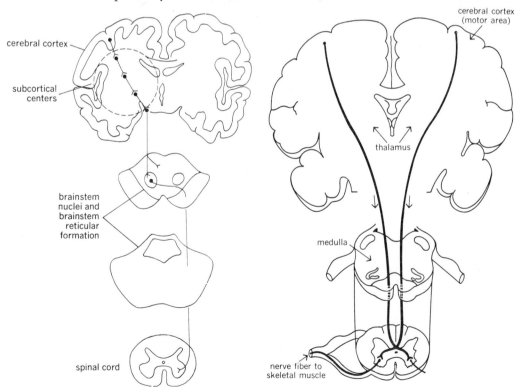

6-24: The direct corticospinal and the multineuronal descending pathways. The multineuronal pathway makes many more connections to more organizational areas, subjecting impulses to many more modifications on their way to the muscles.

require close conscious attention and constantly refined adjustment. When it is severed, actions become clumsier, because the sharp edge of delicate conscious control is missing.

Multineuronal

The second route that motor commands from the cortex can take out towards the muscles is called the *multineuronal pathway*. The initial cell bodies of the pathway are also in the motor cortex, but they immediately synapse to long chains of internuncial neurons descending from the motor cortex to other cortical areas, to the lower brain, and on down the cord. The final links in these chains end not directly upon the neurons of the motor units, but rather upon the spinal internuncial circuits which organize the patterns of stereotyped spinal reflexes.

Thus instead of bypassing the intermediate net, this pathway channels motor commands all the way through it. Impulses take longer to travel from the cortex to the motor unit, since they pass through many more synapses and are influenced by input from many other sources along their way. This arrangement makes possible the second kind of motor control—the setting into motion of the stereotyped reflex units of movement that are organized in the subconscious levels of the lower brain and spinal cord. For instance, the individual circuits and the overall sequential arrangements which control walking or swallowing are in the cord; the multineuronal pathway has only to stimulate these sequences into activity, and then steer their general course without the cortex having to pay attention to the details involved in each separate step.

All coordinated movements require the interaction of both of these pathways. The conscious mind is not competent to direct every single muscular adjustment involved in general movements, and the reflexes are powerless to arrange themselves into new, finely controlled sequences. Successfully integrated, the two of these paths together provide the best aspects of millions of years of repetitive practice and of moment to moment changes in intent or shifts in the surrounding environment. One is basic vocabulary, the other is variable sentence structure; the language of coordinated movement cannot go forward without both working together.

Ascending Sensory Pathways

There is an analogous pair of pathways which carry sensations from the periphery to the sensory cortex, the *dorsal column system* and the *spinothalamic system.*

Dorsal

The dorsal column system runs through the "white matter" of the spinal cord. It is white because its axons are insulated with white, fatty *myelin,* which increases their transmission speed considerably. Their speed—from forty to seventy meters per second—is also enhanced by the fact that there are few synapses to cross from the peripheral sensory ending to the cortex. Like the corticospinal motor pathway, these fibers have a high degree of spatial organization throughout the length of the spinal cord, and like the corticospinal pathway, they faithfully map the relationships of their origins onto the cortex.

This system transmits touch sensations which have precise localizations and fine gradations of intensity, phasic or vibratory sensations, kinesthetic sensations related to body parts in motion, and sensations which have to do with fine distinctions of pressure.

Spinothalamic

The spinothalamic system, on the other hand, runs through the "grey matter" of the cord, so named because it has no white fatty insulation sheaths around its axons. Their spatial orientation is not nearly so carefully preserved at all levels, and they make many more internuncial synaptic junctions on their way up the cord. Their transmission speed is roughly one-fifth of that of the dorsal tract.

This system carries impulses which announce pain; thermal sensations, both hot and cold; crude touch sensations that are not acutely localized; pressure sensations that do not rely upon fine distinctions; kinesthetic sensations having to do with chronic conditions, or the body at rest; tickles and itches; and sexual sensations.

It is a fact of considerable significance to our reflex responses that pain sensations are carried exclusively by the slower spinothalamic pathway. This means that more neutral and at the same time more detailed sensory information will always reach the spinal circuits and the cortex slightly before the stab of pain arrives. This gives us a brief moment to assess the location and the cause of the pain before we react, so that our reflex withdrawal can be more appropriately tailored to the actual source of the pain and more effectively directed; that is, so that we will be able to assess the intensity of the burn, and will be sure to jerk away from the flame rather than towards it, and will arrest our jerk before we crash into the wall.

This time lag gives a special role to general tactile sensations—including body work—when we are in pain. It means that it is possible to bombard the consciousness with more rapidly transmitted and more detailed *touch* sensations which tend to displace the pain response from the foreground. This is why rubbing the spot that hurts, or jumping up and down, or shaking the injured hand are often effective for alleviating pain. This is the principle behind the mother's instinctual rocking and stroking of her hurt child, and it is a principle that can be turned to great advantage in bodywork. If the rest of the body can be inundated with touch sensations, particularly *pleasurable* ones, the part that is in pain can be shifted away from the mind's central focus.

On the other hand, this very same mechanism presents a danger: By keeping ourselves busy, and by forcing our attention onto other matters, it is possible to suppress pain signals which may be very important, possible to bury our awareness of threatening conditions beneath a layer of faster, more acute, but more trivial sensations. The mind's mechanisms of selection and focus can play tricks that are nasty as well as ones that are helpful. One of the principal strengths of bodywork is that it can generate the sensory information—the self awareness—that is necessary for the individual to identify and gain control over conflicting tendencies of this kind.

Descending Sensory Pathways

There is another mechanism which even further selects and distorts incoming sensory information, and which is extremely important to the therapeutic purposes of bodywork. Side-by-side with the ascending dorsal and spinothalamic pathways are *descending* sensory pathways, outgoing tracts from the brain which are *not* efferent, which do *not* contact motor neurons, but which synapse densely with the ascending sensory pathways and exert a centrifugal flow against their incoming sensory information.

These descending sensory pathways originate in the cortex, and their influence can either inhibit or facilitate the sensory input from any area of the body, greatly amplify or suppress altogether any given sensation. Sensory activity within the central nervous system is a Janus head, a two-way stream; signals originating in the brain have just as much to do with conscious sensation as do actual stimulations of the peripheral nerve endings.

In fact, it is the descending paths which determine the *sensitivity* of any particular ascending pathway. The wide varieties of sexual response—among different individuals and within the same individual at different times—provide us with a clear example of this centrifugal influence upon incoming sensations: Depending upon past experiences and the present situation, the same stimulation of the genitals can produce intense ecstasy, bland and neutral sensations, or extreme discomfort. Orgasm can be immediate, or deferred indefinitely. And the imagination alone can produce constant engorgement, utter impotence, and all the degrees of arousal in between. The descending paths can color all kinds of sensory input to this degree.

These pathways allow the mind to determine the active threshold for different sensory signals, and make it possible to focus attention upon a single source of input in the midst of many. It is difficult to imagine what practical use our sensory apparatus would be to us without such a mechanism. We would have no way of selecting a voice from all the other sounds around us, of locating specific objects within the swirl of visual impressions, or of retiring from our senses for contemplation and sleep. This centrifugal principle of *selectivity* is as vital to our appropriate responses to the world as is stimulation itself.

And yet what a potentially dangerous device, what a terrible opportunity to bury valuable, even vital sensory information beneath the fears and prejudices and suppressions of the higher brain! We absolutely must exercise constant discrimination upon the steady barrage of sensations if they are to take on any meaningful form and direct sequential activities; but what bizarre, even ghastly shapes this discrimination is free to invent. Attitudes, moods, neuroses, fixations, and avoidances of all kinds contribute to the sensitivity of the *ascending* sensory pathways themselves, so that minor irritations can be magnified to overwhelming proportions, pleasures can be erased or actually turned into torments, serious internal difficulties can be blotted completely out of consciousness. The principle of selectivity is crucial to organized behavior, but the possibilities for its abuse are enormous. The mind is capable of distorting incoming information to almost any degree, and it can actually construct a body image that has very little to do with the bulk of sensory data which the body is providing.

These two directions of sensory transmission are both occurring all the time, and we cannot say that our idea of reality is more clearly established by one than by the other. Or, if we have to make a choice, we must admit that it is the descending, centrifugal sensory current that is the more important one: We all receive stimulation from the same external world through identical sensory devices, but it is the process of selection and interpretation which makes us respond differently, makes each of us the unique individuals that we are. In this process, discriminating mind descends into and is active in every synapse of the sensory system.

> The two processes of transmitting data through the nervous system and of interpreting it cannot be separated. Information is processed at each synaptic level of the afferent pathways. There is no one point along the afferent pathways or one particular level beneath the central nervous system below which activity cannot be a conscious sensation and above which it is a recognizable, defineable sensory experience. Perception has many levels, and it seems that the many separate stages are arranged in a hierarchy, with the more complex stages receiving input only after they are processed by the more elementary systems.[13]

And the more elementary systems are in turn facilitated or inhibited by the higher, more complex ones. The conclusions towards which these observations push us seems unequivocal. The cognitive, associational processes of the higher brain have just as much to do with our construction of physical reality—both within us and outside of us—as do our sensory devices and their specific stimulations. And remember, it is the perception of this sensory reality which initiates and directs our motor responses, our postures, and our behavior.

If bodywork is to be more significant than just so many pokes and rubs, if it is to effect lasting changes, then it must not merely address the tissues. It must use tactile sensations to reach the mind, the whole mind, from the surface of the skin to the spinal reflexes, to the subconscious responses of the lower brain, to the fields of awareness in the cortex. When this happens, touch is genuinely, profoundly therapeutic.

7

Muscle As Sense Organ

The brain recalls just what the muscles grope for: no more, no less.
— Wm. Faulkner, *Absalom, Absalom!*

Sensationless?

Muscle tissue seems to be almost sensationless. When we look into it, we find far fewer pain endings and pressure sensitive capsules than in the skin, the joints, and the connective tissues, and none of the other sources of tactile sensation that we find in varying numbers in most of the rest of the body's tissues.

The absence of the normal modes of sensation in muscle tissue was clinically observed as long ago as the eighteenth century by Albrecht von Haller, the famed German physiologist. As a physician, von Haller often had the opportunity to work with accidental and surgical wounds in muscle tissue, and since he worked before the discovery of sophisticated anaesthetics, his patients were often fully conscious. He was able to use these circumstances to make many fascinating and important observations concerning the structures and the responses of the nervous system.

By experimenting directly with open wounds in conscious patients, he was able to demonstrate that no degree of touch, pressure, pinching, pulling —even cutting or burning—of muscle tissue produced sensations for the conscious subject. Pain, and any other sensations present, invariably seemed to come from other tissues disturbed by the wound or by the probing: skin, deeper connective tissues, joint surfaces, joint capsules, the periosteum covering the bones, and so on. Prodding red muscle tissue itself yielded no sensory results.

Subsequent experiments and more sophisticated techniques have to a large degree confirmed von Haller's findings. Our muscles, seventy to eighty-five per cent of our physical bulk, deliver to us very little sensation that we can consciously identify. And if we block with a local anaesthetic the sensations arising from the skin and the associated joint capsules, we have no feeling whatever for where our limbs are in space, or what sort of movement they are up to.

Nerves Surrounded by Muscle

So muscle tissue, in and of itself, supplies us with no conscious sensation. Yet it would be patently absurd to imagine that muscle tension and movement play no direct part in our conscious sensory experience.

In the first place, all of my peripheral nerve tissue—everything that is outside of my cranium and my spine—is surrounded by, and in large measure supported by my body's muscles. The trigeminal nerve, which controls movements and creates sensations in my face and scalp, and the brachial plexus, which sends out the axons for the innervation of the arm and hand, are surrounded by the muscles of the neck. The lumbar plexus is surrounded by the bellies of the psoas and the iliacus muscles. The sciatic nerve, which services the muscles and nerve ends of my legs and feet, is literally in the grip of the deep rotators of the hip. These juxtapositions of muscle and nerve are to be found in all parts of the body, and they are responsible for purely mechanical and hydraulic effects through which the tension and movement of the muscles influences the basic health and vitality of the peripheral nervous system as a whole.

These peripheral nerve structures are, after all, living cells which rely upon the free irrigation of the spaces which surround them for their proper nourishment and waste removal. Muscles that have fallen into disuse and flaccidity just don't provide enough pumping action for these intercellular fluids to adequately feed and bathe the nerve cells, and so the general strength of their functions is diminished.

The circulation of the fluids *within* the long, thin tubes of the axons is equally important to the proper conditions of the nerves' membranes and the propagation of action potentials, and inadequate pumping action of the surrounding muscles reduces this *hydraulic* flow as well.

Chronically tense and constricted muscles can complicate things in an even worse fashion. Not only is fluid circulation in and around nerve cells curtailed, but the capillaries which *supply* the nutrition and carry off the waste products are squeezed tightly as well. At the same time, the contracting muscles are producing increased waste products and demanding increased nutrients from capillaries that are less able to supply them. This creates both an oxygen shortage and a waste build-up in the area, both of which are directly toxic to the nerve cells, irritate them, and contribute to even further muscular contractions.

In addition to this, chronically high levels of pressure upon nerve trunks is itself detrimental to their electrical activity, apart from such general circulatory complications. Some researchers have estimated that five pounds of pressure for five minutes on a nerve trunk can reduce its transmission efficiency by as much as forty per cent.

In time, the results of these pressures can be the sharp ache of sciatica generated by the rotator muscles of the hip, numbness or tingling sensations in the hands from the neck muscles clamping down on the brachial plexus, chronic pains in the face and the head from pressure on the trigeminal nerve, and so on. And of course, since the nerve supply to internal organs can be similarly effected, such chronic constrictions can bring along a wide range of organ dysfunctions in its train of events as well—organ dysfunctions that can be extremely difficult to diagnose and treat because no "disease" state exists and no observable damage has been done to specific organ tissues.

Indeed, the complications for circulation and neural transmission which follow in the wake of chronic muscular contraction present some of the gravest potential dangers for the health of the nervous system, and of the body as a whole. Loss of neural efficiency means a less and less vivid reception of the messages that the nerves convey, both from the sensory endings and to the motor units. And areas of the body that are not adequately irrigated stagnate precisely like the choked and swampy backwaters of a sluggish stream, creating septic situations that are ripe for discomfort, disease, and decay.

Nor should we forget the facts that increasingly constricted capillaries require higher and higher blood pressure to make them function at all, and that once they either collapse from the muscles squeezing them or burst from increased blood pressure, they will be replaced with scar tissue and not by new capillaries, thus making the local loss permanent.

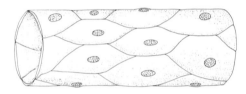

7-1: A section of a capillary. Its structure is very delicate—the walls are only a single cell in thickness. Once a chronic muscular constriction has squeezed it hard enough to collapse its walls, either killing cells or adhering the inner surfaces, the capillary is finished. It will be replaced with scar tissue, and circulation in a local area will be permanently curtailed.

Movement and Sensation

But the contributions of muscle activity to our sensory experience is not by any means limited to these simple mechanical and hydraulic effects upon our internal ecology. Just as every thought and every sensation find expression in our systems through some modification of motor behavior, so every tension and every movement produce a sensory stimulation. Touch receptors in my skin, pressure-sensitive Pacinian corpuscles in my deeper tissues, and Ruffini end organs in my joint capsules all provide sensory responses to every distortion of tissue caused by muscular activity, from a tiny twitch in the corner of my eye to the full mobilization of my body in walking or running. I cannot move without touching myself internally and externally; these internal and external pressures and frictions inform me about my own body and its activities in exactly the same way that my sense of touch informs me about external objects. I am not tactily aware of things, not even my own body parts, unless I push or rub up against them in some way.

The Kinesthetic Sense

All of these tissue distortions and sensory responses taken together are what give me my kinesthetic sense, my feeling for my body's size and shape, the locations of all my joints and limbs, and what all of them are up to. "It is important to remember that in its simplest form, the nervous system is merely a mechanism by which a muscular movement can be initiated by some change in peripheral sensation ..." All of these peripheral sensations initiate some kind of muscular response. What is equally important to remember is that every one of these muscular movements in turn initiates a reciprocal change in peripheral sensation, creates a new chain of sensory and motor responses.

The path of an impulse through the nervous system is not linear, from stimulation to sensation to motor response; it is always circular, each motor response in turn providing stimulation which colors the sensations, which alter the subsequent motor responses, and so on and on and on. My own tissues are among the objects that touch my awareness, and my own muscular responses are continually a part of the creation of

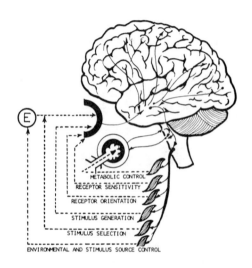

7-2: The many roles of muscular movement in sensory perception. The muscles may interact with the environment, manipulating the objects perceived; they may act to isolate a single source of interest; their own activity creates sensory stimulations which condition our observations of other objects; muscular movements are used to orient the sensory receptors—cocking an ear, for instance, or moving a hand back and forth to test the heat radiating from an object; contractions can increase the sensitivity of a sensory receptor—focussing the eyes on a particular object in the distance, stretching the tympanum of the ear so that it is more acutely responsive (literally "straining" to hear), adjusting the hand so as to perceive the slightest resistance in the turning of a dial, and so on; and muscular activities condition internal metabolic activities, influencing in many ways the manner in which input is processed and the significance of any particular observation.

sensory information about the world that floods my central nervous system. Movement itself is the factor which unites the two halves of my nervous system into a unified relationship of continually mutual reciprosity—"Perception and motor answer are the two sides of the unit behavior."[1]

This is to say that the extent to which I move, and the manner in which I move, have everything to do with what I know about myself and what I know about the outside world.

> Movement leads to a better orientation in relation to our own body. We do not know very much about our body until we move it. Movement is a great uniting factor between the different parts of our body. By movements we come into a definite relation to the outside world and to objects, and only in contact with this outside world are we able to correlate the diverse impressions concerning our own body. *The knowledge of our body is to a great extent dependent upon our actions.*[2]

If I lie motionless for a period of time in a sensory isolation tank, or if I adopt a stress-free and motionless meditative posture, I begin to lose the crisp sense of my physical boundaries, and my mental picture of the spatial relationships of my body parts undergoes bizarre distortions. My feet and hands may seem miles away. My body as a whole may seem incredibly dense and heavy, or large and inflated like a huge balloon. I may feel my physical self dispersing like a gas into the void, or I may feel no connection whatever between my conscious awareness and my living cells.

Such induced hallucinations are often triggered by the disassociation of the sensory and motor elements that are welded together in our normal states of perception. This disassociation can also be accomplished by various drugs, such as the opiates or the hallucinagens, whose pharmaceutical effects disrupt our sensory/motor unity in various ways. In all these situations, the conscious activity of my mind leaves far behind the concrete feed-back loops between my nerves and my muscles, and without the solid underpinning of movement correctly correlated with sensation, my thought processes confuse some of the basic distinctions between internal feeling states and objects. The marvelous or hideous phantasmagoras of my imagination take on as much palpable reality as anything else I am experiencing at the moment. The same hallucinatory effects occur in the muscularly and sensorily suppressed state of sleep.

Movement as the Basis of Perception

So perception is formed on the basis of movement, just as surely and completely as movement is initiated and guided by perception. I have only to begin gently moving my limbs in the water of the isolation tank for the sharp lines of my physical boundaries to leap back into focus; I have only to make a few wriggles and stretches for all of my body parts to reassert their physical relationships in my consciousness.

> We do not feel our body so much when it is at rest; but we get a clearer perception of it when it moves and when new sensations are obtained in contact with reality, that is to say, with objects.[3]

Various degrees of friction, stretching, and impact are the only things which build up our sense of tactile reality, and motion is fundamental to all of them. And this is just as true when I am feeling myself as it is when I am feeling other things.

"It does not appear that we perceive our own body differently from any other object."[4] My body is simply the object, or collection of objects, which has the most intimate proximity and the most constant presence for my consciousness. To consistently ignore it requires enormous repressive energies; and to learn more about it, to use it more efficiently and for more activities, requires constant attention to the creation of new movements and the novel sensations they bring.

Movement and Learning

This sensorimotor interplay, movement and contact, is the basis for learning any new skill. This is what is going on, for instance, when infants are indulging in the endless exercise of vocal movements—spitting, smacking, blowing, cooing—what we call "babbling."

> In the course of these vocal tract gestures, [the infant] produces sound patterns which resemble the vowels and consonants of his parents' language as well as those of many foreign languages. Most importantly however, he is being flooded with sensory feedback. His every unintentional variation of vocal pitch, quality, or loudness, every movement of the tongue or lips in the midst of a vocal sound, produces simultaneous changes in sound and in tactile-proprioceptive sensation. Perhaps one reason why infants can master some simple speech movements, complex as they are, by nine or ten months is simply this fact of inevitable, simultaneous auditory and tactile-proprioceptive feedback from the vocal tract.[5]

Children who have poor sensation in their mouths and lips do not learn to talk normally, no matter how much training they are given. "To learn normal motor patterns for speech a child must not only hear the speech of others, but also he must hear and feel his own speech movements."[6]

It is only the sensations associated with movement that allow me to explore random movements and to begin to select those that approximate the skill I am trying to learn, and it is by means of sensations that I continue to refine my selected movements until my efforts produce the precise effect I want. Thus it is that we call acquiring new skills, or modifying old habits to fit new situations, "getting the feel" of the activity.

In addition to these physical sensations, there is a mental feeling of "rightness" that comes to be associated with the specific manner of movement which produces satisfactory results. This sense of "rightness" is a large part of the pleasure of learning a skill, and is also one of the main reasons why habits become so ingrained, why my behavior takes on such recognizable personal patterns. So much of my sense of psychological and physical continuity, my sense of unity and security, depends upon my ability to repeat appropriate and predictable actions, that this feeling of "rightness" can scarcely be overestimated in its importance as an element of my psychic integration as a whole. Each time I "get the feel" for a new response, I also get a new feel for myself and for my relation to the world of external objects at large. The activities of my muscles play an enormous role in my relationship to internal and external reality.

The Sense of Normalcy

This system of sensorimotor integration is marvelously adaptive, and its simple elements can be manipulated to produce the entire range of complex human skills. Unfortunately, the sense of "rightness" which comes about through repetition may not necessarily correspond to movements and habits that are "optimal" in their efficiency. "Rightness" in this context only means "familiarity," grooves that are well-worn. It is often strongly associated with movements and habits that are merely "satisfactory" in some limited way, or "normal" in the sense of "like before."

We all stand and walk differently, but we all stand and walk with an identical internal sense of "normalcy" associated with our own way of doing it; and this sense of norm has for each of us an equal feeling of "rightness" to it. Yet, in spite of each individual's own sense of "rightness," some of us stand and walk with far more ease and efficiency than others, while some have accustomed themselves to doing it so poorly that their posture and manner of walking undermine the health of the whole system.

Astonishingly enough, once a sense of normalcy has been established in connection with a way of doing something, perceived inefficiency or even pain are usually not enough to alter our behavior. We will tend to continue to do a thing the way we learned it, the way in which we first established our "feel" for it, in spite of the fact that subsequent problems develop as a result.

The security offered by the "normal," the familiar, is so powerful that it typically prevents us from achieving improvements. Often the truth is that if we stopped standing the way we have learned to do over the years, we would be able to stand much more easily. But the constant illusion is that if I stop standing the way I have learned to do, I will not be able to stand at all.

Bodywork and the Sensations of Movement

This conservative tendency inherent in the feeling of normalcy has a great force of inertia. By maintaining learned patterns, it contributes much to my sense of continuity, the stability of my body image, and the firmness of my sense of self. On the other hand, it can prevent constructive changes and refinements just as much as it preserves learned patterns.

The central role played by the tactile sensations in the development and maintenance of this familiarity, this normalcy, provides bodywork and movement therapy with an extremely useful means of confronting the limitations of this protective conservatism.

Just as the individual utilized muscular movements and the sensations they evoked to select and establish his present patterns of behavior, so can the therapist induce new movements and new sensations to begin the selection and establishment of new patterns. If the individual tries to move in new ways on his own, his overwhelming tendency is to favor patterns of movement that feel "normal" to him — movements that he has characteristically used before. But by remaining passive and allowing the therapist to create movements and sensations *for* him, the individual can begin to "get the feel" for patterns of movement that might take laborious weeks and months for him to master on his own, and can experience a new sensory norm that would have been impossible for him to establish as quickly by himself. Then once this *feeling for* a new, more graceful, more efficient gesture is a conscious reality for him, he can move quite rapidly towards establishing for himself the muscle contractions that will reproduce the corresponding sensations.

It is possible for bodywork to be effective, and often to be so rapidly effective, because in fact it utilizes the same principles that guide the learning of motor responses in the first place. It is the motions and pressures of the muscles which create sensations, and it is the selection and repetition of specific sensations which condition the learned patterns of motor activity. It is the task of the body worker to provide movement and sensations which more nearly approximate "optimal," so that the conservatism of what is merely "normal" may be transcended in the learning processes of the individual.

From this perspective, there is certainly nothing mysterious, or illusory, or temporary about the benefits of effective bodywork. Bodywork is simply a method of educating, or re-educating, motor responses, a method which capitalizes upon the central role played by tactile stimulation in the selection and reinforcement of patterns of muscular activity. New ways of posturing and moving cannot be readily learned until the characteristic tension patterns developed by old habits of movement and posture are removed, until the muscles in question are relaxed: "Muscular relaxation...is a prerequisite to any type of successful muscle retraining."[7]

Relaxation does not merely allow the bodyworker to move the client's joints more freely for the moment; it considerably increases the suggestive potency of those freer movements; it creates a neutral ground where the information of new sensations can be introduced into a system normally locked into its old patterns; it produces a palpable hint of what it would be like to respond differently.

Even more importantly, for hours after the session is over, this relaxation and these new sensations give the client the opportunity to experience and practice movements that are relatively free from the habitual sensorimotor patterns that define and dominate his "normal" state. Under these conditions, a great deal of relearning can take place in a short period of time, and the sense of conscious self-control—so crucial to our competence and well-being—can be tremendously enhanced.

Muscular Sensations

Von Haller in the eighteenth century, we have remarked, could produce no clear, conscious sensations by prodding exposed muscle tissue. It was reasonable to conclude that muscle contained no sensory elements. Generations of researchers that followed turned up little evidence to the contrary, and so it came to be regarded as a truth that muscle was largely insentient.

But in our own century, the development and use of far more technically improved microscopes yielded visual evidence that displaced this old truth. When we peer into the fine structures of their tissue, we discover that our muscles

turn out to be full of sense organs, and very fine sense organs at that. The principle kind, the muscle spindles, are the most elaborate sensory structures in the body outside the eyes and ears.[8]

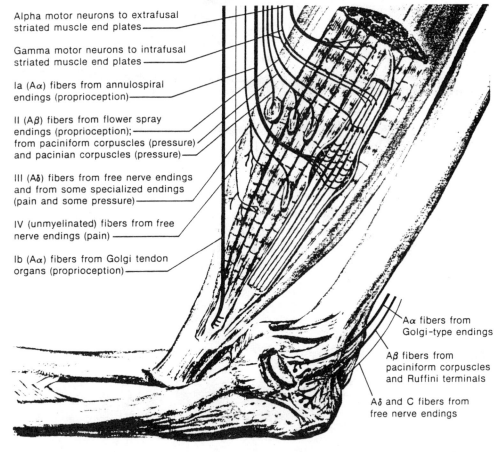

Alpha motor neurons to extrafusal striated muscle end plates

Gamma motor neurons to intrafusal striated muscle end plates

Ia (Aα) fibers from annulospiral endings (proprioception)

II (Aβ) fibers from flower spray endings (proprioception); from paciniform corpuscles (pressure) and pacinian corpuscles (pressure)

III (Aδ) fibers from free nerve endings and from some specialized endings (pain and some pressure)

IV (unmyelinated) fibers from free nerve endings (pain)

Ib (Aα) fibers from Golgi tendon organs (proprioception)

Aα fibers from Golgi-type endings

Aβ fibers from paciniform corpuscles and Ruffini terminals

Aδ and C fibers from free nerve endings

7-3: The sensory devices found in muscle; they are all shown much larger than actual scale. Note that there are a few pressure and pain endings found in the muscle belly. Perhaps von Haller's subjects did not report sensations from them because there are fewer of them than in other surrounding tissues, and because the shock connected with the subjects' own open wounds may have clouded the lesser input coming from the fewer numbers of sensory endings within the muscle itself.

Whether or not we "feel" anything in our muscles, these sensory elements are there, and there in great numbers. Their discovery has been one of the greatest challenges in recent history to our understanding of muscular activity, and what we have learned about their functions has greatly complicated and greatly enriched our concept of motor coordination.

Their activity is proving to be an indispensable link in our understanding of the phenomena of tonus, posture, inherited reflexes such as sucking and swallowing, learned skills such as walking, writing, or driving a car, and a host of other muscular responses. They help to explain both how I can train myself to acquire the exquisite motor control of a dancer, a painter, or a watchmaker, and

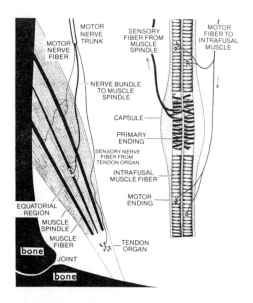

7-4: Intrafusal (spindle) muscle fibers within a muscle belly showing both the motor nerve and the sensory element which serve them.

how it is that I can slowly lose that control to become stiffened and awkward, straight-jacketed by the very muscles that should give me movement and freedom.

The Muscle Spindle

A *muscle spindle* is a bundle of three to ten specialized muscle fibers, smaller in diameter and much shorter than the larger skeletal muscle cells surrounding them. It is called a "spindle" because the overall shape of each bundle resembles that of a textile spindle—cylindrical, thick in the center portion, and tapering to

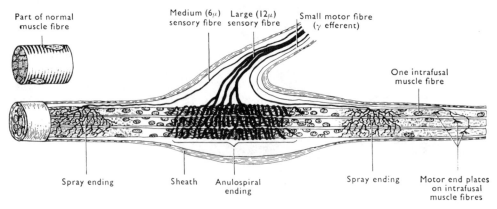

7-5: A detail of the central region of a spindle, showing the anulospiral receptor coiled around the intrafusal fiber.

points on both ends. At their pointed ends, the bundles are gathered together by connective tissue and attached to the connective tissue septa that form the small compartments holding together the skeletal muscle cells. These spindle fibers, also called *intrafusal fibers,* have throughout most of their length a striated appearance like the larger skeletal fibers, and even though they are too delicate to contribute much to the work load, they lengthen and shorten or assume a fixed length by means of the same kind of telescoping strands of myosin and actin that power the skeletal muscles.

Unique to these intrafusal muscle fibers is an absence of the striations—due to an absence of the overlapping contractile proteins—in the central portion of the cells' length. This central zone does not actively lengthen and shorten like the rest of the fiber, but is passively stretched and slackened by the contraction and release of the striated, muscularly active parts of the cell.

Around this central region is wrapped a long sensory ending, the *anulospiral receptor.* Because of its spiral arrangement around the cylindrical intrafusal

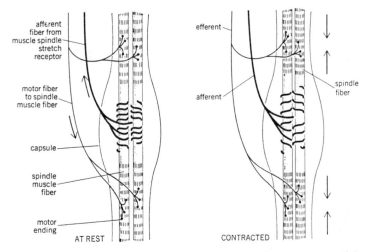

7-6: Contraction widens the coils of the anulospiral receptor, registering the exact amount of shortening of the fiber as a whole.

fiber, it can register quite precisely the degree of lengthening and shortening of the central zone: The gaps between the coils are either widened or narrowed to the exact degree of the expansion or contraction of the fiber as a whole.

Since the spindles are attached to the surrounding septa in the muscle belly, this anulospiral receptor can be stretched and compressed either by the shortening or lengthening of the intrafusal fiber itself, or by the shortening and lengthening of the larger skeletal muscle fibers that surround it.

Movement and Sensation Unified

The muscle spindles are the locations where the two halves of the nervous system, the sensory and the motor, have their closest physiological association, where movement and sensation are joined directly together in a firm embrace, where no intervening barriers exist and no intermediary messengers are necessary. The anulospiral receptor does not feel the distortion of surrounding tissues when a muscle produces a movement; it feels the movement of the muscle itself. It feels exactly what length the muscle was before it lengthened or shortened, it feels just how far and how fast the change occurs, and it feels the exact length again when it stops. Spindles are motor units that can feel *themselves,* a combination that is unique to them.

Spindle Reflex Arcs

The sensory axon that ends in the anulospiral receptor reaches out from its cell body located in the spinal cord. This cell body synapses with its own spinal sensory tracts which carry the spindles' sensory information up each segment of the spinal column and finally to the brain, much like the orderly, parallel spinal tracts for the skin receptors, the joint receptors, and so on.

But in addition to joining together in its own sensory stream like all other sensory nerves headed for the brain, the cell bodies of the anulospiral receptors make another interesting connection within the spinal column. They synapse directly to the body of a *motor* nerve as well, and to precisely the motor nerve which stimulates the skeletal muscle cells that surround the corresponding spindle. This means that the terminal motor nerve, the one which directly excites the muscle cells of the skeletal motor unit, can be excited not only by *motor* com-

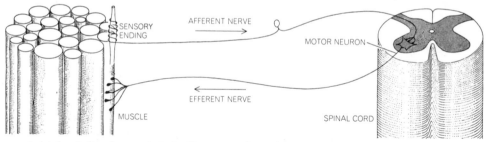

7-7: A simple spindle reflex arc. A single afferent nerve forms the anulospiral receptor at one end and synapses directly to a motor nerve at the other end, in the spinal column. This motor nerve in turn synapses to muscle cells in the immediate vicinity of the spindle, creating a very sensitive local feedback loop.

mands from the brain, but can also be excited by a *sensory* signal from the muscle spindle surrounded by the muscle cells of the *same* skeletal motor unit.

This sensory-to-motor synapse in the spinal cord forms a *reflex arc,* the most direct linkage we have between local sensory events and local motor response. Activity in specific muscle cells creates a local sensory impulse which directly effects the subsequent activity of the same muscle cells. Thus the reflex arc constitutes a feedback loop which both keeps my muscles themselves constantly informed as to what they are up to, and constantly modifies their efforts. And most of this feedback takes place in the spinal cord, far below my levels of conscious awareness, and far more rapidly than I could consciously command it.

The simplest kind of control that is afforded by this feedback loop is the so-called stretch reflex, which is illustrated by the knee-jerk test which most of us have experienced in the doctor's office. The common tendon of the quadriceps is tapped with a rubber mallet just below the knee-cap, and this tap has the effect of a sharp tug on the muscle fibers of the thigh. This sudden change in length of the thigh muscles is registered by the anulospiral receptors, which in turn stimulate in the spinal cord the motor neurons which power the thigh, causing a brief contraction of the thigh muscles which makes the foot jerk forward in a healthy reflex. This automatic contraction has the effect of keeping the thigh muscles at a constant length regardless of outside forces acting upon it, and makes it possible to maintain erect posture in spite of external disturbances.

The components of this arc are like a miniature or primitive nervous system, a microcosm of our nervous system as a whole, a system which "in its simplest form is merely a mechanism by which a muscular movement can be initiated by some change in peripheral sensation." My spindles and their reflex arcs are tiny neural units that monitor and influence motor events that are so continuous and so numerous that they would totally overwhelm my conscious mind. Even if I could keep track of the changing lengths of every one of my millions of muscle cells, I certainly would have no room left to think about anything else.

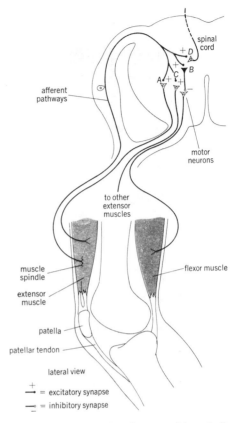

7-8: The simple stretch reflex created by spindle reflex arcs. When the patellar tendon is tapped with a mallet, it has the effect of sharply tugging upon the attached extensor muscles, the quadriceps. This sharp tug fires the anulospiral receptors, which in turn fire the motor neurons in the spinal cord associated with the quadriceps, causing them to contract quickly. As a result, the knee jerks, indicating a healthy reflex.

The Gamma Motor System

The contractile portions of the intrafusal fibers are innervated by motor neurons in the same fashion as are the larger skeletal muscle cells. The intrafusal membrane is contacted by a motor end plate, which stimulates the spindle cell into contracting or fixing its length, or ceases stimulation to allow the cell to be lengthened passively. But the motor neurons which control the intrafusals are altogether separate from the ones which control the skeletal muscles.

The skeletal motor neurons are larger in diameter, with their own vertical tracts through the length of the spinal cord, and end their paths near the summit of the brain, in the motor cortex. They are called *alpha* motor neurons. The motor neurons for the intrafusal fibers, in contrast, are smaller in diameter, have their

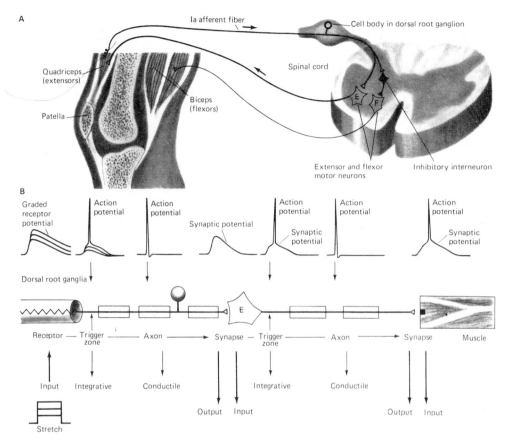

7-9: A detail of a stretch reflex involved in a knee jerk. The patellar tendon is tapped ("stretch"), and the resulting action potential follows the series illustrated from left to right. Note that the anulospiral receptor's afferent fiber also must contact an inhibitory neuron connected to the antagonistic muscles in the hamstrings, so that they will release enough to allow the knee to jerk forward in the reflex reaction.

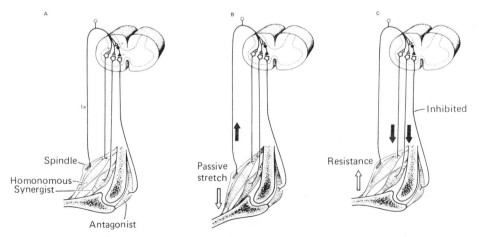

7-10: The stretched spindle may also recruit other synergistic motor units in different muscles, to strengthen the contractile reflex response. Observe again the necessary reflex inhibition of antagonist muscles.

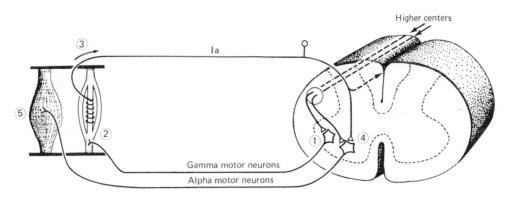

7-11: Gamma motor neurons stimulate the intrafusal muscle fibers, and alpha motor neurons stimulate the skeletal muscle fibers. They are each a completely different motor system; the ascending gamma paths terminate in cell bodies in the brain stem, while alpha paths ascend all the way to cell bodies in the motor area of the cerebral cortex. One third of our motor neurons are gamma.

own discrete pathways up the spinal cord, and end in collections of cell bodies, or ganglia, deep within the brain, in the brain stem. These are called the *gamma* neurons.

We have, then, two separate motor systems within us. One of them, the alpha system, originates in the cortex, is closely associated with conscious sensations in the sensory cortex, operates the skeletal muscles, and is responsive to my conscious commands to lift an arm, take a step, and so on. The other system, the gamma, originates deep inside the older part of the brain, is associated with lower sensory centers that produce no conscious sensations, controls the lengths of my intrafusal fibers, and functions primarily beneath the levels of my conscious awareness.

These two systems are linked together at their peripheries by the anulospiral receptor, which wraps the middle of the intrafusal fiber and synapses in the spinal column with its alpha partner. And because of these spinal reflex arcs, any impulse and movement initiated by one of the systems necessarily triggers an immediate reciprocal impulse and movement in the other, since the anulospiral receptor is stretched or compressed in either case.

The significance of these two motor systems and their sensory interconnections will become clearer as we proceed to examine their functional relationships. The fact that this significance is considerable is strongly hinted at by the sheer size of the "hidden" gamma system: As it turns out, fully one third of the motor neurons in the human body are gamma.

Golgi Tendon Organs

In addition to the muscle spindle, there is another sensory device, also present in large numbers, which is intimately associated with the activities of individual motor units of the skeletal muscles. The *Golgi tendon organs* are found among the collagen bundles of the tendons, in the border zone where the muscle fibers are attached to the tendons. Although they are located in the connective tissue of the tendon rather than in the midst of the muscle cells, they are, like the spindles, minute gauges for the efforts of the alpha muscle fibers.

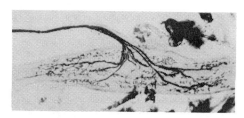

7-12: A Golgi tendon organ embedded in a slip of tendon. The collagen fibers of the tendon are stacked in zig-zag forms, so that as the tendon receives weight, the zig-zags straighten out. This movement is registered precisely by the tendon organ.

The collagen fibers of the tendons are not arranged in straight parallel lines, but in parallel ripples, or zig-zags (see Fig. 3-13). This gives the tendons a small degree of elasticity as the ripples are pulled taut and straight and then allowed to recoil back into their zig-zag resting pattern. The Golgi organs themselves are multi-branched type endings of sensory axons, which are woven among the collagen fibers near the muscle cells, and which are stimulated by the straightening and recoiling of the tendon.

As is the case with the muscle spindles, the stimulation of a single tendon organ is highly specific: Each particular organ is most directly affected by the lengthening and contracting of the few alpha muscle fibers which attach to the collagen bundles containing that tendon organ, so that each Golgi is responsive to the activities of only ten to fifteen alpha motor units.

And as with the sensory elements of the motor spindles, the pathways carrying the sensory information from the Golgis are culminated in the ganglia of the brain stem, with very few direct connections to the conscious cortical areas. We normally receive no more conscious sensation from the Golgis than we do from the muscle spindles; their information is processed and responded to primarily in the brain stem and the spinal cord, beneath our level of conscious awareness.

The Golgi Organ/Muscle Spindle Relationship

As a sensory device, the tendon organ is a close partner to the muscle spindle in the assessment of the specific activity of every one of my alpha motor units. The anulospiral element of the spindle measures the *length* of a muscle's fibers, and the speed with which that length is changing. Adding to this information, the Golgi tendon organs measure the *tensions* that are developed as a result of these changing lengths. The degree of distortion in the parallel zig-zag collagen bundles is a precise gauge of the force with which a muscle is actually pulling on the bone to which it is attached.

Such a gauge is really necessary in order to fully and accurately assess the net amount of work force actually being delivered by a muscle, as opposed to merely knowing now much and how fast it is lengthening or shortening. I can shorten my bicep exactly the same distance at exactly the same speed, whether there is a book in my hand or not, and my spindles will register identical information in either case. It is only the differing stress placed upon the tendon organ during the gesture which announces and evaluates the added weight of the book.

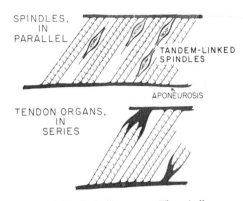

7-13: Golgi and spindle partners. The spindles measure length and speed of changing length, while the tendon organs measure the amount of pull generated upon the tendon as a result of these changing lengths and speeds.

There are many variable factors which help to determine just how much work a muscle is doing as it is contracting, and information coming only from the muscle spindles can be as confusing as it is helpful with regard to measuring this real work. For instance, a muscle cell is "weaker" when it is stretched out to its full length, because the myosin and actin chains do not overlap very much and therefore have fewer cross-bridges to ratchet; hence contractions at this end of the muscles range do not have as much force behind them as they do when the muscle shortens by about half and the myosin and actin filaments are overlapping deeply and creating many cross-bridges. Or, increasing fatigue can make a muscle *feel* as though it is working harder and harder, in spite of the fact that it is actually contracting with less and less force. Even changes in my mood can significantly alter my sense of ease or effort during any given contraction.

So the Golgis add an indispensable quantum of information to the spindles' measure of changing muscle lengths: The Golgis assess the exact amount of *resistance* which is overcome in order to contract a given distance in a given time.

Mass

As a team, then, the Golgis and the spindles produce a sensory impression that is very different in kind than the impressions of color, texture, odor, or sound produced by our more conscious sense. Instead of measuring any of these surface qualities, the muscle and tendon organs assess the pure *mass* of an object.

Now mass is an invisible thing. We have only to contemplate the surprises offered by a tennis ball filled with lead, or a large "rock" made of styrofoam in a movie studio, to remind ourselves how easily deceived our other sense organs can be with regard to mass. Mass has nothing to do with surface qualities; it is the measure of an object's *resistance to movement,* and I can have no idea of its value until I am actively engaged in *moving* the object.

Nor are the sensory cues relating to mass at all constant with regard to the object. They vary continually, as a function of inertia, according to the *speed* with which I move the object, or the relative suddenness with which I attempt to change the direction of movement or stop the object. A five pound bucket "feels" much heavier if I swing it rapidly in a circle over my head—that is, I have to brace myself much more forcefully in order to resist its pull. It is the precise value of this resistance which is measured by the Golgi tendon organs, and when their information is correlated with the spindles' measurement of the exact speed and distance of movement, I can arrive at an accurate estimate of mass, that invisible yet crucial property of all matter.

The Golgi Reflex Arc

Like the spindles, the Golgis are much more than just sensory organs; they also are strategic links between our sensory and our motor functions, a part of the exquisitely sensitive system we have for the fine local control and overall coordination of our posture and our movement.

Like the spindle receptors, the Golgi axons do not only ascend in orderly pathways to their terminal sensory ganglia in the brainstem. They also synapse directly onto motor neurons in the spinal cord, and as with the spindle receptors, the motor neuron contacted by the Golgi receptor is the very same motor neuron which activates the alpha muscle cells which pull directly upon the tendon fibers in which the Golgi is located. Thus another kind of local sensor/motor feedback loop is created.

A Complimentary Opposition

But the action of the tendon organ's synapse onto its corresponding motor neuron is not the same as that of the spindle; it is its complimentary opposite, The action of the anulospiral receptor upon its motor nerve is *excitatory:* When the spindle is suddenly stretched beyond a pre-determined "normal" resting length—as in the knee jerk reflex test—it excites the alpha motor nerve so that a contraction immediately follows which quickly re-establishes the desired "normal" resting length.

The tendon organ, on the other hand, has an *inhibitory* effect upon its alpha motor nerve: When the tension developed upon a tendon exceeds a pre-set "normal" limit, the Golgi inhibits the motor nerve, reducing its level of stimu-

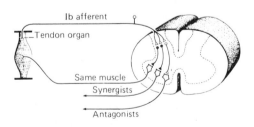

7-14: A Golgi reflex arc. In the spinal cord, its effect upon the alpha motor neurons is the opposite of that of the spindles: The Golgi afferent impulse inhibits the muscle fibers associated with it, and excites antagonists. The two kinds of arcs form complementary reflex devices.

latory firing and thus relaxing the tension back down to its "normal" resting value.

The simplest and most basic function of this inhibitory reflex arc is to prevent the contractile power of the muscles from damaging the tendons and the bones. Many of our muscles are capable of generating enough pull to rip themselves loose from their own moorings, and even the smaller ones which do not have such brute power are in danger of being torn by the uncontrolled pulling of the larger muscles around them. When the Golgi organ senses, due to the increasing tensional distortion of the tendon's fibers, that a strain or a tear is imminent, its signal becomes powerful enough to inhibit the alpha motor neurons that are stimulating the contraction. Tension is reduced instantly, and the damage is avoided.

Higher Brain Influences

This simple inhibitory reflex has a wide range of applications and functions, because the "normal" tension setting which activates it is not limited to that level which announces "imminent danger of injury." Through practice, the tension values controlled by the Golgi reflex can be adjusted to fit almost any job, and the muscles are automatically inhibited from producing more effort than is necessary for the work in hand. This fine degree of tensional adjustment and virtually instantaneous local control is absolutely necessary for the accomplishment of fine motor movements; it is the major reflex mechanism that keeps our efforts appropriately attuned to our desired ends. For example, the fine gradations of sound volume controlled by the force of the strike of the pianist's fingers against the keys is a function of such a precise tension setting in the tendon organ reflexes of his fingers and arms, accurately established by long practice.

Thousands of such discriminating tasks confront us daily, jobs which require a steady force to be applied smoothly, or an identical force to be applied repeatedly, regardless of changes in local muscle lengths, our changes in overall body position, emotional excitement, depression or boredom, growing fatigue, or a host of other variables extraneous to the exact force required.

I can bring a glass of water smoothly to my lips because practice has taught me just how much contractile effort and speed is necessary to lift it and carry it through the air without either dropping it or throwing the water towards the ceiling. This familiar feel for the resistance of the glass of water, and for the appropriate muscular effort to both overcome that resistance and remain in constant control, are functions of the variable settings of the inhibitory response of the Golgi tendon organs. And I use this reflex mechanism every time I use a screwdriver or a wrench, row a boat, push a car, do a push-up or a deep knee bend, pick up an object—in short, every time I need a specific amount of effort delivered in order to accomplish a specific task—any time "too much" is just as mistaken as "too little." This includes, of course, almost all the controlled uses to which I put my muscles.

Now in order to be helpful in all situations, this variable setting of the tension values which trigger the reflex must be capable of both a wide range of adjustment and rapid shifts. Objects that we need to manipulate with carefully controlled efforts may be small or large, light or heavy. Building a rock wall can require just as much finesse and balance as building a house of playing cards, but the levels of tension which require equally sensitive monitoring are very different in each case.

Since these relative tension values can be altered rapidly at will, and are refined with practice, it seems evident that they can be controlled by higher brain centers. This is presumably done through descending neural pathways which can generate impulses that either facilitate or inhibit the action of the Golgi/motor neuron synapses.

> In this way, control signals from higher nervous centers could auomatically set the level of tension at which the muscle would be maintained. If the required tension is high, then the muscle tension would be set by the servo-feedback mechanism to this high level of tension. On the other hand, if the desired tension level is low, the muscle tension would be set this level.[9]

"Required Tension"

So it is necessary that we have a means of monitoring the tension developed by muscular activity, and equally necessary that the threshold of response for the inhibitory function of that monitor be a variable threshold that can be readily adjusted to suit many purposes, from preventing tissue damage due to overload, to providing a smooth and delicate twist of the tuning knob of a sensitive short-wave receiver. And such a marvelously adaptable tension-feedback system we do have in our Golgi tendon organs, reflex arcs which connect the sensory events in a stretching tendon directly to the motor events which control that degree of stretch, neural feed-back loops whose degree of sensory and motor stimulation may be widely altered according to our intent, our conscious training, and our unconscious habits.

This ingenious device does, however, contain a singular danger, a danger unfortunately inherent in the very features of the Golgi reflex which are the cleverest, and the most indispensable to its proper function. The degree of facilitation of the feed-back loop, which sets the threshold value for the "required tension," is controlled by descending impulses from higher brain centers down into the loop's internuncial network in the brain stem and the spinal cord. In this way, conscious judgements and the fruits of practice are translated into precise neuromuscular values. But judgement and practice are not the only factors that can be involved in this facilitating higher brain activity. Relative levels of overall arousal, our attitudes towards our past experience, the quality of our present mood, neurotic avoidances and compulsions of all kinds, emotional associations from all quarters—any of these things can color descending messages, and do in fact cause considerable alterations in the Golgi's threshold values.

It is possible, for instance, to be so emotionally involved in an effort—either through panic or through exhilaration—that we do not even notice that our exertions have torn us internally until the excitement has receded, leaving the painful injury behind to surprise us. Or acute anxiety may drive the value of the "required tension" so high that our knuckles whiten as we grip the steering wheel, the pencil suddenly snaps in our fingers, or the glass shatters as we set it with too much force onto the table. On the other hand, timidity or the fear of being rejected can so sap us of "required tension" that it is difficult for us to produce a loud, clear knock upon a door that we tremble to enter. We can be "up tight"—so agitated that a high value of "required tension" is suffused throughout our muscular system, hampering some movements and exaggerating others. Or we can be "down"—so enervated at the prospect of a dreaded ordeal that we cannot even muster the "required tension" to get ourselves out of bed.

The Silent Sounds

As we can now appreciate, muscle tissue is anything but insentient. The muscle spindles and the Gogi tendon organs are extremely sensitive monitors, and between the two of them our central nervous system is kept constantly informed about the activities of every individual motor unit: how long they are, how rapidly they are changing this length, and how much tension they are delivering to the tendons which anchor them to the bones. And there is no sensory activity from any other source which more directly influences my motor behavior; local Golgi and spindle sensations are instantly translated into local muscular responses through their direct synapses onto the alpha neurons in the spinal cord.

These muscular responses in turn create more spindle and Golgi sensations, which further modulate muscle responses, which again change the nature of the sensations, and so on and on, in a mutual interchange whose continuous immediacy creates an intimacy that is more refined than that of any other sensory/motor link, an interchange which by-passes most of the intermediate steps in transmission which modify virtually every other type of sensory input. Furthermore, it is not only the alpha motor system that is so closely linked to these sensory devices, but also the entire gamma system as well, a second motor system which has its own distinct pathways and its own distinct control centers in the brain stem.

These direct internuncial circuits in the spinal cord and this brainstem-directed gamma motor system create an astonishing condition in the numerous and elaborate sensory feedback loops of the spindles and the Golgis: We are consciously aware of almost none of their constant activities.

> Signals transmitted to the central nervous system from these two receptors operate *entirely at a subconsious level*, causing no sensory perception at all.

Instead, they transmit tremendous amounts of information from the muscles and the tendons to 1) the motor control systems in the spinal cord [and the brain stem], and 2) the motor control systems of the cerebellum.[10]

The large extent of the Golgi-spindle system, and the minute complexity of its wiring, provide the apparata for enormous numbers of impulses and responses, and if we probe the feedback loops or the ascending gamma pathways with electrodes we do indeed find their activity is normally quite brisk. Yet they are almost utterly silent to our conscious awareness. It is as though the eyes were working perfectly, and yet we saw nothing, as though our ears were functioning and yet we heard no sound. What is it that can be accomplished for us by all this unfelt sensory information and all this unconscious muscular response?

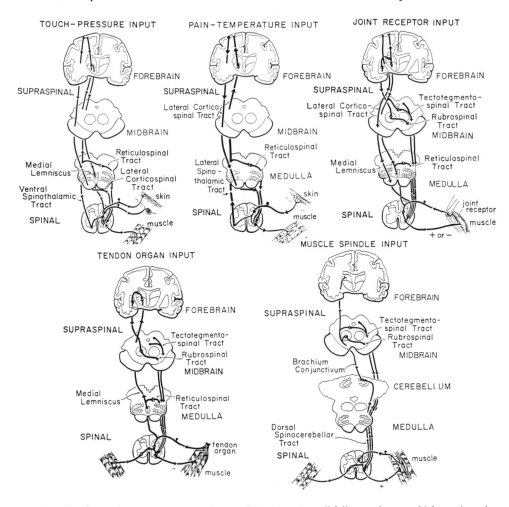

7-15: Impulses from pain, pressure, temperature, and joint receptors all follow pathways which terminate in the sensory cortex, and create conscious sensations. But impulses from tendon organs and spindles terminate in the brainstem, never reaching the cortex to register as conscious sensations.

Simple Reflexes

We have already had occasion to look at the basic "stretch reflex," illustrated by the quick jerk of the knee in response to a blow from a rubber mallet upon the common tendon of the quadriceps. This same jerking phenomenon can be elicited by similarly tapping the tendons at the elbow, the ankle, the wrist, and so on, and it is used diagnostically to assure us that at least our spindles and their spinal reflex arcs are intact. A proper jerk is the benchmark of the healthy function of these basic apparata of neuromuscular coordination.

The stretch reflex is the activity of the simplest reflex arc found in man. It employs only a single sensory element—the anulospiral receptor—and a single muscular element—the motor unit of the corresponding alpha neuron (although we must remember that in the knee jerk response many spindles and motor units are involved, not just one of each). It is initiated by a sudden stretch applied to a muscle by an external force, and its net effect is always to oppose that stretching force and maintain the current length of the muscle.

Gravity is by far the most common of these external forces, whether it affects us through the weight of an object or through the weight of our own tissues. For the purpose of initiating a stretch reflex, the weight of my own arm is just as "external" to the spindles and muscles of my shoulder as is a bucket of water held by my hand. Thus the role of my shoulder spindles is to keep my shoulder in its proper—or more precisely, its "normal"—place, whether it is only the mass of my arm or the additional mass of the bucket which tends to stretch it downward.

Needless to say, there is a wide variety of such "external" forces, both inside and outside of my body, the accurate tap of a mallet being only the most isolated possible clinical example. Every physical encounter with an object generates such forces, as does the weight of every limb and organ of my own body. And the directions and degrees of stretch change every time I shift positions. Consequently, there is a wide latitude in the strength of the reflex response. A steady, continuous stretch produces an equally steady and continual contractile response to balance it. The weight of my body parts is constant, and my muscles apply a constant force to counteract them and hold me erect. This is called a "static" stretch reflex. A sudden shift in my center of gravity, or a sudden addition of weight on a limb, produces a stronger and more sudden reflex response, like the knee jerk; this is called a "dynamic" stretch reflex.

Resting Tension

Interestingly, the actual strength and suddenness of the external force are not the only factors for determining the strength of the static or dynamic reflex responses. After all, a quick tug on a slack wire does not produce the same physical results as an equal tug on a taut wire. The higher the resting tension values are in the muscles of my thigh, the more violent will be the knee jerk in response to the tap of the mallet. Conversely, a flaccid thigh may produce little or no jerk, in spite of the fact that the spindle mechanisms are basically intact.

We have noted earlier that it is states of mind in the higher brain centers which often have the decisive influence in establishing these basic tension values. The spindles and Golgis are, in the end, just mechanisms that react when their stimulation threshold is crossed; it is the mind which sets and changes these thresholds.

Our attitudes and emotions are tugging on our muscles just as effectively as are any external forces, setting the stage for our reflex responses just as surely as a blow from a mallet or a shift in center of gravity. To state it figuratively, objects pull on my muscles from the outside, and my emotions pull on my muscles from the inside; a pull from either vantage point increases the tension on the muscle.

Stress and anxiety are notorious for their effect of driving up the resting tension values of our musculature, and we can readily observe the exaggerations they lend to simple reflex responses by popping a balloon behind a person who is poised upon tightened coils of anxiety. These emotional states are themselves like "weights" which drag on our muscles just like any physical object, and can in fact profoundly alter the strength of our responses to the forces exerted by physical objects or gravity. Here we are standing at the edge of a large vicious circle, where mental strain and physical tension can reinforce one another around and around the spinal reflex arcs until fatigue, pain, and dysfunction are the results.

It is obvious, then, why bodywork cannot use mere *force* to make a muscle relax and lengthen. It is one of the fundamental jobs of the muscles' sensory system to *resist* sudden change due to external forces, and this resistance is enforced by an automatic reflex that cannot be bullied or argued with. No matter what specific bodywork technique is being applied, only slow, patient, unthreatening pressures and stretches can avoid triggering more contractile responses through these reflex arcs, and only a growing trust and surrender in the mind of the client can succeed in calming the mental turmoil which has established increased tension settings and exaggerated responses.

Pleasure and Tension Settings

The physical and emotional pleasures that accompany such a soothing manner of working are not by any means incidental extras to the process of relaxing and lengthening muscle tissue. Pleasure is one of the principal tools of the bodyworker, no matter what sort of manipulations he or she has been trained to do, because soothing pleasure is one of the most potent means available to us to defuse the exaggerated responses of the body's reflex defense systems.

We feel this truth instinctively when we must calm a crying baby, or comfort an injured child, or quiet a nervous animal. And controlled testing has confirmed that it is impossible for laboratory animals to sustain states of fear or rage while the pleasure centers in their brains are being artificially stimulated; they immediately cease their aggressive or defensive behavior and focus upon the pleasure.

This appears to be a response that is just as obligatory as the reflex contractions we have been examining, and as such it is one of our most useful means of interrupting self-perpetuating cycles of excess tension. Pain and discomfort, on the other hand, can only be counterproductive to this process, because local pain acts just as surely as a sudden tug to initiate another contractile reflex to which we will now turn our attention.

The Withdrawal Reflex

A sudden, unexpected change of length in a muscle is one way to stimulate a reflex contraction; a sharp, unexpected pain is another. Every time we unwittingly touch our hand against a hot object we feel for ourselves a demonstration of this sort of reflex response. A feeling something like an electric shock shoots through the whole arm, and the hand is instantly jerked away from the painful stimulus.

This shock and the jerk which follows often occur so fast that the actual sensation of pain does not register in the mind until after the hand has been withdrawn. This is because the pain stimulus, like the sudden-stretch stimulus, synapses with internuncial groups in the spinal cord which feed the message directly into the appropriate alpha and gamma motor circuits, causing the withdrawal contractions before the pain impulse has time to travel all the way up the spine to the cortex, where it becomes a conscious sensation.

Quite often it happens that our jerk of withdrawal is more dramatic than it really needed to be. When we examine our hand, we find no actual injury, even though we have yanked it away from the hot object as though it were burning us severely. This is not neces-

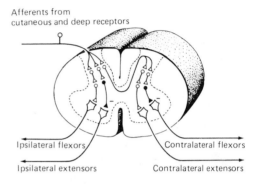

Afferents from cutaneous and deep receptors

Ipsilateral flexors — Contralateral flexors
Ipsilateral extensors — Contralateral extensors

7-16: Peripheral sensory nerves—including those for pain—also connect with alpha motor neurons in the spinal column. This means that peripheral sensations can also trigger the reflex contraction of the muscles, as well as Golgi and spindle input.

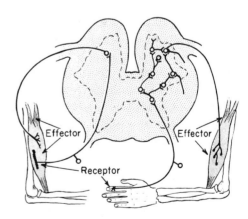

Effector — Effector
Receptor

7-17: The circuits of a simple stretch reflex and of a withdrawal reflex. The withdrawal reflex is more complex, since a sensation from one part of a limb must activate muscles in another part of the limb, rather than just the motor units in the immediate vicinity of the sensory receptor. This requires more internuncial connections.

sarily over-reaction, but is rather the *successful functioning* of this reflex: The point is to get our hand away from the hot object *before* actual damage has occurred. This is the reason for having a direct spinal reflex response to pain in the first place.

But even so, both the level of the threshold which triggers the response and the magnitude of the response are to a large degree determined by influences from the higher brain, just as with the stretch reflex. Our attitudes about pain and our relative levels of anxiety play a large role in the strength of the withdrawal reflex as well. We can be so conditioned to pain, or so numbed to it, that the effect of the defensive reflex is markedly dampened, even though actual damage is occurring. Or on the other hand, we can be so sensitized and fearful of pain

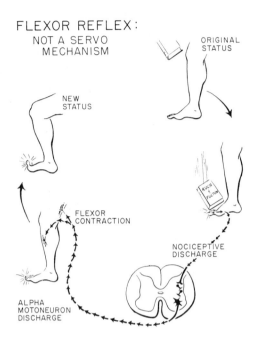

FLEXOR REFLEX: NOT A SERVO MECHANISM

ORIGINAL STATUS

NEW STATUS

FLEXOR CONTRACTION

NOCICEPTIVE DISCHARGE

ALPHA MOTONEURON DISCHARGE

7-18: A schematic example of a flexor (withdrawal) reflex.

that the slightest hint of injury sends a limb into spasmodic contractions. This is yet another avenue through which stress, anxiety, or depression can begin to amplify reflex muscular contractions until "normal" tension settings are seriously distorted. Local pain, or even mere discomfort, can initiate a vicious circle of contraction, until the contractions *themselves* are strong enough to cause the very pain that keeps the cycle going.

Once again we can see the relevance of these spinal reflex responses to the manner with which bodyworkers manipulate tissues. The very stress and anxieties which tend to produce muscular tension are the same influences which lower the thresholds and exaggerate the responses to additional pain or discomfort. Thus anything we do to cause local discomfort, or often even the *hint* of possible discomfort, reflexively tightens the contractions we are seeking to release.

Patience, sensitivity, and careful timing will reward the bodyworker far more than mere force, no matter what structural or functional end he wished to achieve. The client's patterns of reflex contractions cannot be bullied into submission. In fact, the very act of bullying can only ignite more stretch and withdrawal reflexes.

The Crossed-Extensor Reflex

An isolated stretch of reflex may only involve a few spindles and a few alpha neurons and muscle cells. The withdrawal reflex typically affects an entire limb. In the next level of this reflex organization of behavior—the crossed-extensor reflex—we can see how reactions to local stimuli can begin to involve the entire body in contractile responses.

The withdrawal reflex causes the *flexors* of the affected limb to contract, pulling the limb away from the painful stimulus. If the stimulus or the reaction is strong enough, the crossed-extensor reflex is triggered as well. Less than a second after the flexing of the injured limb begins, the extensors of the opposite limb are activated by the strong pain impulses, and they move to extend the opposite limb to push the whole body away from the source of the pain. Indeed, if the source of the pain is intense enough, most of the muscles of the body may be reflexively called into play to escape it: Rather than just jerking my hand away, I twist my torso violently and jump back with both legs.

This spreading of the basic withdrawal reflex throughout the body is what makes a person "jumpy" when it is aggravated to a high degree. And remember that it is not just the strength of the stimulus which affects this spreading process, but also the underlying tension settings in the body's musculature as a whole. This is how stress, anxiety, fearful attitudes, and the like tend to amplify local reactions and diffuse them into distortions of the entire body.

The "optimal" response to pain or discomfort, dictated by the natural efficiency of the reflexes, would be just enough contraction to remove a single body part from the source of pain. However, a person's "normal" response to a relatively minor pain—a response dictated by past experiences, emotional tendencies, levels of anxiety, and so on—may be a violent jerk of the injured

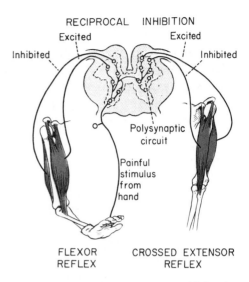

7-19: When a crossed extensor reflex is added to the flexor reflex, an even more complicated exchange occurs in the internuncial circuits of the spinal cord. Several spinal segments may be involved, and several distinct peripheral sensations may guide the reflex contractions and inhibitions.

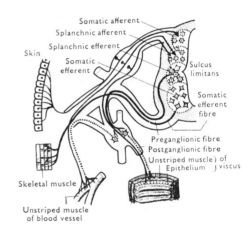

7-20: Through the involvement of various internuncial connections in the spinal cord, sensory stimulations may cause reflex responses in any of the body's muscles, including those of the internal organs and the blood vessels. Thus pain from skin may cause contractions of the stomach or the colon as well as of the skeletal muscles. Or conversely, pain in internal organs can stimulate spasms in the skeletal muscles. Many organ dysfunctions stem from this mechanism and the vicious circles it can generate.

On the other hand, this mechanism can be of great use to body work. Internal organs can be reflexly influenced by positive sensations coming from the periphery.

part, a twist of the torso, a jump with the legs, distortions of the face, a yell, or a holding of the breath, a ducking of the head, and a general cringing posture that persists long after the stimulus has been effectively escaped. These exaggerated reflex responses, facilitated by both physical and mental tension, can begin to force the whole body to over-react to local discomforts, waste large amounts of energy, and distort our normal posture for periods of time much longer than are necessary merely to escape a specific painful stimulus.

Learned Reflexes

Up to now we have considered only two kinds of stimulation in the activation of reflex muscular responses—the pull of gravity and pain. Fortunately for us, these are not the only two sensations that call my reflex circuits into play, otherwise my behavior would amount to little more than stabilizing myself against gravity and avoiding pain.

This bracing and avoidance are primary reactions which are fundamental to the continued life and health of our organisms, and so we accordingly find that they often play a very large defensive role in our behavior, and that the sense of alarm they entail is usually capable of over-riding all other considerations.

But we are motivated not only by defensiveness. Our reflexes respond to all sorts of experiences of pleasure, objects of interest, novel situations, curiosities. We are given a few reflex reactions that help preserve our lives; we are free to develop any number of additional ones that can help to direct and enrich them.

Our Two Earliest Reflexes

At the prenatal age of seven and a half weeks, the very first foetal response is the turning away of the head in response to stroking the cheek. For the next four or five weeks of intrauterine life, this withdrawal response strengthens, until the whole body is engaged in turning away from the stimulation on the cheek.

But by the time of birth, stroking the infant's cheek causes an *opposite* reaction: The body turns *toward* the stroked cheek, and begins exploring with its mouth for a nipple. A stimulus that was originally associated with "potential danger" is later associated with "breast" and "nourishment", and instead of a withdrawal, a reaching out is the result. In other words, our sensations and reflex responses—in conjunction with the higher brain—are equally capable of leading us in two different directions —away from harm or towards pleasure, nourishment, interest, novelty. There are reflexes for opening and experiencing, as well as for closing and avoiding.

The balance between these two tendencies is constantly being worked out as we grow from infancy to adulthood to old age. As we learn to crawl, to toddle, to talk, to walk and run, to dance, to ride a bicycle, to play a musical instrument, or to perform job skills, our sensorimotor systems are constantly feeling out the differences between new dangers and new advantages, new pains and new pleasures.

As we progress in this groping and tasting, our accumulating experiences, our attitudes, and our emotions begin to direct the overall patterns of our responses one way or the other. Some of us drift towards closed and defensive behavior, and some of us towards more open and experimental behavior. Some develop more skills and graces than others. And the two basic reflex responses—to withdraw and avoid, or to extend and explore—become cumulative, so that these inner tendencies are eventually faithfully mirrored in stiff, braced, defensive bodies or relaxed, supple, open ones.

Conditioning Reflexes

There is nothing that is inevitable about either of these two avenues of development. One experience simply leads us to the next, so that all past experiences tend to color present ones, which in turn tend to select and color future ones. In this way our basic reflexes become "conditioned," both with regard to the skills we learn to perform and the manner in which we perform them. This conditioning of the reflexes into new patterns is the essence of all motor learning, and the process of conditioning is always dominated by *feeling*—both in our sensory organs and in our higher brains.

It is by the direct manipulation of these feelings that bodywork can begin to alter the conditioning that has shaped our learned patterns of reflex response. And it is because these feeling states—both sensory and emotional—are such formative influences upon our behavioral conditioning in the first place that changes in them can have such dramatic results in changing defensive muscular patterns, self-perpetuating tendencies towards withdrawal and rigidity.

By having tactile pleasure generated in his tissues, an individual may be quickly reminded that his body can be the source of a wide range of pleasant and interesting feelings, not just painful or gravitational ones. By having the range of motion in a stiffened limb slowly increased while these pleasurable sensations are continued, he is allowed to discover that many of the movements of which he was fearful are not really dangerous or painful to him, and that a good deal of his self-restraint was unnecessary, was even the source of much of the discomfort that he imagined he was avoiding.

And as these positive sensations accumulate, the individual can begin to make new *emotional* and *intellectual* associations with his body as well: It can be trusted and used to the full extent of its possibilities, not just suspected and defended; it is the source of movement and pleasure, not just limitation and discomfort; it provides many kinds of sensory information upon which a wide variety of choices may be based, not just one or two sensations which trigger one or two inevitable responses.

Feelings of heightened interest and an urge to explore these new sensations can begin to loom larger than the old feelings of fear and defensiveness, and the exploration of these new sensations can lead to new patterns of behavior. The body can start to *move* differently because the mind has *felt* something different. New conditioned responses can arise because they are supported by new sensations and facilitated by new attitudes. Negative tendencies can be shifted in their course, and the reflex gestures of extension and opening can be reinforced, just as the reflex gestures of contraction and closing were reinforced, and reinforced through exactly the same learning medium — appropriate sensory input.

Alpha-Gamma Integration

The large alpha skeletal muscle cells are the ones which do the actual work of supporting my body parts, moving them through space, and dealing with external objects. It seems natural to regard them as the primary source of this support and movement, and to regard the spindles and Golgis as sensory devices which function mainly to monitor the changing lengths and tensions inherent in this muscular activity.

And this is indeed the model of motor behavior that was held for a long time: Motor commands issue from the motor cortex in the form of specifically coded bursts of action potentials that are carried down the alpha motor neurons to the skeletal muscles, which then perform the command; the actual details of changing lengths and tensions of this performance is tracked by the spindles and Golgis, generating the sensory information which then assists the motor

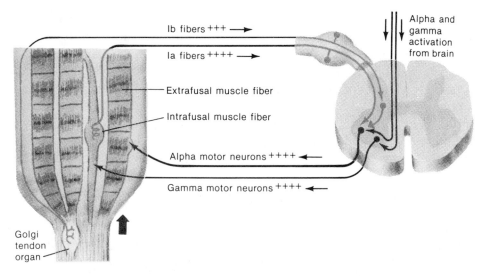

Ib fibers +++ ⟶

Ia fibers ++++ ⟶

Alpha and gamma activation from brain

— Extrafusal muscle fiber

— Intrafusal muscle fiber

Alpha motor neurons ++++ ⟵

Gamma motor neurons ++++ ⟵

Golgi tendon organ ⟍

7-21: If the gamma motor system does not activate the intrafusal fibers at the same time that the cortex is activating the alpha fibers, misalignments will occur in the lengths of these two types of fibers. This will set off reflex responses that will stymie smooth synergistic/antagonistic coordination.

cortex in making fine adjustments in its commands, improving the precision and accuracy of the commanded movement.

However, this logical and relatively simple view of the matter does not account for a couple of important factors. First of all, the spindles are not just sensory endings; they are direct links to spinal reflex arcs which are built to resist any sudden, unexpected change in muscle length. Any command sent to the skeletal muscles would necessarily be stymied by this reflex locking action unless the reflex arcs from the anulospiral receptors to the alpha motor neurons were somehow neutralized, disengaged. No muscle can lengthen unless the stretch reflex allows it, and hence while the reflex is active, the shortening of any muscle

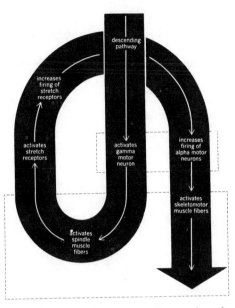

7-22: The gamma loop. The small rectangle at the center indicates those effects occurring in the brainstem or spinal cord; the large rectangle, the effects in the muscle.

would be blocked by the stretch reflex of its antagonists.

Secondly, the spindles are not passive, stimulated only by the movement of the muscle tissue around them; they are muscle cells in their own right, with the power to lengthen and contract independently of the alpha motor units. Moreover, when these gamma motor cells of the spindles change their lengths, their spinal reflex arcs oblige the surrounding alpha cells to change with them. The stretch reflex is activated any time that the length of the alpha muscle cells does not coincide with the length of the gamma muscle cells.

It is clear, then, that the muscle spindle complexes are more than just measuring devices. They constitute an active motor system, with its own contractile cells, its own motor neurons, its own spinal pathways, and its own areas of organization in the brain—a motor system which must move in tandem with the commands from the alpha motor cortex to avoid throwing on the brakes of the stretch reflex, a motor system which does not merely track the movements of the larger alpha system, but which to a large degree either stimulates or inhibits alpha movements. The role of the muscle spindles is not just to maintain singular "normal" resting lengths, but to maintain a sense of "normalcy" throughout every phase of the active use of my muscles.

The Gamma Loop

We now know that a movement may be initiated by either of these two motor systems. The motor cortex can initiate a voluntary movement without being blocked by the stretch reflex, because when I *know* that I am going to move in a specific way, then the critical "unexpected" quality of stretching muscle lengths is neutralized. The unconscious gamma command centers in my brain stem can mimic a move directed by my conscious mind, lengthening and shortening its intrafusal cells in concert with the alpha cells around them so that the anulospiral sensory element is not stretched or collapsed during the movement. In this instance, the gamma system follows the lead of the alpha, with the anulospiral ending's reflex arc silenced as long as the two are synchronized—that is, as long as the alpha movements correspond to "expected" limits that are successfully mimicked by gamma movements.

A movement may be initiated by the gamma motor system as well. In this case, the command signals are organized in the terminal gamma ganglia in the brain stem (the gamma system's counterpart for the alpha's cerebral cortex). These signals are then sent through a complicated path known as the *gamma loop:* They descend through gamma motor neurons out to the intrafusal fibers. These small spindle cells are not strong enough to move a limb, but they are strong enough to stretch their own anulospiral receptors. This stretch automatically fires the spinal reflex arcs connected with the receptors, and the larger alpha motor cells are immediately stimulated to match the contractions of the gamma fibers. As soon as the desired muscle length has been reached, the commands from the brain stem cease, and the spindles hold their new resting length. When the alpha fibers catch up to this new resting length (a matter of a fraction of a second), the anulospiral element is quieted, and contraction ceases.

Space and Time

Thus two different kinds of muscle contractions are possible, each using a different sort of patterning principle to control movements. The motor cortex and the alpha neurons directly cause contractions of the skeletal muscles, contractions which continue as long as the command signals are being sent. The brain stem and the gamma neurons, on the other hand, cause contractions that are mediated through the spindle system, and their commands cease once a predetermined length has been achieved, at which point the stretch reflex arrests further effort. The alpha system organizes its commands in terms of the duration of neural bursts; the gamma system organizes its commands in terms of the starting and stopping lengths of the muscle fibers.

In other words, the stimulation patterns which organize the alpha contractions are coded as a function of *time*—the duration of the bursts coming from the motor cortex. And the patterns which organize the gamma contractions are coded as a function of *space,* stimulation ceasing when a predetermined length is achieved in the anulospiral gauges.

"New" Brain and "Old" Brain

So in our thinking about the ways in which we feel and control our myriad and constant muscular responses, we must always regard not just one motor system, but two: The alpha system, which originates in the motor cortex and terminates at the motor end-plates that contact the individual cells of the skeletal motor units; and the gamma system, which originates in the brain stem and terminates at the motor end-plates that contact the smaller intrafusal motor fibers distributed as spindles among the skeletal muscle cells. These two systems are firmly united through the millions of spindle reflex arcs in the spinal column, and through vertical connections between the cortex and the brain stem, but they each add distinctly different influences to the performance of the skeletal muscles.

The alpha system is rooted in the surface of the more recently developed cerebral cortex; its control is more accessible to my conscious awareness, and is the avenue through which I direct voluntary commands to my muscles, such as "raise right arm. The gamma system, on the other hand, is rooted in the deeper, more ancient strata of the brain stem that are not normally accessible to my conscious awareness, and this is the avenue through which I direct largely unconscious impulses which adjust the position and tone of my body as a whole in such a way as to correctly support my conscious, voluntary movements. For instance, "raise right arm" cannot be smoothly executed unless the unconscious commands "brace right leg and muscles to the left of spine" are not automatically triggered by the stretches initiated by the pulling imbalance of the extending arm.

We have already come up against so many unanswered questions concerning even the simplest sensorimotor transmissions, and have seen that the internuncial network through which they pass is so vast in its numbers of interconnections and so microscopic in its complexities, that it would seem pure folly to complicate the issue further. But the whole neuromuscular system works all together, or it does not work well at all. Assembling pieces of a jigsaw puzzle is not made easier by putting half of them away so that there are fewer to contend with. An individual's motor behavior is such a rich and minutely interpolated blending of the influences of the alpha and the gamma systems that we must either try to come to grips with their interaction or give up inquiring how gesture, posture, and habit are achieved.

Old Brain

The brainstem, where the cell bodies of the gamma motor system are most densely organized, is both phylogenetically and ontogenetically the oldest part of the brain, the primitive cephalic ganglion. It is often referred to as the "reptilian brain," because its principal enduring features were developed during the evolutionary age of reptiles. It was (and still is) the reptile's highest level of sensory integration and motor command.

And it is reproduced with remarkable fidelity to its ancient design in the growing human foetus. Further outgrowths of the evolving central nervous system, and of the growing human brain—principally the thalamus, hypothalamus cerebellum, and cortex—are offshoots of this older core structure, offshoots which proliferate more and more interconnections, which add more and more numerous and sophisticated levels of integration and modification to the individual's conscious awareness and motor behavior.

Although this older portion of the brain is complex in its own right, it has only a fraction as many cell bodies and synapses as does our cortex, and specific anatomical pathways followed by specific kinds of impulses are much easier to discern in the stem. It has been developed by millions of generations, both before and after the Earth's age of reptiles, and this long usage has produced neural forms and patterns of transmission which are remarkably constant within each species.

Like the spinal cord, many of the brain stem's interconnections are "hard-wired," and their stimulation initiates obligatory responses that are not unlike those of the spinal reflex arcs. It is these relatively fixed pathways and responses which control the range of behavior and style of movement that are so characteristic of each species; a cat and a small dog have pretty much the same skeletal and muscular structure, yet each moves this structure about in ways which clearly identify it as canine or feline. These distinctly different styles of moving similar physical frames are the result of different patterns of integrating sensory information and of organizing motor commands, primarily in the spinal cord and in the older, "reptilian" portion of the brain— that is, the centers of gamma motor control.

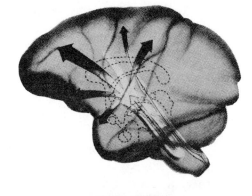

In the human being, this gamma system accounts for fully one third of an individual's motor neurons. In most other mammals, this fraction is even larger, since their cortexes are much smaller relative to the size of their brain stems than is man's.

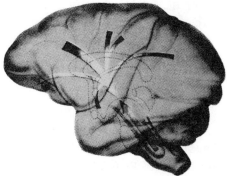

The Reticular Formation

Much of the bulk of the brain stem is made up of a large area called the *reticular formation*. The relative amounts

7-23: The reticular formation. Levels of arousal may be influenced either by peripheral sensations or by activities in the higher brain centers.

of neural activity in this region corresponds to our general levels of arousal, both mental and physical. When we are in deep sleep, the measurable activity in the reticular formation is low; when we are awake and fully alert, reticular activity is relatively high. In fact, going to sleep and waking up are so consistently characterized by markedly decreasing and increasing levels of neural activity in this area that one is tempted to say that sleep is caused by a sharp falling off of neural transmissions in the reticular formation, and awakening is caused by their resurgence.

Normally, these levels of activity follow a fairly regular cycle of waking and sleeping, but a wide variety of conditions can obviously exert their influence. Some stimulants, such as amphetamine, target this region and accelerate its activity, producing a feverishly alert state of consciousness, and tense, twitching muscles. Conversely, some muscle relaxers, tranquilizers, and depressants lower normal levels of activity in the reticular formation, producing relaxation, grogginess, loss of motor coordination, or even general anaesthesia as their dosages increase.

In our seventy-year-old surgical patient, we saw that this deep anaesthesia not only rendered him totally unconscious, it also interrupted his muscles' ability to hold his joints together. The gradual return of his former stiffness during his recovery period was the return of his reticular formation's normal waking levels of activity, once again stimulating his brainstem, his cortex, and all of his muscles to their characteristic degrees and patterns of contraction.

Apart from the physical necessities of a wake/sleep cycle and the predictable effects of certain drugs, it is difficult to pinpoint precisely what things influence the activity of the reticular formation, and exactly what the nature of that influence will be. The absolute silence that one person requires in order to fall asleep may be so unnerving to another that sleep is impossible. My father could never sleep without his radio on, but its faint squeaks would keep me stark-staringly awake two rooms away. A person may sleep blissfully through dozens of surrounding noises, yet awaken instantly to a familiar voice, or even a familiar step on the stairs.

Clearly sensory inputs effect our levels of conscious arousal, but there are wide individual differences between what sort of input and how much, and what kind of specific patterns they will arouse. And when we reflect for a moment upon the enormous effects such variables as past experiences, specific training, and emotional reactions can have on our mental and physical states, we can indeed appreciate that "arousal" is a thing that comes in many sizes, shapes, and colors. A distant screech of a tire may throw a mother who knows her child is near the street into a panicky wave of anxiety and tension, while a shell exploding in the next bunker may scarcely cause a combat-seasoned soldier to flinch.

The level of activity in this reticular formation reflects an individual's general state of arousal. Artificial over-stimulation of the entire area does not result in limb-flailing or the exaggeration of particular gestures; instead, it causes a stiffened tetany in all the muscles of the body simultaneously. Everything locks rigidly into place and cannot be moved until stimulation recedes. Conversely, blocking stimulation from reaching the whole area results in a general loss of muscle tone throughout the body, as in our anaesthetized patient. That is, activity in this area as a whole does not command our muscles to produce any particular gesture or assume any particular posture. Rather, general activity here provides the conditions of general muscle tone, sensory awareness, and mental alertness which will support and color whatever postures and gestures are made.

The reticular formation cannot issue the command "raise right arm," but it does help to establish the trembling tension, the calm readiness, or the sluggishness which will characterize how I raise my arm in response to a situation. To direct these general levels of arousal into particular movements requires the next level of the "old" brain, the basal ganglia—the highest level of sensory and motor organization of the gamma motor system.

The Basal Ganglia

If a person is lying asleep, quick, firm pressure applied to the soles of his feet is very likely to wake him up. This is the result of a sensory stimulation activating the reticular formation and arousing the system into general readiness to act. But if the pressure on the soles of the feet is quick enough and strong enough, something much more specific than general arousal takes place: The muscles of the body—and particularly those of the feet and legs—contract and brace themselves at the same lengths and tension values appropriate for maintaining a standing posture.

The sensory stimulation of the quick pressure on the soles simulates the stimulation that occurs when the person suddenly stands on them, and his musculature responds by instantly and unconsciously bracing against the gravitational forces that are normally associated with that pressure. His contracting and stiffening in the patterns which years of experience of standing and walking have established as appropriate to prevent a fall.

This highly coordinated motor response is distinct from merely "waking up," and it is an illustration of how the basal ganglia begin to orchestrate the body's specific muscular responses to sensory stimulation. If I stand up quickly, I do not have nearly enough time to consciously direct all of the contractions that will stabilize me. How do my legs know instantly and accurately what to do?

The Orchestration of Muscular Response

Any general movement is made up of hundreds of small contractile movements, each one arranged in a closely timed sequence to contribute its increment to a smooth and controlled gesture. These small local contractions are generated by the stimulation of the alpha motor neurons in the spinal cord which connect to their individual motor units. It appears to be the job of the basal ganglia to orchestrate the basic selection of the appropriate motor neurons, initiate their stimulations in the proper sequence, and direct their precise timing.

Some of these movement patterns, such as swallowing, are fully established in the basal ganglia at birth; others, such as walking, are the result of long years of practice. Each of these ganglia seem to add a specific quality to any general movement, qualities that are notably absent or exaggerated when the activity of one of the ganglia is out of balance with the others. A few examples will help to indicate how each

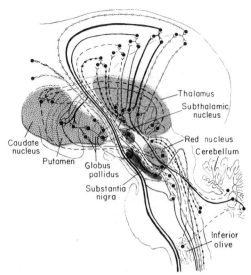

7-24: Pathways through the basal ganglia. The organization of each ganglia adds a specific quality to the final muscular expression of motor commands.

ganglia adds its organizational component to a successfully controlled movement.

Substantia Nigra

The *substantia nigra* is a ganglia that is particularly responsible for the interpretation and coordination of the overall sensory information coming from the muscle spindles and the tendon organs. The totality of this information is crucial to assessing and controlling the changes in lengths of the body's muscles, the speed with which a movement is occurring, and the actual work-load that is being handled.

This, remember, is sensory information that we do not "feel" in the normal sense, but when it is incomplete or scrambled, fine control of movement becomes seriously hampered. The step becomes unsteady, the reach inaccurate; motions become jerky; the limbs may tremble or even oscillate grossly in their attempt to find the correct lengths and speeds. These aberrations are among the symptoms of Parkinson's disease, a syndrome which appears when the substantia nigra is damaged in any way.

Thus the substantia nigra adds to the brain's overall arousal the specific messages coming from all of the spindles and tendon organs: Which exact motor units are lengthening or shortening, how fast are these changes in length taking place, and how much force is being developed?

Globus Pallidus

In the *globus pallidus*, all of this sensory information is sifted and selected so that we can appropriately *brace* some parts of our bodies in order to support the desired *movement* in other parts. For instance, the back, shoulder, and arm must be suitably braced in order for the fingers to successfully manipulate a pencil on a page. These phenomena of "fixation" seem to be the elements added to coordinated movement by the globus pallidus. If a laboratory animal is electrically stimulated in this ganglion, it will become fixed in whatever position it had achieved at the onset of the stimulation, and it will remain fixed in that position until the stimulation ceases.

The possible varieties of these fixations are endless, enabling us to brace all parts of our bodies in any shape the anatomy will allow. Without them, the violinist could not hold his instrument, the marksman could not steady his rifle, the artist could not hold his arm straight to daub the canvas, the host could not pour coffee from the pot.

Typically, of course, we do not continue to explore the innumerable fixations of which we are capable. Our habits, our jobs, our social situations, our general dispositions all tend to urge us to prefer certain fixed positions over others. For each of us, our characteristic posture is nothing but a particular fixation to which we return again and again, until the idea that we might in fact stand up in a different way passes out of our conscious consideration.

Over the years, it is the habitual repetition of these preferred fixations which creates the individualized tension patterns in our musculatures, and eventually even alters the thickness of our fascia and the shape of our bones in order to more efficiently accommodate a limited number of positions. As we select postural fixations and become more and more attached to them, their increasing familiarity begins to give us a comforting sensory and psychological stability, a constant norm to which we return as to a favorite jacket or an old friend.

Indeed, my favorite fixed positions eventually cease to be *something I am doing* and become to a large degree what I am. The fixation becomes dominant, and the release more difficult; person, posture, and point of view become firmly welded together, unfortunately limiting all three. And what was a familiar old friend can become an increasingly tormenting millstone around the neck. I find that the position which held me up comfortably for a while cannot do so indefinitely. In fact, there simply is no single position that will support me for indefinite periods of time without producing areas of fatigue, pain, and eventual dysfunction. I need a large *repertoire* of fixations, so that I am not trapped in the discomforts inherent in any single position.

Releasing these compelling fixations is of course one of the principal jobs of bodywork. By manipulating the body so that other positions are concretely experienced, the bodyworker can remind a stiffened back that other positions are in fact possible, that other muscles can take over for a while, that the limitations previously experienced are not *anatomical* ones. And it is extremely important to remember while manipulating these stiff muscles that the fixation is not in the tissues under my hands, but is deep in the unconscious processes of the mind. My physical contact with the local tissues is merely a means of generating new sensory input into the sensorimotor process; it is the mind that is coordinating this process which must release its hold upon a fixed position.

Striate Body

Just as the globus pallidus fixes various body parts in particular positions, so does the *striate body* initiate and monitor many *stereotyped movements*. Cats and dogs and horses and pigs all graze and chew, prick up their ears at a new sound, coordinate various gaits, and so on. Humans also share a wide range of stereotyped movements, similar in their features because they are designed to accomplish the same things for each individual.

And further, we have noted that although both dogs and cats do many similar things—sitting, walking, drinking, jumping, grooming, and the like—they each do them in distinctly canine or feline ways. Every species has a way of doing the normal tasks of living, a manner of movement that is peculiar to it. A good mime can represent "cat" or "mouse," or "horse," or "ape" with a brief imitation of these animals' *manner* of movement just as effectively as he could with an elaborate costume. These too are stereotypes of movement.

The striate body seems to control a wide range of such movements—individual movements that have common utility, movements which continually correct our balance, movements which are the synchronized background motions that necessarily accompany the use of a limb, or movements which establish such standard communications as sexual arousal, docility, fear, anger, or defensiveness. As with fixed positions, in the human being both the repertoire of stereotyped movements and the stereotyped manner in which all movements are done may markedly display habitual preferences built up by compulsions, training, job requirements, and dispositions.

And as with chronic fixations, there is the tendency over long periods of repetition to confuse how I *do* things with who I *am*. My most common movements, designed to be controlled by my unconscious mind so that I can freely direct my attention elsewhere, become more than stereotypes; they become straight jackets, and I find myself the prisoner of the very unconscious processes which are supposed to protect and liberate me. Re-establishing for the individual the sense of a wide array of equally possible movements is the real significance behind the work of freeing a person from limited neuromuscular patterns.

Controlling Tone

These two primary reflex arcs—the spindle and the Golgi—are the principal sensory devices which the nervous system uses for the enormously complicated task of maintaining and adjusting the appropriate levels of muscle tone throughout the body.

The normal tone of a muscle is dependent upon the simple stretch reflex, through which the the sensory endings in a muscle, stimulated by even the slightest stretching of the muscle, initiate a segmental reflex increasing muscle tone.[11]

The muscle spindle, whose associated reflex arc tends to excite alpha motor neurons and their motor units, is complimented by the Golgi tendon organ, whose reflex arc tends to inhibit the same alpha neurons and motor units. Between the two of them, they produce a summation of excitation and inhibition on the alpha neurons which keeps the active muscle fibers within a narrow range of tensional forces—just the right amount to stand, to lift a book, to hold a glass.

Now the problem of maintaining this precision is such a complex one not only because there are so many muscle cells in the body to monitor, but also because proper muscle tone must accomplish so many different things. It must be able to shift its various tensional values in the various parts of the musculature back and forth so rapidly in order to do all of my muscular tasks competently.

First of all, the tone of my muscle cells must hold my skeleton together so that it neither collapses in upon my organs nor dislocates at its joints. It is *tone,* just as much as it is connective tissues or bone, that is responsible for my basic structural shape and integrity. Secondly, my muscle tone must superimpose upon its own stability the steady, rhythmical expansion and contraction of respiration. Third, it must support my overall structure in one position or another— lying, sitting standing, and so on. Finally, it must be able to brace and

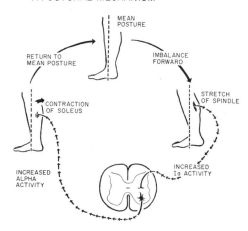

THE STRETCH REFLEX
A POSTURAL MECHANISM

7-25: Using the simple stretch reflex to control basic postural tone.

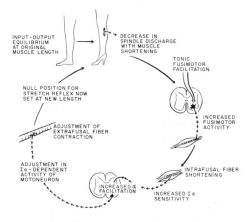

RESETTING
THE STRETCH REFLEX

7-26: The stretch reflexes controlling postural tone can be reset to accommodate new circumstances. However, settings maintained for long periods of time become postural and tonal habits, which may resist resetting if they become entrenched enough. This is how simple postural aberrations may eventually create limitations and force compensations throughout the musculature.

release any part of the body in relation to the whole, and to do this with spontaneity and split-second timing, so that graceful, purposeful action may be added to my stability, my posture, and my rhythmic respiration.

It is no wonder we find that such large portions of our nervous systems are so continually engaged in controlling the maintenance and adjustments of this tone. The entire system of spindle cells, with both their contractile parts and their anulospiral receptors, the Golgi tendon organs, the reflex arcs, much of the internuncial circuitry of the spinal column, and most of the oldest portion of our brains—including the reticular formation and the basal ganglia—all work together to orchestrate this complex phenomenon. We have, as it were, a brain within our brain and a muscle system within our muscle system to monitor the constantly shifting values of background tonus, to provide a stable yet flexible framework which we are free to use how we will.

Nor is it a wonder that these elements and processes are normally controlled below my level of consciousness—if this were not the case, walking across the room to get a glass of water would require more diversified and minute attention than my conscious awareness could possibly muster.

It is the old brain, along with the even more ancient spinal cord, that are given the bulk of this task, because they have had so many more generations in which to grapple with the problems and refine the solutions. Millions upon millions of trials and errors have resulted in genetically constant motor circuits and sensory feedback loops which handle the fundamental life-supporting jobs of muscle tone for me automatically. Firm structure, posture, respiratory rhythms, swallowing, elimination, grasping, withdrawing, tracking with the eyes—all these intact and fully functional activities and more are given to each of us as new-born infants, the legacy of the development of our ancestors.

The evolution of hundreds of thousands of generations has established the necessary muscle lengths and tension loads in order to keep my joints together and to support my structure against the pull of gravity. This information appears to be stored in some form in the brainstem—particularly in the basal ganglia—and transferred genetically from one generation to the next. The whole gamma motor system of the individual then uses this information to adjust the lengths of his muscle spindles appropriately, which in turn adjust the lengths and tensions of his skeletal muscles throughout the reflex arcs.

Getting the Feel For It

The sensory roles of the anulospiral receptor and the tendon organ are absolutely central in this process of exerting and adjusting muscle tone. These devices establish the "feel" for length and tension, and it is this *feeling* which is maintained by the gamma motor system, the reflex arcs, and the alpha skeletal muscles. All of my muscle cells—both alpha and gamma—are continually *felt* by the mind as they work, whether most of these "feelings" ever reach my conscious awareness or not. And it is primarily these muscular feelings which supply my central nervous system with the constant information necessary to successfully combine the demands of free motion with those of basic structural stability.

The sophistication required for this maintenance of structure and flexibility can be appreciated if we remember that almost any simple motion—such as raising the arm out to the side—changes either the length or the tension values in most of the body's muscle cells. If one is to avoid tipping towards the extended arm, then the feet, the legs, the hips, the back, the neck, and the opposite arm all must participate in a new distribution of balance created by the "isolated" movement of raising the arm. The difficulties experienced by every child learning to sit, to stand erect, and to walk with an even gait attest to the complexity of the demands which these shifts in balance and tone make upon us. The entire musculature must learn to participate in the motion of any of its parts. And to do this, the entire musculature must *feel* its own activity, fully and in rich detail. Competent posture and movement are among the chief points of sensory self-awareness. The purpose of bodywork is to heighten and focus this awareness.

It is the child's task during this early motor training to experiment by trial and error, and to set the precise lengths and tensions—and *changes* in length and tension—in all his muscle fibers for these basic skills of standing and walking. This is the education of the basal ganglia and the gamma motor system, as learned reflex responses are added to our inherited ones. The lengths and rates of change of the spindle fibers are set at values which experience has confirmed to be appropriate for the movement desired, and then the sensorimotor reflex arcs of the spindles and the Golgis command the alpha motor nerves, and hence the skeletal muscles, to respond exactly to those specifications that have been established in the gamma system by previous trial and error.

And this chain of events holds true not only for the actual limb being moved, but for all other parts of the musculature that must brace, or shift, or compensate in any way. In this complicated process, the child is guided primarily by *sensory cues* which become more consistent and more predictable with every repetition of his efforts.

Adding the Old to the New

Many of our muscular responses have to be virtually automatic. When we are born, we do not have time to practice holding our joints together securely, or breathing, or swallowing, or eliminating, or withdrawing from painful stimuli. We must be ready very quickly to carry out these activities, or we will die. And we must *continue* doing them, no matter what else we may get involved in as we develop.

The basic life-supporting patterns of muscular activity which respond to the sensory messages signalling collapse, dislocation, lack of oxygen, thirst, hunger, full bladder or bowels, and the like, are already established for us by the time of our birth, programmed into the unconscious, older parts of our nervous system, the spinal cord and the basal ganglia. But for our subsequent development as individual human beings, much depends upon a great many of our muscular responses being voluntary, a matter of choice, and not automatic. One of the marvelous things about humans is that they are *not* confined to a repertoire of genetically given reflex responses; we are free to add almost limitless variations in the uses to which we can put our muscles.

The patterns of muscular contractions for these additional skills are not genetic givens. They must be learned. They are the products of experimentation, trial and error, conscious repetition, painstaking refinement. They often do not have any particular survival value. In fact, often enough the learning process involved in some of these skills can be downright dangerous, and we must actually overcome protective reflexes that would withdraw us from the situation. These sorts of skill do not spring from our unconscious genetic inheritance, but from our curiosities, our imaginations, our conscious desires. The neurological arena of their initial formation is not the older reptilian brain, but the newer cortex.

Both the old brain and the new brain can initiate muscular effort—the cortex through direct stimulation of the alpha neurons and their associated skeletal muscle fibers, and the brain stem through the indirect route of adjusting the lengths of the gamma fibers and relying upon the spindle reflex arcs to trigger the same skeletal muscles—the gamma loop. These two control centers are quite different in their origins and in their neurological organization, and the quality of the motor commands that come from each of them reflects these differences. They may be said to preside over two complementary, not to say conflicting, tendencies in human behavior.

Conservatism vs. Radical Experimentation

The older control center is fundamentally conservative in its operations. It preserves and perpetuates those elements of our behavior that have proven to have survival value, both for the individual and for the species. Its hallmarks are obligatory reflex response and stereotyped movements. It is the cellular wisdom of our ancestors guiding us through problems that would quickly defeat us if we were left to the limited resources of our own personal experience. It is the source of the muscular efforts that maintain a constant base tonus, that keep our structure intact, and that coordinate the large menu of fixations and standardized movements which are the bases for any posture we choose to adopt and any gesture we choose to make. It is what makes a cat always act like a cat, and a dog always act like a dog.

On the other hand, the tendency of the newer cortical control center is radically experimental. Always fascinated by novelty and diversity, restless and inquisitive, our conscious attention pulls us this way and that. We explore new environments, try new foods, learn new skills, adopt new modes of behavior.

Old limitations are challenged and transcended, old reflex responses are modified, or even extinguished. New responses are cultivated, and these in turn lead to new explorations, new novelties, new skills, and the transcendence of previous limitations. In addition to our conservative compulsion to stick faithfully to the "tried and true," we have also the compulsion to "try the new". In addition to a strong tendency to preserve life by means of stable habits, there is another strong tendency to make it more interesting, more exciting, to disrupt those habits.

Dangers

There are dangers lurking in both of these tendencies. Unlike the tried and true, new situations and new behaviors may have unpredictable effects. Often it happens that the first attempts to experiment with a novelty are costly, even disastrous. On the other hand, strict adherence to the conservative tendency can precipitate disasters as well—museums of natural history are filled with the remains of species that perished because they could not change and adapt.

Almost no other animal even comes close to man with regard to the large size of our newer cortex relative to our older brain stem. This is why we are free to adopt or reject such a wide variety of attitudes and behavior patterns. This is also why man, more than the others, is a species that has the potential to be so deeply divided against itself, to be neurotic, psychotic, suicidal.

Ideally, the harmonious combination of conservatism and free experimentation may provide the balance that is necessary in order to preserve life while at the same time developing and improving it. But the potential conflict between the two just as often produces extremely uncomfortable states, and even lethal ones. No species has adapted itself to so many different and hostile environments, or developed as many skills, or produced such remarkable differences among individuals. Yet no other species has indulged in so much behavior which actively undermines its own relationship to the environment, its own health, its own future.

Acquiring Reflexes

Throughout our lives, our inquisitiveness and our attraction to novelty lead us to discover new things to do and new ways of doing them. We are given at birth a few stock forms of behavior to help insure our survival, such as breathing, sucking, swallowing, eliminating, or withdrawing from painful stimuli, and we very quickly begin to expand our motor skills until the activities we have learned outnumber by far those which came to us as our inheritance. And one of the things that is remarkable about this process is that much of what we learn tends to become almost as automatic and as consistent as these more primitive reflexes. I can walk or drive my car with almost as little conscious concern for how I am doing it as I have for how I am chewing or swallowing. A virtuoso musician fingers his instrument with a degree of familiarity and thoughtless ease that we call "second nature."

Repetition

It is those actions which I have repeated over and over the most number of times which most come to resemble my more primitive reflexes in their automaticity and their regularity. It was, after all, countless repetitions by millions of generations which produced the genetic patterns in the brain that control those primitive reflexes; and it is the same sort of amalgamation of repetitions—albeit on a lesser scale—which produce the patterns for the many skills and reflexes I learn within my own lifetime. And the more I do a particular thing in a particular way, the more stable and predictable that action becomes. In this way, all novelties in which I persist end up becoming habits, all successful experimentation tends to move towards conservative norms.

No one has been able to describe exactly how this happens, yet this process is the basis for virtually every learned skill. I begin with a conscious desire to do something which I have never done before. My conscious motor commands from my motor cortex serve to guide me through my initial awkward attempts, largely through my direct neuronal path, while my conscious visual, auditory, or tactile sensations inform me of the accuracy or error in each trial. Given enough trials, these conscious sensations guide my conscious motor commands until my actions begin to consistently produce the particular sensations that announce successful repetitions: I "get the feel" for the desired movement. Once this has happened, my speed and accuracy begin to increase much more rapidly.

Eventually, if I practice consistently enough, this conscious "feel" for the new activity begins increasingly to be translated into the specific muscle lengths, rates of change, and tension loads required to produce the desired results. At this point, my control over the new skill takes on a new level of refinement, aided now by the addition of my practiced conditioning of the muscle spindles, the Golgi tendon organs, their reflex arcs, and ultimately the basal ganglia—the multineuronal path, including the entire gamma system—to the conscious sensory cues and motor commands that guided me previously.

And at the same time, this new skill begins to resemble in some respects an inherited reflex: The sensory and motor coding which directs it is taken over increasingly by unconscious centers in the older parts of the brain, and it requires less and less conscious vigilance on my part to repeat the action.

To the degree that practice and refinement continue, this resemblance deepens. Learned motor patterns can become so ingrained that they need almost no conscious direction to start them up and guide them on their way. Indeed, an expert typist can physically punch out a line faster than it is possible to "think" the same series of individual letters. I can develop a response to the degree that it is impossible for me not to do it once the appropriate stimulus is presented. And I can develop a skill to the degree that the muscle lengths and tension loads required for it become a part of my "norm," and they begin to condition the ways in which I do everything else, including the older established patterns of breathing, swallowing, and eliminating.

In the Groove

Perhaps because the cortex sits physically above the brain stem, or perhaps because we associate conscious experience with visible surfaces and the unconscious with hidden interiors, the metaphor which seems most readily to fit this process of developing a skill is one of "sinking in," or "dropping" from one level to a "deeper" one.

It does in fact appear that the earlier stages of learned responses are directed by the conscious mnd, and are more associated with the sensory and motor cortexes, while later stages take on the more fixed and automatic properties of unconscious responses associated with the inherited patterns of activity in the brain stem. Thus the apparent "control center" for a learned activity may literally "drop down" into deeper neural strata as we repeat it over and over.

No doubt we are here running the danger of over-simplification. Probably any skill requires simultaneous activity from every part of the brain during *all* phases of its development, so that the very idea of a specific "control center" that is "sinking to lower levels" becomes problematical in actual fact. We just don't know enough about the process to visualize it accurately in this way. Nevertheless, we are all familiar with the impression that new skills, awkward at first, settle "into the groove," and that habits become "entrenched," at which point they become conditioned reflexes, similar in some respects to the reflexes with which we were born.

Baby and Colt

The differences between the motor development of a colt and a human infant suggest some of the consequences for us of this process of turning new behaviors into acquired reflexes. A colt is born with nearly all of the locomotor patterns that it will use all of its life intact in a highly developed brain stem. Within hours it can stand, walk, trot, canter, and run, and apart from some improvements in grace and strength, these patterns will not alter appreciably as the colt matures. Soon after birth it enjoys something like the same competence it will have as an adult.

For the human infant, on the other hand, the level of competency at birth is a rather sorry affair. It is born with far fewer patterns of reflex movement established in its basal ganglia, and its initial efforts are utterly ineffectual for sitting and standing, let alone walking or running. The human requires years of rolling, rocking, crawling, pulling up, and toddling before it can begin to approximate the control displayed by the colt in its first day of life.

Humans need to acquire the reflexes associated with locomotion, a long and arduous process of trial, error, and endless repetition of each partial success until it becomes ingrained in the sensorimotor system. Because this is the case with us, it is not only the collective wisdom of my ancestors that determines how I will stand and walk; all the accidents, the felicities and traumas, the personal attitudes and emotions, all the thoughts and experiences that I have while I am learning become a part of my developing patterns of movement as well.

Consequently, my locomotor patterns become individualized by my unique combination of experiences. Unlike colts, no two of us humans learn to walk in an identical fashion. One child may be carried so much for so long that it is late in building the strength and balance necessary. Another may be coached through its early learning phases so rapidly that it never gets a chance to master the fundamentals properly. One may always have to practice with a large wad of diaper between its legs, or another may be arrested by shrieks of "No, no!" whenever it seeks to venture. Injuries of one kind or another will almost certainly occur during the years of learning to walk, retarding the process and forcing various compensations while they are healing. Casts or braces may be prescribed, which completely interrupt the child's practice and development.

In short, the human child can encounter any number of a large variety of encouragements and set-backs during the time it is learning to walk, and every one of them will affect the manner of his walking. It is no wonder that we so often admire the precociousness of the colt and the natural grace and power of the horse. None of us will ever run so early, and very few of us will ever run with such ease and purity of style.

However, this longer and more laborious process of motor learning does not necessarily imply a disadvantage to the human. It may take us longer, but we have the opportunity to make more of it. The colt learns to stand, walk, trot, canter, and gallop so quickly precisely because the neural stimulus patterns for these movements are fixed permanently in its nervous system. It can immediately do these things superbly well, but it will never be able to change them significantly, or to add more than a few variations.

But in the long learning period of the human, a period dominated not by inherited patterns but by voluntary trial and error, there are many opportunities for individual experiments, and variations. Just because our patterns are not fixed, we have the means to develop innumerable new movements and individual styles. A horse will never dance ballet, or ice-skate, and it takes years of rigorous training even to teach one to pull off a stiff curtsy. The famed Lipezaners may be remarkable *horses,* but they do not approach *human* levels of skill in acrobatics or dance. Those of us who feel relatively awkward in our gaits can take heart from the reflection that after all it is not a difficult thing for most of us to stand on one leg, climb a ladder, jump rope, or walk a narrow plank, simple adaptations that would befuddle the cleverest horse.

Individual Stereotypes

As our cortex has mushroomed larger and larger during the evolution of our nervous system, the process of encephalization has carried forward more and more of the neural activities of sensory analysis, decision making, and motor command to the newer, more conscious—and more flexible—areas. As with the visual tectum of the frog, the older and deeper structures of the brain have gradually given up their inherited reflex control of various functions, in favor of the greater sophistication, the wider variation and the more conscious selection possible in the vastly more complex cortical structures.

And yet we have not eliminated stereotyped behavior. Rather, we have exchanged one sort of stereotype for another. Each of us begins life with a relatively small number of *obligatory* stereotypes generated by genetically fixed neural patterns in our brain stems. These species stereotypes, established in animals like the colt by millenia of natural selection and inheritance, are supplanted by more highly *individualized* stereotypes, established in humans by the particular experiences, selections, and practice of every new member of the species.

So the burgeoning of the cortex and the simultaneous encephalization has opened up seemingly infinite variations in the behavior of a single species. This is indeed the crux of the advantage that humans enjoy over most other creatures: We are capable of mastering any number of new skills, and of radically modifying our behavior as conditions demand. In a word, almost everything that is not *given* to us can be *learned*. And with regard to behavioral patterns, every human individual becomes to a large extent a species unto itself.

Limitations

It is of extreme importance for us to realize, however, that this great freedom cannot be extended to mean that I can do anything I want any time I want. Stable patterns of behavior, consistency, and competence have to be built up by means of the repetition of specific movements, whether this is done over the course of development of a species or of an individual. And the more stable a behavior becomes—for a species or an individual—the more difficult it is to change it or stop it.

Ingrained habit and obligatory reflex have much in common. Any novelty that is repeated enough becomes a stereotype. All radical experiments tend towards conservative norms, which guide further selections and developments. As I practice and repeat, I involve deeper and deeper associations, until I engage the basal ganglia, the gamma motor system, and the muscle spindles in the control of a skill. I have then *acquired* a reflex, a bit of behavior that is to a large degree autonomous, initiated and directed unconsciously.

I can, in other words, grow out of the limited fixations of my biological ancestors, but I cannot avoid the force of my own habits. I am free to create a new stable pattern, but once it is established, I am not free to dismiss it with a snap of my fingers. Experimentation becomes gesture, gesture becomes posture, and posture becomes structure. These are things that are well to keep in mind as I make my choices and pursue my developments.

In the context of habitual postures and acquired reflex responses, bodywork can be an invaluable addition to the other factors involved in motor learning, because it can focus an individual's awareness upon the sensory information which defines his present condition, and can introduce a more acute self-consciousness into the making of choices concerning experiments, practice, and habits. The exploration of tactile sensations can both reveal the counterproductive response, and provide the cues that suggest activities that will lead the way out of the vicious circle.

Dangers

The elements of even my simplest movements are so numerous and complicated that I have to rely to a large extent upon these reflex responses, either inherited or acquired, in order to coordinate them properly. It is the organization of the basal ganglia that is primarily responsible for the orchestration of my reflexes, arranging them in sequence and timing them so that they flow smoothly to produce a purposeful movement. These organizational patterns directly stimulate the gamma motor neurons, which activate the intrafusal muscle fibers in the spindles, either fixing them at precise lengths or changing their lengths at precise rates. In turn, these relatively weak impulses of the small spindle muscle fibers stretch or collapse the anulospiral sensory elements, whose spinal reflex arcs stimulate the larger, stronger alpha muscle cells to exactly mimic their actions, producing the desired patterned movement of the limb.

While marvelling at the complexity and ingenuity of this sensorimotor, alpha/gamma feedback network, we must also recognize that the feature which makes it so useful is also the quality which most exposes it to abuse: It is for the most part completely unconscious. If these myriad events were not processed at an unconscious level, we would have no time to think of anything else, and our conscious direction would never attain the speed and acuity of our reflexes anyway. But there is a danger inherent in this advantage. I do not have to consciously think about what to do with all of muscles; on the other hand, my muscles are not necessarily doing what I consciously think.

If the *only* things affecting these unconscious sensorimotor processes were purely mechanical considerations, such as muscle length, rate of change, weight of the limb or object to be moved, forces of inertia, and so on, then we would probably all learn to stand and move in a manner very like one another; practice would lead us all to refine any movement to the same point, that is, the most efficient use of the anatomical structures and the required energy to produce a particular result. Our major differences in behavior would stem from anatomical variations, relative strength, and possible anomalies in our nervous systems.

Reflex Body Language

But the fact is that the unconscious levels of our mental activity are filled with all manner of influences, only some of which have to do with direct perceptions of the body in relation to gravity, objects, and space. The rapid coming and going of our emotional states, slowly developed attitudes, biases, and prejudices, lingering reactions to physical or emotional traumas, the quality of our various trainings, our perceptual errors, depressions, elations, nightmares, daydreams—all these things and more commingle freely beneath the surface of our conscious awareness, sometimes harmonizing together and sometimes jostling sharply against one another, and each perpetually coloring the others as they circulate and develop. And since all mental events are eventually expressed in some form of motor response, all of these activities are interpolated into our mechanisms of muscular control as much as any objective data concerning physical forces.

Often these internal states are directly expressed in small, involuntary movements, such as anxious twitching around the eyes or mouth, a slight shrinking back of the chest in fear, a clenching of the jaw in anger, an embarrassed ducking of the head. These muscular reactions constitute a silent "body language" that often speaks more clearly than words or voluntary actions, and for those who observe them closely they provide many concrete clues about the trains of thought and the emotional reactions occurring in the mind of another.

Body langauge which expresses the contents of the unconscious are of course not limited to these kinds of small involuntary gestures. The shape of my overall posture, poses which I characteristically repeat or maintain for long periods, the general tightness or flaccidity of my musculature as a whole, chronic local tensions or weaknesses, the particular individual style with which I perform all of my movements—all of these things are manifestations of the dominant themes of the development of my mental life, broadly coloring my physical activities with values such as "straight" or "stooped," "sprightly" or "phlegmatic," "dull" or "nervous."

These typical postures and general characteristics are formed in each individual in exactly the same way that I turn a new skill into an acquired habit; By means of repetitive practice. Mental states that occur over and over, or that are intense and sustained, begin to evoke the same muscular responses again and again, until the sensorimotor patterns which initiate and guide them more and more resemble the relatively fixed organization of the basal ganglia, the gamma motor system, and the mechanisms of reflex response.

An often repeated mental *event* becomes a tendency, a *tendency* followed long enough becomes a *habit,* and a habit exercised long enough becomes a bit of *personal identity.* This process parallels that of repeated *gestures* becoming *postures,* and sustained postures becoming *structures,* and these two processes continually interact over the course of our lives, producing our general appearance and styles of behavior. A twitch about the eye caused by a particular anxiety develops into a chronic tic that can be generated by almost any anxiety; habitual anger produces the permanently frozen grimace of the choleric; frequent embarrassed duckings of the head produces the cringing stance of the servile; a chest shrunk back often enough in fear produces the slumped shoulders and sunken rib cage of defeat.

Tone and Mental State

Whether I am practicing a skill until it is reliable, or slowly adopting a postural distortion until it is chronic, I am engaging the same neural mechanisms, mechanisms which are responsible for establishing all the stable patterns of my behavior. The physiological events that are central to both learning processes are 1) the adjustments of the lengths of the muscle spindle fibers, and 2) the adjustments of the tension loads on the Golgi tendon organs.

The generation of the appropriate amount of muscle tonus, whether it is for sustaining postural norms or for executing a smooth and accurate gesture, depends directly upon the information supplied by these sensory devices. Their primary task is to sense the precise amount and duration of effort required to make a particular movement or maintain a position, and then to transfer the corresponding stimulations via the spinal reflex arcs to the skeletal muscles. Through these millions of feedback loops, sensations and movements are inextricably woven together into a single unit, behavior.

What is crucial to stable postures is the central nervous system's ability to set all of these lengths and tension values. And what is crucial to reliable skills is its ability to rapidly *alter* these settings to the appropriate lengths and loads established by experience. In this way, just the right amounts of effort are distributed to the right areas for any task at hand. What complicates this elegant process for humans is the fact that these settings can be influenced either locally or generally by many factors that have nothing to do with the precise amount of muscular effort required to most efficiently do a particular job.

Anxiety, for instance, typically drives these settings higher and higher, regardless of what I am actually doing. Conversely, depression can set them lower and lower. My emotional tendencies enter into my neuromuscular calculations just as surely as do physical forces, and these feelings, attitudes, and reactions produce all kinds of distortions in the actual levels of effort generated by my skeletal muscles.

SERVO-CONTROLLED MOVEMENT

REESTABLISHMENT OF Ia INPUT — α OUTPUT EQUILIBRIUM IN FOLLOW-UP OF CHANGE IN FUSIMOTOR ACTIVITY

INCREASED γ ACTIVITY

FURTHER INPUT-OUTPUT ADJUSTMENT

'BIAS'

EXTRAFUSAL SHORTENING

INCREASED Ia ACTIVITY

FACILITATION OF ALPHA ACTIVITY

INCREASED ALPHA ACTIVITY

7-27: This is how the unconscious mind, acting under a wide array of influences, can increase or decrease the basic tonal values of the musculature through the gamma loop. The mechanism is designed to make possible necessary adjustments to changing circumstances; but it can also raise or drop tonal values in response to attitudes and emotions that have little or nothing to do with the actual task at hand. In this way it is possible for such a mechanism to become extremely counterproductive, making us literally "uptight," or "unstringing" us.

That is, the settings of my spindles and tendon organs are not merely responsive to weight, motion, and repetition; they are also responsive to every sort of mental state. They are not only physically efficient, they are emotionally expressive as well. This is why we so often exhaust ourselves and distort ourselves, why we are so often so wasteful, awkward, and uncomfortable in spite of the elaborate neural devices we have with which to monitor our every effort. Our "sense of effort" is mental and emotional as much as it is physical and mechanical, and this fact leads us into muscular exertions that are inappropriate for the work we want to do, and into postures which distort the efficient arrangement of our bodies' weights in the field of gravity.

The Best of Two Worlds

What we have at our disposal are two very different sensorimotor systems, each with its own muscle cells, its own neural pathways, its own areas in the brain where its terminal cell bodies are clustered, and its own principles of organization.

The gamma system and its terminating basal ganglia are primarily responsible for preserving the levels of tonus and the changes in muscle length which are the basis for my inherited reflex responses, those species-constant units of posture and behavior which generations of usage have proven to have reliable survival value. These responses are fixed and obligatory, so that there is no danger of my "forgetting" them.

The alpha system and its terminating motor cortex, on the other hand, are primarily responsible for modifying reflexes to suit different situations, and for exploring new movements, units of behavior for which there are no inherited programs, gestures and skills that often have no precedent in the history of the species.

The gamma system preserves for us the muscular wisdom of our ancestors, while the alpha system provides us with a doorway out of the confining circle of our primitive reflexes—the ability to try novel experiments, learn new skills, and continually expand our repertoire of ways in which to cope with changing conditions.

These two systems are admirably designed to give us the best of two worlds—a conservative one which obliges us to maintain behavioral norms that are inherited, and a radically experimental one which seems to be capable of learning everything that is not included in our inheritance. The two working together make it possible for us to preserve our lives while at the same time expanding their possibilities, a truly ideal combination.

The only catch is that given its radically experimental nature, it is very possible for the cortex and the alpha system to learn new patterns of behavior that are not necessarily beneficial to the organism. I can cultivate learned habits that are downright destructive to my health. Further than this—and here is the real problem for us—the more I practice bad habits that have begun innocently enough as experimentations or temporary compensations, the more deeply they involve the unconscious reflex mechanisms of the gamma system, until they end up becoming acquired reflexes.

Once enough repetitions have occurred for this to have taken place, then the staunchly conservative nature of the gamma system will maintain the bad habit almost as tenaciously as it preserves inherited patterns. In spite of the fact that my way of standing and walking has become stiff, wasteful of energy, and chronically uncomfortable, my gamma system will not relinquish the patterns it has been trained to reproduce. Nor can any amount of intellectual knowledge alter this situation. The pattern of the bad habit is no longer voluntary or experimental; it is unconscious, physiological, cellular.

It is the business of body work to constructively modify these harmful acquired reflexes and the vicious circles that they generate. This is not accomplished by giving the subject new "ideas"; mere ideas cannot appreciably alter reflex response. Nor is it done by simply telling the subject how to move better, because any solo attempts at new movements will have the strong tendency to be conditioned by the same older habits we wish to modify or eliminate.

It is done by giving the subject new sensations in his flesh and new feelings in his mind, so that these *sensations* and *feelings* can begin to redirect his musculature in new movement patterns in exactly the same way that older sensations and feelings built up the existing patterns. It is not the muscles themselves that must be changed, but the internal "sense of effort" which guides them.

The Cerebellum

While we are exploring our two motor systems and their mutual integration with the input from our various sensory receptors, we must take some account of the activities of one of the areas of the brain about which we know the least—the cerebellum.

We have seen how sensation and motor response are fused into a single mechanism in the muscle spindle, and how they are inseparable elements in the functioning of the spinal reflex arcs of the spindles and the Golgi tendon organs; we have seen how ascending sensory pathways, descending motor pathways, and descending sensory pathways all mingle their interconnections throughout the length of the spinal cord, influencing one another's streams of neural impulses at every level; and we have seen how two separate motor systems—the alpha and the gamma—each process the various sensory streams and use them to program our patterns of movement in their very different ways. The topmost of these many layers of sensorimotor integration appears to be the cerebellum, where the activities of the two motor systems and all the streams of sensation—both conscious and unconscious—are continually blended into coherent experience and behavior.

There can be little doubt that the specific communications occurring in the cerebellum are of central importance to the kinds of changes that can be effected by bodywork. It is here where all of the disparate neural threads we have been taking up one by one are ultimately woven together into one fabric, where sensory experience, cognitive reflection, and muscular response are thoroughly homogenized and translated into highly individualized bodies and lives.

This is not to say that the bodyworker cannot be effective until he can get his hands on the cerebellum, or that his successes are doomed to be random because we understand so little about it. On the contrary, this highest level of integration

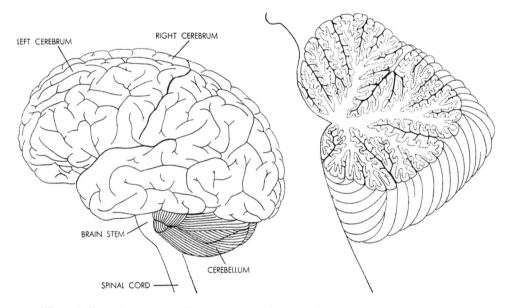

7-28: The cerebellum. The cutaway at the right shows the deep fissures and invaginations of the surface, which give the cerebellum a cortical surface area that is very large compared to its diameter. This gives it the means to accommodate a large number of cortical cell bodies, and thus to handle a large amount of information at the same time.

is one of the bodyworker's most valuable tools, even if he can't touch it: If acquired reflex responses, muscle tone, posture, and manner of movement are being altered by the practitioner's manipulations, then we can be all but certain that activities in the cerebellum are altering as well, and to the same degree. And if new content can be added to the associations occurring in the cerebellum, then all of the individual's future feeling, thought, and action may be affected as well. It may even be the case that bodywork and the complex responses that it can elicit will prove to have an important role to play in future discoveries about the functions of the cerebellum, and offer some clues for the ways in which it integrates sensations, attitudes, and responses.

A Black Box

The cerebellum was for a long time regarded as a mysterious "black box," a "silent area" of the brain. Because it is so large, neurologists assumed that it must have some important function, but their research was frustrated; tools and procedures that had been developed to explore the other areas of the brain yielded no results in the cerebellum. No matter where it was probed with electrodes, no spontaneous sensations, movements, or emotions could be elicited. And the cerebellum's own electrical activity is of such a high frequency (more than ten times as fast as the cerebrum) and of such a low amplitude (less than one tenth of that of the cerebrum) that it was only after the relatively recent development of exceedingly sensitive instruments that it could even be detected.[12] Added to these difficulties was the fact that large portions of the cerebellum could be destroyed without producing obvious functional losses, rendering it very nearly impossible to make simple deductions about specific functions in specific areas of its cortex.

It had been known since the turn of this century that this "black box" was a center for some sort of muscular control. Luigi Luciani, an Italian physiologist, experimented with cerebellar destruction in cats, and established that his cerebellized animals retained their gross motor functions, but that they suffered from trembling, loss of motor tone, lack of coordination, and severe disturbances in their equilibrium, their posture, and their gait.[13] Since Luciani's experiments, this resulting "drunken gait" has been the clinical symptom of cerebellar damage. But it has taken much longer to explain how this occurs, and why such large portions must be destroyed in order to produce severe and lasting impairment of coordination.

Sensory Connections

The more we learn about the operations of this part of our brain, the more it seems to confirm the primary importance of sensory information for the stimulation, the patterning, and the continuous control of muscular efforts.

Anatomically, it has developed as an elaboration of the area of the brain stem that serves to process sensations from the various sense organs that are particularly concerned with the orientation of the body in space and the orientation of body parts to the whole.[14] The major sensory tracts which branch into the cerebellum include those from the eyes, the ears, and the touch receptors from all over the body's surface; the vestibular system (the organs of balance in the middle ear), the proprioceptors of all the joints and deep tissues, the Golgi tendon organs, and the entire system of muscle spindles.

The cerebellum has become the largest and the most sophisticated in those animals who have the greatest need to constantly assess their body positions with respect to gravity, and it has reached its fullest flower in the erect human, whose relationship to gravity is the most problematical of all.

Motor Connections

In addition to these sensory tracts, several major motor pathways feed into the cerebellum as well, linking it directly with the reticular formation (which controls our overall levels of arousal), the basal ganglia (where our primitive spinal reflexes are organized into coherent patterns), the subthalamus and the thalamus (where these reflex patterns are arranged into meaningful sequences), and the motor cortex (where our voluntary motor commands are initiated). Coming out the opposite direction of neural flow, motor tracts exit the cerebellum and reach directly back out to the reticular formation, the basal ganglia, the subthalamus and thalamus, the motor cortex, and the internuncial reflex circuits of the whole spinal column.

These features of the cerebellum's developmental origins and its anatomical connections, in conjunction with the symptoms of severe disturbances of equilibrium and coordination which result from extensive cerebellar damage, strongly suggest that the cerebellum has emerged as a center for monitoring and controlling the body's muscular efforts in the light of the sum total of all the sensory information regarding the position of its parts and its stability in the field of gravity.

Maps

As is the case with the arrangement of cells in the sensory and motor cortexes, the terminal cell bodies of these incoming and outgoing tracts are mapped quite precisely on the surface of the cerebellum, so that signals coming from adjacent areas in the periphery are sent to cell bodies that are adjacent on the surface of the cerebellum. These arrangements of terminal cell bodies form small "homunculi," projections of the body's spatial relationships, very like those formed by the sensory and motor cortexes.

Moreover, the sensory, the motor, and the cerebellar homunculi are interconnected in such a way that the axons coming from the sensory cortical area which corresponds to "right hand" reach out to contact the "right hand" area of the motor homunculus, which in turn sends out axons that contact the "right hand" area on the surface of the cerebellum. The associated motor tracts are similarly arranged, so that all three of these "maps" mutually correspond to every part of the body, limb for limb, digit for digit, muscle for muscle. In this way, the cerebellum, along with the sensory and motor cortexes, "know" from what part of the body a signal comes, or to what part it is going.

The details of this spatial mapping in the cerebellum were far more difficult to establish than they were for the mapping of the sensory and motor cortexes. This was true because sensory impressions are not formed in the cerebellum, nor are motor commands initiated there. The cerebellum only monitors and modifies signals which begin and end somewhere else. It is the control box for a vast circuitry, but it is never the source for any of that circuitry's activities. This is why probing its surface does not produce sensations, emotions, or movements.

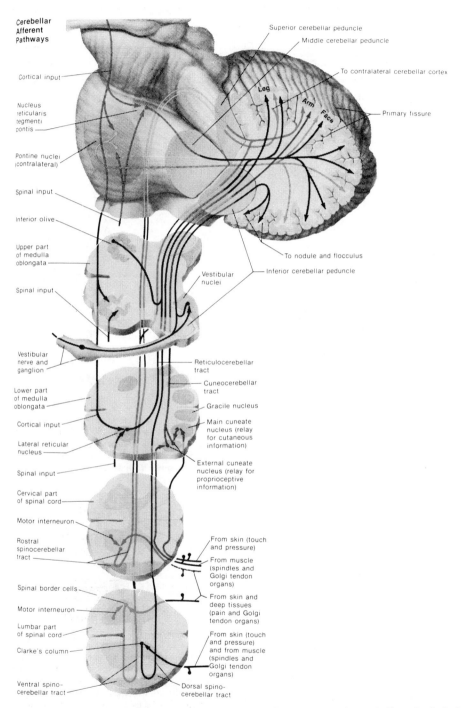

7-29: Sensory pathways of all kinds and from all over the body converge in the cerebellum. Particularly prominent are the pathways carrying impulses from various sense organs concerned with the orientation of the body in space, and with the orientation of body parts to one another, making the cerebellum the "head ganglion of the proprioceptive system." It performs the highest level of sensory, alpha, and gamma integration.

Cerebellar mapping, then, had to be accomplished by stimulating either areas of the body or known areas of the sensory or motor cortexes, and then probing the cerebellar surface with an extremely sensitive electrode for the tiny area that was activated by the sensorimotor stimulations—at first an exceedingly random, and later an exceedingly painstaking process. By these means, the "silence" of the cerebellum was finally broken.

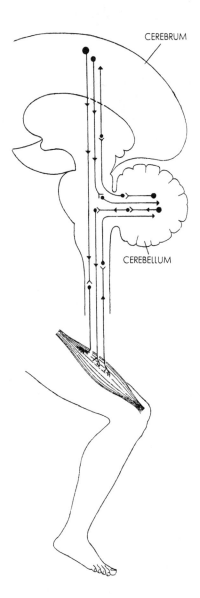

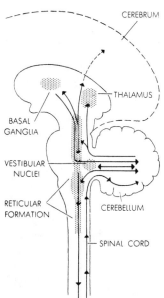

7-30: The basal ganglia—where the terminal cell bodies of the spindles, the tendon organs, and the gamma motor system are organized—have rich connections with the cerebellum, connections which flow in both directions.

7-32: Two body shapes are mapped onto the cerebellar cortex, similar in their organization to the sensory and motor homunculi on the cerebral cortex (although the two cerebellar homunculi are not strictly divided into "sensory" and "motor" areas). The actual hand, the sensory cortex "hand," the motor cortex "hand," and the cerebellar "hands" are all linked by corresponding pathways, as are all other body parts.

7-31: The alpha motor system also routes most of its impulses through the cerebellum, where voluntary actions are integrated with information coming from the special senses and the proprioceptors.

Cerebellar Activity

Through such exploration, we have learned that it is very far from being silent. There is no part of the nervous system where activity is more brisk. There is not a sensory event in the body that is not evaluated, nor a motor event that is not monitored by the cerebellum. No signals originate here, but almost all pass through at one stage or another on their way to becoming behavior.

A Typical Sequence

A typical sequence of cerebellar activity probably goes something like this: The desire to do a particular thing, such as "raise right arm," is conceived in the conscious mind; this desire is converted to commands that are sent through both the alpha and the gamma motor systems, which select and initiate the appropriate bracings and movements to begin the gesture. At the same time, an identical copy of the command is sent to the cerebellum.

As soon as the movement begins, it sets off a continuous stream of sensory stimulations from the skin, the joint receptors, the Golgi tendon organs, the muscle spindles, and also from the eyes. All of this sensory feedback is immediately forwarded to the cerebellum, just a split-second behind the copy of the motor command. Thus the "arm" area of the cerebellum receives two different sets of information—the motor command expressing the desired movement, and the sensory report—coming from many quarters—of the movement *actually taking place*. The cerebellum minutely compares these two sets of information, identifies discrepancies, and sends out to both the alpha and the gamma systems whatever modifications in the motor impulses are necessary to keep the actual movement in line with the specifications of the desired movement.

Thus a large measure of refinement is added to the efforts of both the alpha system, which activates the muscles directly from the motor cortex, and of the gamma system, which engages the genetically patterned responses of the basal ganglia and the spinal column's internuncial circuits. This constant checking of command against performance guides all of our movements, smoothing out the small errors even as they are being made, and providing the standard against which we judge our own mastery of any repeated motion. Without the guidance of these comparing and correcting functions of the cerebellum, our attempted gestures would be jerky and inaccurate, constantly oscillating between too much and not enough.

Equilibrium

Now as the arm is raised, the body's center of gravity is shifted significantly, causing another feature of cerebellar function to assert itself—the maintenance of equilibrium. For this function, which operates at all times, sensory information from the entire body must be constantly correlated and compared to the sensations generated by the tipping back and forth of the fluid in the semi-circular canals in the skull. These data are compared, and signals for muscular modifications of tone, length, and tension load are dispatched simultaneously throughout the feet, the legs, and the trunk to create compensatory contractions for any shift which threatens the body's upright stability. Again, a great deal of flexibility and refinement is hereby added to the simple "holding" mechanism of

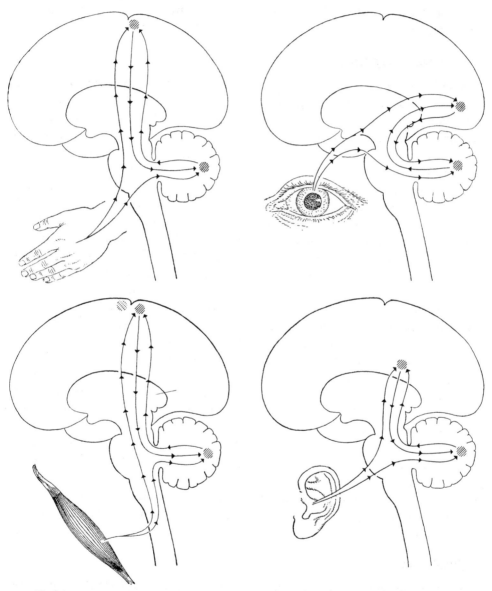

7-33: All of the sensory sources send signals both to their centers in the cerebrum and in the cerebellum. The cerebrum and the cerebellum then send these signals back and forth to each other, creating reverberatory feedback loops between the two. It is thought that these feedback loops help to continually modify ongoing motor commands—which are also sent through the cerebellum—constantly refining them in the light of current sensory information. It is here that the sensations evoked by body work may in fact have their most potent effect upon patterns of motor behavior.

the stretch reflex, refinement which allows us to maintain our balance while in motion, as well as while we are in a fixed posture.

Eyes and Ears

So with every motion of the body or any of its parts, the cerebellum must monitor two very different and very complex sensorimotor concerns. First, it must function as an "automatic pilot" for every gesture, comparing command with performance; this it does by correlating the activities in the same specific area in each of the sensory, the motor, and the cerebellar homunculi. Second, it maintains the body's general equilibrium, either at rest or in spite of whatever motions might be involved; this it does by comparing sensory cues from all over the body to the stream of signals coming from the vestibular organs of balance.

In addition to all of the somatic senses, the cerebellum is particularly closely associated with two other special senses, sight and hearing. Our eyes and our ears are responsible for a great deal of the information that tells us about our orientation in space and our relationship to objects around us, and the cerebellum considerably refines our muscular responses to the information coming from these senses as well.

Even though the mechanisms of sight and hearing are essentially unimpaired in animals who have had their cerebellums destroyed, these subjects show a poor ability to pinpoint the source of a sound, or to accurately estimate distances to objects. They have even been known to run into a wall with enough force to dash their brains out, because they could not judge its distance or the speed of their approach.[15]

A Feedback Control Box

Generally speaking, then, the cerebellum is an accessory control system which bases its motor modifications upon the sensory information generated by all parts of the body.

> In engineering parlance, it is the control box in the feedback circuitry of the nervous system. Such a control box does not originate the input of commands to the system, but is informed of them by the command center, for example, by the setting of a dial.[16]

It is the highest level of integration for all incoming sensory information and all outgoing motor commands; and it is responsible for the split-second coordination necessary between the alpha and the gamma motor systems. By means of its functions, the widely diverse streams of information from the special senses, the proprioceptors, the reflexes, and voluntary commands are blended smoothly together into competent movement.

It is like an orchestra conductor, silent yet absolutely necessary to produce harmonious music from a huge collection of instruments. Stimulation of the internuncial spinal circuits can elicit all possible isolated movements; the brainstem and basal ganglia can organize these isolated units into rudimentary patterns; the thalamus and subthalamus can arrange these rudimentary patterns into various useful sequences; and the conscious cortex can select commands which utilize any combination of these elements at will. But without the intervening cerebellum, none of our resulting actions will have reliable accuracy, fine control, or smooth transitions.

It is here in this neural organ, "the head ganglion of the proprioceptive system,"[17] where much of our enormously complicated sensorimotor coordination is accomplished, coordination which is necessary for the same musculature to maintain the tonus of a more or less rigid framework while at the same time superimposing almost any degree of movement upon that framework. And once again, we find that physical sensations—the same sensations that are systematically stimulated by bodywork—are crucial to this coordination.

8

The Sense of Effort

Our most sacred convictions, the unchanging elements of our supreme values, are judgements of our muscles.
—Nietzsche, *The Will To Power*

It is clear that the body's smooth and accurate response to motor commands depends to a very high degree upon the guidance of sensory feedback. This guidance does not come from any one sort of nerve end or sense organ, but rather from the full array of tactile sensory devices all over the body. It is comprehensive and constant, reporting and adjusting the slightest movement of a finger directing a pencil, as well as the motions of the whole body during violent exercise.

Without this sensory guidance and refinement of motor activity, we would be like the pathetic monsters of a Dr. Frankenstein, our walk staggering and jerky, our reach clumsy and inaccurate, our reflexes spasmodic; we would career about our environment, our every movement a source of danger to ourselves and to those around us.

This interdependence of sensory and motor functions is reflected in the structure of the nervous system at almost every level. They are so perfectly blended in the structure of the muscle spindles that it is arbitrary whether we regard these devices as sensory endings that contract and lengthen, or muscle cells that feel. These two functions are only separated slightly more in the Golgi tendon organ. And every spinal nerve trunk carries both sensory and motor neurons, terminating in sensory endings and motor end plates that intermingle in every nook and cranny of the musculature.

Throughout the length of the spinal cord, all of these sensory nerves synapse richly to the internuncial circuits and their associated motor nerves, densely linking these two halves of behavior together with reflex arcs at every segmental level. The final integrative functions of the cerebellum are only the highest level of the blending of sensory and motor elements, elements that in fact mingle with one another at innumerable junctions on their journey from the periphery to the brain and back out again.

Truly, it is only while we are dissecting corpses or while we are drawing theoretical circuitries of the nervous system based upon man-made electrical models that we can preserve for ourselves a clear distinction between sensory and motor impulses. In the behavior of the living creature, the two either combine inextricably at all levels to produce smoothly controlled movement, or various degrees of impairment result.

The Numb Thumb

If, for example, we anaesthetize the sensory receptors in the skin and joints of a subject's thumb and then blindfold him, he cannot tell when we move his thumb back and forth for him; he has no sense whatever of where we put it in space. Secondly, if we ask him to move his own thumb back and forth one inch, it is impossible for him to gauge the distance accurately, or even move his thumb the same distance with every trial. Moreover, while he is attempting this task, he cannot tell whether he is moving his thumb freely, or whether we are sometimes blocking his efforts. The efforts to move the thumb are crude to the observer and confusing to the subject without the necessary sensory accompaniment.

If we now stop manipulating our subject, and allow him to take off his blindfold to try out the normal use of his partially numbed hand, we will continue to observe this muscular incompetence in the absence of sensory feedback. He will drop things because his thumb grasps with too little pressure, or he will break fragile things because it grasps with too much. He cannot place small objects on a table with accuracy. His handwriting turns into a scrawl, and he begins to suffer from emotional discomfort because of his hand's unreliability.

> Thus normal dexterity in man is not simply the result of a hand with a wide range of movement, but depends upon the fact that it is also an important sensory organ. Loss of this sensation removes the ability to use the hand properly, even though the muscles, tendons, and joints which are essential to the production of the fine movements are in no way impaired.[1]

Strong muscles are just not enough. Without a constant sensory progress report concerning all of the variables at each stage of any motion, our conscious mind and our alpha motor system alone can only make a guess as to how many motor units to recruit for it, and can only roughly approximate the correct bursts of action potentials required for its accurate execution. In order for our movements to be effective, we must constantly feel our way along.

Gauging Effort

The tactile sensations of the skin and the joints play a crucial role in the use of the hand, or of any other part of the body. They are necessary in order to judge how far I have moved a finger or a limb, what the dimensions of objects are, how much pressure my contractions develop against surfaces I grasp, how soft or hard those surfaces are.

What sort of sense do the muscle spindles and the tendon organs add to this picture? How does their sensory information assist the motor commands and the reflex responses in the orchestration of smooth and accurate movements? They are delicate in their structures, acute in their responses, and vast in number, yet they are just as silent as the cerebellum to our familiar conscious sensory modes. Almost all of their activities are processed by reflex arcs in the spinal cord and by unconscious centers in the brain. It is as though they were always "anaesthetized"; they give us no qualities like those we receive from other sensory sources — color, texture, odor, taste, pitch, volume, equilibrium, limb position, pressure, and so on. Our muscles are filled with sensory organs which produce no conscious sensation, and it is not readily apparent just what sort of "sense" their obscure activity might add to our picture of our selves and our surroundings.

Eye Effort

The movements of the eye offer us a special case in which to isolate this "sensationless sensitivity" of muscle tissue. The muscles of the eye are more densley supplied with spindles and tendon organs than any others in the body. But unlike the thumb, neither the eyeball nor its socket has any touch or pressure receptors of any kind to tell us how far the eye has moved, or in what direction it is pointing.

Yet we are certainly aware of this degree of movement and direction all the time. It would seem logical to assume that it is the stability of the objects themselves in my field of vision which provides the reference points that tell us the angle of our gaze. However, it is easy to demonstrate that this is not really the case. If while holding my eye open, I rapidly press and release one corner of it with a fingertip, so that it moves back and forth in its socket, it appears that it is the "stable" scene before me which jiggles back and forth, and not my eyeball. This illusion disappears the moment I cease manipulating my eye with my fingers and begin moving it back and forth with its own muscles.

In other words, if the eye is moved voluntarily with its own muscles, the scene before me remains fixed and the angle of my gaze sweeps back and forth across it. But if the passive eye is moved with the finger, the scene begins to jump around and we interpret what we see as if the eye were stationary.

These easily confirmed experimental results were noted as long ago as 1867 by Helmholz in his *Handbook of Physiological Optics,* but it was not until the middle of the twentieth century — after the discoveries of the functions of the muscle spindles, the gamma motor system, and the cerebellum—that they were repeated with particular emphasis upon what these ambiguities of the eye might reveal about the sensory properties of muscle.

In these more recent experiments, the skin around the rim of the eye was locally anaesthetized, so that there was no possibility of tactile cues to indicate the eye's position or movement, and the object of vision was reduced to a single spot of light on a neutral surface. Once again it was confirmed that the manipulation of the passive eyeball produced the visual sensation of a moving object, while the normal use of the eye muscles gave the experimental subject the sense of a stable dot on the wall and allowed him to accurately assess the movements of his eye in relation to it. "Hence," says our more recent researcher,

> we only know in which direction our eyes are pointing when we move them voluntarily, and this must be because we make an unconscious estimate of the effort put into moving them. Other sense organs in the area do not tell us which way the eyes are pointing, because if they are moved passively we do not seem to know that they have moved.[2]

And to press the point even further, if both the tactile and the visual sensations are removed, by local anaesthetic around the eye socket and a black cap placed over the cornea, the subject can easily direct his own eyes any direction we command, but he experiences no motion in his eye whatever no matter how it is passively manipulated.

> The unequivocal conclusion of all these experiments is that we have no sense organs in the eye muscles or near them that tell us which way our eyes are pointing. We normally know which way we are looking, but only because an internal "sense of effort" gives us an estimate of how much we have exerted our eye muscles. If voluntary movements are artificially impeded, we absolutely do not know what is going on—unless we can see and reason back from the visual impressions we receive.[3]

Here then is the "sense" that is provided by the muscle spindles and the tendon organs—the *sense of effort*. It is the tendon organ which records the tension developed by any contraction, and it is the spindle which records how far and how fast the contraction proceeds; and it is their combined feedback which is summed to determine how many motor units and what degree of contraction are necessary to roll the eye a particular distance at a particular speed. This sense is normally so sensationless, processed so thoroughly by the brainstem and the cerebellum, that we are scarcely even aware of its operation until its absence throws our other senses into confusion, as in the case of the manipulation of the passive eye.

Physical Effort

This "sense of effort" constitutes an additional special sense, with as complicated apparatus and as rich and constant an input as those of smell, taste, sight, hearing, or touch. It gives us assistance that is just as indispensable for managing our own bodies and for manipulating external objects as is that from any other sensory source. The "organ" of this extra sense is nothing less than the body's entire musculature, our most abundant tissue, seventy to eighty-five per cent of our selves. We are aware of our muscles as workhorses, because we continually observe their effects; we are surprisingly unaware of them as sense organs, because the data they provide are handled almost exclusively by centers in the brain that are normally unconscious.

Effort And Space

Yet no sense contributes more to our conception of material reality. My idea of an object's dimensions is based upon precisely how much effort is required to trace its boundaries, with my fingers, my arms, my paces, or with the sweep of my eyes. Likewise, distances between objects have little concrete meaning apart from an experienced estimate of the effort necessary to traverse them.

> We are able to say that a table is four feet away or a piano ten feet, not, as one might suppose, because of any special quality of vision or a specific reaction to light, but chiefly by means of the muscular sense. The ability to differentiate distances is built up from a multitude of impressions conveyed from the eye muscles used in focusing the visual image upon the retina. The degree of muscular adjustment that is necessary varies with the distance and size of the object, and also with its shape. Impressions of the work done by the eye muscles as they move are reported to the brain, where they are interpreted as indicating a certain distance in space.... This is due to memories of former experiences—movements of the eye, movements of the body going towards objects and reaching them, together with tactile impressions of their shapes, extent, and textures derived from handling them. This spatial perception built up from past experience of the body muscles is remembered in conjunction with the impressions from adjusting the eye muscles.[4]

This is to say that our very concept of space itself is primarily muscular. Space is, after all, nothing but the volume that is occupied by objects and the distances that separate them, values that are given concrete significance only by means of so many gradations of "this far"—quanta that are stored in the mind as known increments of muscular effort. "Extension" acquires no meaning unless we have extended. And infinity begins for each of us at exactly that point where the distance conceived goes beyond our own ability to imagine the effort required to traverse it.

Effort and Time

Our notion of the regular passage of time is probably built up from the same muscular materials. To be sure, the cycles of the days and nights and seasons have always been there to observe, but these cycles are so long and include so many variables that the time frame we deduce from them alone is rather vague. The accurate accounting of hours and minutes and seconds seem to be much more readily associated with the steady beat of the heart, the constant oscillations that underlie tonus, the steady swing of the arms and legs as we walk (which, incidentally, the pendulum of the clock imitates). It is surely these shorter physiological oscillations that have sophisticated our sense of the irregularities of the solar cycles, and not the other way around.

It is of particular interest to note that hallucinatory disturbances of our normal senses of space and time occur when we are disassociated from our muscular sensations, such as during meditation, while we are motionless and lost in thought, while we sleep, while we are under the influences of drugs, or suspended in a sensory isolation tank.

Effort and Gravity

We cannot see or smell or taste or touch gravity, but there is nothing in our existence that is more continually present for us. All material things have mass. There is no part of our bodies or of any other object we know of that does not share this attraction to the ground. There is no physical force that more conditions our state of being and our activities. All movement is a question of a balance between surrender and resistance to this downward pull, and in the movement of my own body it is my muscular sensations which provide the mechanisms used by the mind to sense and control this balance.

The whole notion of mass and weight is nothing more than a measure of the effort necessary to resist this force of gravity. Mass, inertia, acceleration, and momentum—the primary physical qualities shared by all material objects—have no meaning apart from the amount of effort required to make the objects which possess them move, or to stop them from moving. Our muscle spindles and tendon organs are the scales with which we continually assess these invisible forces.

Our other sense organs tell us about the additional properties of matter, such as texture, temperature, color, odor, taste, and so on. In philosophical language, these are called "secondary," or "surface," or "accidental" qualities. Our muscles give us our sense of matter's most primary property, mass, that which underlies all other qualities which an object may possibly have. This is not the perception of this matter or that matter; this is the perception of materialness itself, upon which all other judgements must be based in our dealings with physical reality. Our muscles measure the essence of the relationships between ourselves and objects.

It is not surprising that questions concerning "essence" and "primary qualities" should have fretted so many thinkers for so many centuries. When we sit still and contemplate the problem, the sensory information which might enlighten us is largely silenced; and when we are active, this information is processed largely by our unconscious. This is perhaps why philosophical obscurity seems to reign so often in these sorts of questions, why perfectly sound minds can lead themselves into perplexing doubts about the "reality" of the physical world, and why such speculations appear so absurd to the working man.

Psychological Effort

When I support or move my body, it is perceived by me just like I perceive any other object with mass. I assess the effort required to lift and move my arm in exactly the same way that I assess the effort used in lifting a book, or a stone. In this regard, my body is simply the mass (or collection of masses) with which I most often interact.

It would seem only natural, then, that my repetitious familiarity with this object would eventually establish for me a perfect economy of effort in my movements and postures; the masses of my head, limbs, and torso would be such well known values that I would never exert more or less effort than was necessary to do what I want with them. This ideal is indeed possible.

Tone and Weight

However, the actual physical masses of our body parts are not the only values with which we have to contend here. It must be remembered that the assessment of resistance and weight are dependent upon the amount of tension placed on the muscles and tendons, and the accompanying sensations of pressure. We have already seen how the "normal resting tone" of our muscles can fluctuate widely, influenced by a long list of local injuries or discomforts, general attitudes, depressions, elations, anxieties, lassitudes, interests, boredom, and so on.

When the value of this normal resting tone is raised, whether by short-term emotional reactions or by long-term habits, tendencies, or attitudes, the degree of tension enforced by the basic stretch reflex is increased. Conversely, a lowering of normal resting tone results in a decrease in the degree of tension that is maintained by the reflexes.

> Every pull of the tone will therefore be either added or subtracted from the muscular tension and will influence the perception of gravity in the limb and in a weight put on the limb. Every extremity which rises is lighter, and every extremity which sinks is heavier. Every lighter extremity underrates and every heavier extremity overrates weight.[5]

My perceptions of mass, weight, and effort can all be changed merely by increasing my resting tonus, without the slightest amount of actual "work" being added to my burden.

> When we move an outside heavy mass, we have to exert our muscles. The more force we have to use, the heavier the mass seems to be. When we keep a part of our body in its position by muscular effort, this part immediately feels heavier.[6]

And the higher the level of tone maintained by the muscles while holding our body parts in their positions, the heavier they will appear to feel. One can readily experience this phenomenon by tensing all the muscles in one arm; this arm will immediately feel heavier and more substantial than the relaxed one, and the longer I hold this tension the more pronounced my sense of added weight will become. "The impression of heaviness in our bodies varies therefore according to the muscular strength that we apply to it."[7]

The Weight of the World

These variations in tonus can have enormous significance for our notions of the amount of effort it takes to sit, stand, mow the lawn, go to work, make love, or what have you. The "weight of the world" can literally be placed on someone's shoulders, not by means of any external object, but simply by means of increasing the pull exerted by the muscles of the back and shoulders. The sensations of tension and pressure will mimic exactly those produced by the addition of weight, and will just as surely increase our sense of the efforts we are making to "bear up" under the "burden".

> When one breathes in deeply and keeps the chest in a position of maximum inspiration, one immediately feels one's chest as a heavy mass. This seems paradoxical, since the specific weight of the chest is, of course, diminished in deep inspiration. But here we come to an important addition to our insight. The chest feels heavier and like an almost solid heavy substance, *because a muscular effort has taken place.*[8]

And of course these muscles which maintain this increased tension are literally efforting more. No movement or work may be taking place, but they are pulling harder, raising their metabolism, consuming more energy, producing more wastes, and creating more fatigue.

It is as though gravity were pulling on our muscle cells from one end, and our internal mental states were pulling on the other. It is the net effect of these two forces which determines the stress load placed on the muscles, not only the actual mass of physical objects. And an increase on the pull from either end constitutes an increase in our experience of effort.

Habitual perseverance of these internal increases in effort, for whatever reason, will eventually result in higher values for the normal resting tone maintained by the stretch reflexes, which in turn results in distortions of posture and movement. And once a new norm is established, all of the body's reflexes will strive to preserve it in spite of the fact that uncomfortable symptoms crop up to tell us how far our "normal" has drifted away from "optimal."

> All these tones pull the limb into a position which is unknown [i.e, unconscious] to the individual. The normal posture then becomes the posture into which the tone would pull the limb. From this posture, every other posture is judged.[9]

This *persistence of tone*

> is present in every muscle of the body. It is also present for every single posture of the body. We are dealing therefore with a phenomenon of general significance.[10]

Effort, then, may be something I am feeling, rather than something I am doing. But no matter what its source, the physiological consequences of the burden are the same: The muscles increase their tension to support the added "weight." More tension is placed on the connective tissues, and more compression is placed upon the joints; more fuel is burned, depleting the body's supply of energy, and more wastes are produced, while at the same time circulation to the contracted area is reduced; discomforts, and eventually dysfunctions, appear; and, in fact, it requires more and more real effort to get around.

The subjective experience of this vicious cycle, and even its outward manifestations, can feel and look so much like inherent weakness, chronic disease, or natural aging that we often simply pursue the struggle with an air of heroic acceptance of our plight, an attitude which in itself only serves to further increase our sense of the efforts which life demands of us. In our muscles as much as in our ideas, we are trapped by the tyranny of what we come to regard as "normal."

The Aesthetics of Effort

The various pulls on our muscles, and the sense of effort which they generate, underlies every interpretation we make of all our other sensory cues. All activities are judged to be difficult or easy, possible or impossible, attractive or unattractive, by virtue of the effort which we perceive as being inherent in them.

Some of us enjoy a challenge, while others collapse at the first sign of resistance. The difference between these two responses very often has little to do with the objective difficulties of the task, and even little to do with the actual strength we possess to meet it. Much more to the point is our *sense of the effort* it will cost us to do the thing, a sense that has every bit as much psychology as it has physics behind it. But because so much unconscious habituation is involved in our assessments of efforts, we are seldom aware of their decisive role not only in our physical activities but also in our emotional preferences, our intellectual pleasures and repugnances, our speculations, our judgements of all kinds. These subjective assessments of efforts are highly individualized, and they mold our personal responses to the world around us in far-reaching and surprising ways.

Sporting Effort

Some competitive sports, like weight-lifting or football, require brute strength. Others, like track or swimming, demand speed and endurance. Still others, golf or billiards, for example, place much more emphasis upon fine control than upon strength, speed or endurance. But success in any of these kinds of sports requires the ability to focus intently upon a muscular performance, and to achieve a high degree of development in the motor reflexes.

Now apart from the salaries of individual professional athletes, no great material consequences are decided by either a win or a loss; the contest is, after all, a game. Yet during this game, whose outcome will not be fateful for the spectators in any case, the excitement of the fans can approach hysteria, and the final score can produce hours, even days and weeks, of euphoria or depression for them. What could be the source of so much emotional turmoil around an event whose impact is generally so trivial upon the practical concerns of those fans' lives?

The answer is that sports are celebrations of muscularity itself, and the fans' excitement over them stems from a spontaneous admiration of uncommon muscular feats. The essential content of the jubilant cry "Our team won!" is "Our players can use their muscles with greater skill and to greater effect than your players." A part of this is undoubtedly the unconscious recognition of the survival value of great strength and its refinement; the prototype is combat, where the outcome for superior physical development matters very much. But much of the excitement and admiration has to do with a delight in muscularity for its own sake. The solo mountaineer or the lone fly-fisherman can excite just as much admiration for the exercise of their skill as does the team player in the midst of the crowded stadium.

This phenomenon of the sports fan is extremely interesting with regard to our sense of effort. Babe Ruth unquestionably derived a great deal of pleasure from his ability to bring his bat into contact with a fast ball with great force and precision; but what is equally as remarkable as his legendary skill is that tens of thousands of baseball fans also derived a great deal of pleasure from merely watching him do it.

For Babe, the pleasure came from the exercise of his own skill; for a fan, it came from the appreciation of the power and finesse displayed in the skill of another. For both, the pleasure was grounded in a finely discriminating sense of effort.

Observing highly trained athletes allows us all to develop a rich *vicarious* sense of effort, giving us a chance to experience in some measure muscular feats of which we would otherwise be ignorant. Fans intent upon an event usually squirm visibly along with the action. Electromyographic studies (the use of electrodes to measure and record precise muscular exertions) confirm that very often this squirming consists of subdued imitations of the motions of the athletes—the bobbing of the boxer's fists, the lunge out of the starting blocks, the recoil from a hard tackle, and the like. The fans *are watching with their bodies* as much as with their eyes; they are *participating*, not just watching.

This kind of vicarious muscular experience gives us concrete examples of strength and skill beyond our own, and greatly enlarges our notions of human capabilities—enhancing for us our appreciation of the species to which we each belong. For these reasons, athletic games are far from being trivial. Whether it is watching a hurdler's lunging stride, a high diver's meticulously controlled contortions, the perfect putt, or the precarious leaning of the motorcyclist into a fast turn, it is our own insight into the fine muscular control necessary for such perfomances, and the effort of discipline required to develop it, which take our breath away. And it is our ability to comprehend—even if we cannot duplicate—the skill involved which makes our astonishment so keenly pleasurable that we cannot help but cheer when we see victory or perfection.

Dancing Effort

In the appreciation of the arts, this response to the demonstration of muscular skill reaches a very high degree of refinement, and in most cases our sense of aesthetics cannot be separated from our sense of effort. This is particularly obvious with the performing arts. Applause for the dancer is quite simply homage to his or her strength, flexibility, and precision. Even further, the medium through which dancers express their art, the actual artifact that is created, and their own musculature are all one. And if, in addition to thrilling us with the sheer difficulty and grace of their muscular efforts, they can manipulate these efforts in such a way as to evoke narrative, specific emotions, conflicts, resolutions, insights, then our admiration for their skill knows no bounds—the exercise of their musculature becomes a sublimely communicative thing.

Dance raises the coordination of efforts made by the muscles to an extreme, and it is this which engages our enthusiastic appreciation. Moreover, the various possible qualities of this coordination and expression give us an appreciation not only of dancers in general, but many different appreciations for many kinds of dancing. Because we understand something of the devilish difficulties and uncompromising efforts in the obligatory repertoire of steps, turns, leaps, and gestures in classical ballet, we can form one whole kind of critical appreciation—one that reveres the dancer's ability to overcome his or her individual idiosyncrasies in order to reproduce painstakingly standardized purities of form.

Quite a different sort of appreciation is evoked by various kinds of extemporaneous dancing. Here the effort is not that of recreating the purity of an established form, but quite the opposite—the moment by moment discovery and execution of clear and concise movements and gestures that are expressive of inner states unique to the particular performer, to the passing mood, even to the fleeting instant. What we admire in this instance is the inventiveness and spontaneity of the dancer, the ability to embody an intensely personal vision, while in ballet what struck us as so noble, so fine, was the contrary ability to sublimate the merely personal in the execution of forms which are comfortingly august and unchanging.

Underlying our sense of the beauty of ballet is our appreciation of a musculature that is able to reproduce these exacting standards; on the other hand, the beauty of extemporaneous dance springs for us from the ability of that same basic musculature to studiously avoid imitating past forms in its expression of feelings that are ever imminent, ever changing.

Style of Effort

So it is not merely the *degree* of effort that evokes in us a particular level of response; it is the entire *manner* of the effort as well. The sense of effort also carries with it a sense of *style*. What is often most engaging in the performance of dancers from parts of the world unfamiliar to us is not how well they reproduce standardized forms (for we have no historical knowledge of their art forms against which to measure this), nor what personal impulses they express uniquely and spontaneously (for we do not know enough about their cultural norms to identify "spontaneity" with any certainty), but rather an exotic sense of style, so different from what we are used to seeing, so filled with ideas, feelings, and associations that are unfamiliar and mysterious to us, which open up whole new modes of aesthetic standards, fresh perspectives from which to appreciate body movement and human experience.

Nor are "traditional," "spontaneous," or "exotic" the only kinds of categories which characterize "style" in the dance. Within every mode there is a way of doing things which illustrates not only a particular kind of effort, but *economy* of effort as well, a practiced grace and ease of execution which nearly conceals the difficulty of what is being done. During the Renaissance, this grace and economy of effort was actively cultivated and openly admired in all forms of conduct to a very high degree, and was even regarded as the essence of gentlemanly or ladylike behavior. They had a special word for it: *sprezzatura*—doing the difficult with apparent effortlessness.

It was not enough to be able to boast of numerous accomplishments. The ideal in all accomplishments was strainless ease. The Renaissance Man, with his famous diversity of skills and learning, was not merely a Jack-of-all-trades, but rather the man who could read, address a public assembly, dash off a thoughtfully passionate sonnet, play the lyre, make love, ride a horse, wield the sword and lance, govern his serfs, and manage foreign diplomacy all with equal aplomb.

There is something almost god-like about individuals who can do much as though it were little, because great difficulties do not seem to tax their resources in the slightest. They give us the illusion of the wave of the hand which parts the clouds, a regal control of effort that is far removed from our familiar sense of the gritty bumps and grinds of muscles and masses. This is very often the sort of thing we mean when we say that a person "has style," or does a thing "with style." It is a delicious appreciation of the least amount of effort yielding the greatest amount of result.

And the more aware we are of the actual difficulties, the more intensely we admire the apparent ease. This is not a matter of illusion; rather, *sprezzatura* bespeaks not only skill but also long intimacy with that skill, not mere ability but also comfort in that ability, not mere strength but also great refinement of that strength.

Musical Effort

In dance, the aesthetic object is the shape and movement of the whole musculature itself. In music, we seem to be in quite a different aesthetic realm, where appreciation is triggered by the qualities of pitches, tones, harmonies, discords—phenomena of airwaves, invisible and intangible.

Music is often described as the most ethereal, the most disembodied art. Yet it is not mere pitch, tone, and harmony in which the beauties of music reside; music is the *manipulation* of those elements. There is no note or chord whose charm would not quickly fade without others to surround it, to provide it with a meaningful context. The punctuation of the tone into varying time values, raising it and lowering it to produce a melodic line, combining it with other tones to create harmonic modulations, adding the variations, the trills, the pianissimos and the emphases—these are the effects without which making tones may not truly be called music, and all these practical matters of music are the product of an exceedingly fine degree of *muscular* control.

This is most obvious in the singer, whose own respiratory tract constitutes the instrument, but it is no less the case with the violinist, the pianist, or the trumpet player. We do not have the activity of the whole body to admire, as with the dancer, but rather small, subtle motions of the fingers, the arms, the diaphragm, the larynx, the embouchure. The nature of the *artifact* has changed, but the difficulty of the muscular demands required to produce it have not diminished in the slightest. While the number of muscles and the range of their movement have decreased, the need for precision in placement, pressure, and timing has risen proportionately, and it is our understanding of this refinement—especially if we have attempted to play an instrument ourselves—that adds much to the depth of our appreciation of the performance of any musical piece. Music, like dance, is nothing if it is not actual performance, and performance is nothing if it is not muscular. The muscles are the Muses.

It is clear that our sense of these muscular components of music have as much to do with our aesthetic responses to it as do the purely sensory qualities produced in our ears. Why is it, for instance, that the organist is not categorically regarded as a superior musician to the guitarist, even though the organist can produce far greater volume, numbers of notes, and varieties of tone qualities? It is because the guitarist is equally able to exhibit mastery with his six strings as is the organist with his hundreds of keys, pedals, and stops. It is this mastery which moves us, whatever the formal or mechanical confines within which it operates, and this mastery is a question of a muscular relationship with the instrument. It is, after all, more pleasing to hear a simple air played with exquisite control on a solo flute than to listen to the clumsy botching of a far more complicated concerto with a full orchestra.

Or again, why is a live performance always more engaging than a recording of the same artist? It is because we have the physical facts of the sound's production visible before us, because we can confirm with our eyes the efforts and fine distinctions which our ears alone can only lead us to imagine. And to take this abstraction imposed by recording one step further, why is a computer-produced version of a Bach fugue doomed to remain vastly inferior even to a recorded version performed by a living artist? The computer simply cannot include those minute fluctuations of tone, those almost unnoticeable liberties with tempo, those subtle variations in the phrasing, all of which announce to our ears the muscular aliveness of the source of the sound.

It is even to the computer's disadvantage that its version may be utterly "flawless"—devoid, that is, of those tiny evidences of strain and error which are too small to mar our pleasure, but which on the contrary give us eloquent evidence of the difficulty of the piece, and a delicious sense of the skillful efforts involved in the crises and resolutions of its most demanding passages.

Some musicians even express a dissatisfaction with electronic instruments in general for a very similar reason. George Shearing observes that

> the problem with electronic keyboards—and I've played a few notes here and there—is that there doesn't seem to be a great deal of room for the tender loving care that goes into some pianistic touches, the feeling that you have caressed the depression of this note to such a degree that it will sing rather than shout at you. Generally, this individuality of touch, and being able to recognize one pianist from another, is all but absent in computerized instruments. It's "OK, you want an A?" (Knob). "A B?" (Knob.) You can't dismiss the marvelous variety of tone colors available on the electronic instrument, but tone colors are one thing; degrees of sound and touch, and all the inconsistencies and therefore unpredictableness that go into a human addressing something as sensitive as the piano are another thing.[11]

Indeed, many passages of music—and some whole genres—inspire our enthusiasm with the difficulties inherent in their execution just as much as with the particular qualities of their melodies, their harmonies, their rhythms. The etude is a case in point.

> It is a short piece in which the musical interest is derived almost entirely from a single technical problem. A mechanical difficulty directly produces the music, its charm and its pathos. Beauty and technique are united, but the creative stimulus is the hand of the executant with its arrangement of muscles

and tendons, its idiosyncratic shape. In the Etudes of Chopin, the moment of greatest emotional tension is generally the one that stretches the pianist's hands most painfully, so that the muscular sensation becomes—even without the sound—a mimesis of passion.[12]

In short, it is to a large degree our knowledge of the limitations of the instrument and the difficulty of the piece, together with our perceptions of how well the artist succeeds in coping with these conditions, which afford us aesthetic pleasure, just as much as any values and associations intrinsic to particular melodic and harmonic modes. The popular response to various modes, after all, comes and goes with the fashions, but the muscular skill of the musician remains as the medium in which they all have their day. No matter how ethereal, how disembodied music may appear to be, its basic underpinning could not be more physical. Rhythm, phrasing, pitch, and tone are all direct manifestations of muscular control, and our aesthetic response to them is solidly linked with our perception of the sophistication of these efforts.

Plastic Arts

Our aesthetic responses to painting, sculpture, and architecture are also based in large measure upon our active sense of effort, no matter how exclusively formal and visual their appeal may seem to be. One of the most common reactions of the visitor who enters the Sistine Chapel and stares upward at the painted ceiling is, "Unbelievable! Two years flat on his back clear up there, and with the paint dripping in his eyes!" The power of Michelangelo's imaginative conception, the bold colors, the arresting poses and marvelous modelling of the figures—and the exhaustive effort required to stick them up there—all roll over the observer with equal force.

And analogous to recorded music, prints of paintings and photographs of sculptures have a diminished impact, largely because they erase many of the tell-tale signs of the quality of the activity involved in producing the original. The bronzed thumbprints of Rodin on the surface of a sculptured cheek preserve for us the spontaneity and excitement of the moment when the clay was still wet, still apt to change, still responding to the thrust and caress of the artist's hands. The frenzied brushwork of Frans Hals gives to many of his portraits a vitality and a tension that are wholly absent in the actual attitude of his sitter. And as a matter of fact, a close scrutiny of his strokes and dashes, or of Rembrandt's contrasting meticulous layering of tints and varnishes, is as revealing of the individual artist's state of mind and communicative intent as are the two men's subject matters, their composition, their choice of colors.

The nature of the prints left by the brush in the paint, which preserve for us the exact muscular qualities of each stroke, has in itself much to do with the icy classicism of David, the smoothness of Ingre, the tumultuousness of Van Gogh, or the ambiguities of Monet, and for the art lover these fossils of the act of creation are just as clear evidence of the originality and genius of the artist as are any other considerations. Many artists can skillfully copy the colors and forms of these masters, but a close examination of their brushwork quickly exposes their copies as "fakes."

Architectural structures as well are as filled with the labors of the builders as with the vision of the designer. In his essay "Stones of Venice," John Ruskin utilized a description of the visual splendors of Europe's Gothic cathedrals primarily as a reference point from which to admire the particular efforts of their countless nameless stone masons, and to explore the vast efforts of social and spiritual organization that it took for all of them to sustain generations of efforts and complete their aggregate tasks.

In a similar vein, books celebrating the Golden Gate Bridge contain far fewer views of the finished artifact than of the efforts of the construction workers at various stages of its completion. The more intimately we are aware of these sweaty efforts and their technical difficulties, the more splendid seems the bridge.

Precision of Effort

In forming an understanding of what our sense of effort contributes to our experience of reality or our response to art, it is very important to recognize that it is not simply the amount of effort that is critical, in the sense that more effort necessarily means a more vivid or a more positive effect. The crucial factor, whether assessing the mass of an object or appreciating levels of aesthetic accomplishment, is not the ability to generate *more* effort, but rather the ability to generate precisely the *correct amount* of effort.

It is often infinitely more useful to know exactly how heavy a thing is than merely to know whether we are strong enough to lift it. Too much effort distorts a pirouette or a note or a brushstroke just as surely as does too little. In other words it is the sense of the *appropriateness* of the effort to the desired effect which is so admirable in sports and in the arts, and so indispensable in our own daily affairs. It is a case of "just right" having far more value than does "more than enough".

How "hard" we are working is the crudest sort of measurement we derive from our sense of effort, not to be confused in any way with an awareness of the exact effort required to execute a particular motion. The former is a matter of raw strength and exertion, while the latter requires the refinement and control without which all the raw strength in the world is clumsy and ineffective.

This precision in our sense of effort not only enriches our feeling for art, it fleshes out our concepts of reality as a whole; for this reason it has enormous practical—even survival—value. The mastery that can be attained by the devotees of oriental martial arts provides us with astonishing examples of the practical usefulness of the fine discernment and memory of quanta of muscular effort.

I myself once witnessed a Japanese swordsman demonstrate this kind of muscular refinement in the following manner: His assistant lay flat on his back on the ground, with a sheet of paper on his stomach and a watermelon resting on the sheet of paper. The swordsman was led ten or fifteen yards away, blindfolded, and turned around repeatedly. After a moment of silent concentration, he then bounded forward, cleared the distance to his assistant in four or five leaps, and brought his razor-sharp sword down on the watermelon with great force. The melon was split cleanly in half, and there remained a faint crease in the sheet of paper from the arrested blade of the sword.

How could there possibly be a "trick" to such a performance? How could such a thing be accomplished, except by the training of the musculature and the mind to an intense degree of sensitization, until the swordsman's intimacy with his finest increments of effort becomes developed to the point that the movements of his limbs estimate distance and direction for him as accurately as do his eyes? Once he surveyed his surroundings visually, he could function blindfolded because he was able to exactly match his efforts and movements to his memory of the placements of objects around him. It may be difficult for us to imagine a sense of effort sophisticated to this degree, and yet this is how every blind person navigates his environment, and it is how all of us find the knob of our bathroom door on the far side of the bedroom in the middle of a dark night.

Effort and Value

We should not pass from this subject of the aesthetics of effort without making some note of the quality that is imparted to an object when we say that it has been "hand-made". As automated manufacturing has advanced, the term has become one of increasing distinction. If two functional or decorative items look and perform very much alike, it will usually be the item actually turned out by the hands of the craftsman that is valued more highly than the one produced by mold or stamp or computerized lathe.

Just as the archeological study of ancient hand tools and products can speak volumes to us concerning the interests, the skills, and even the characters of our ancestors, so do the marks from the hands of the contemporary craftsman give his works an element that will never be found in any object that is mechanically produced. This is the human element, the knowledge that it was personal skill and patience that created the functionality or the beauty of the thing, the sense that many hours filled with innumerable small, attentive efforts are contained within it.

And again, it is not flawlessness which imparts this value, but rather the many small imperfections which attest to the difficulties of the task and to the actual efforts of the craftsman. Indeed, it has been one of the most pernicious effects of mechanized mass production that we have lost the opportunity for much of our daily appreciation of these eloquent imperfections; our houses, our clothing, our dishes, our furniture, our utensils stare back at us mute in their flawlessness and their uniformity, void of any of the marks of humanity that would give their appearance personality and add the feelings of muscular intimacy and admiration to the convenience of their use.

The brisk market for "handmade" items—in spite of their greater cost—is not just nostalgia or ostentatious fashion; it is as often as not a sincere search for some endearing marks of humanity—muscular, skilled, creative humanity—upon our shelters, our tools, our decorations, our incidental bric-a-brac.

9

Sensory Engrams

The features of our face are hardly more than gestures which force of habit has made permanent. Nature, like the destruction of Pompey, like the metamorphosis of a nymph into a tree, has arrested us in an accustomed movement.
—Marcel Proust, *The Remembrance of Things Past*

Signatures

We have commented upon the unique pattern of brushstrokes with which every accomplished painter applies his paints to the canvas. Unlike shapes and colors, these patterns of strokes, and the *manner* of working which they imply, are very nearly impossible for another to duplicate, and for the expert they provide reliable evidence to confirm or deny the source of a picture in question.

Few of us will ever develop so individualized an artistic style, but we all participate in this same phenomenon whenever we take up a pen and sign our name. The individuality of every person's signature is a simple and universal example of the ways in which our patterns of motor behavior become invested with our unique experiences and responses as we learn to use our muscles—our personal style.

The signature, unlike fingerprints, clearly has little to do with an inevitable unfolding of genetic factors, yet it is used with almost the same legal authority to identify its owner as are fingerprints. The idiosyncrasies of the signature are, as it were, the fossil remains of the individual's process of getting to know and learning to command his musculature, a process so complicated and so protracted that even though the pen, the paper, and the alphabet are the same for each of us, our use of them displays our unique individual divergencies just as much as it affirms our successful imitation of letters and words.

An experienced analyst might even be able to make accurate surmises about certain personality traits revealed in a sample of handwriting. This practice of handwriting analysis has no more of the occult about it than do the impressions we receive from a person as he shakes our hand or as we observe him walking or dancing; all of these actions can be faithful revelations of the state of mind which produces their specific neuromuscular patterns.

Thus it is that we put our "signature" upon every gesture we make, and cannot normally do otherwise. To change these motor patterns in any but the most self-conscious, superficial, and temporary kinds of ways requires significant changes in our mind's relationship to the body as a whole, a relationship that has all of our training and experience behind it. When Mr. X. signs his name on a piece of paper, we have an artifact of his personal history.

This uniquely personal quality of the signature is a neuromuscular fact of great importance. It not only indicates to us the almost infinite range of differences possible in basic motor patterns, it also affords us a valuable clue concerning the underlying organization of the motor system as a whole. Consider a peculiar fact regarding the signature: If we now give Mr. X. not a paper and pen, but a piece of chalk and a large blackboard, he will again produce a signature that has the same recognizable unique qualities. The muscles used to produce the signature on the blackboard are entirely different, but the individuality of the handwriting remains.

The juxtaposition of the two signatures may not tell us any more about Mr. X.'s personal history, but it can lead us closer towards an understanding of how Mr. X. learns to control his muscles in the first place: What mechanism transfers the motor pattern of the signature from the fingers guiding the pen on the paper to the whole arm, the legs, and the torso guiding the chalk on the blackboard?

Organizing Motor Performance

What is immediately evident from our observation of Mr. X. executing these two signatures is that wherever the learning takes place that makes such a skill possible, it is definitely not in the muscle tissue itself. Mr. X. has spent years developing this skill in his fingers, and they can now accurately reproduce the recognizable pattern by dint of countless repetitions; but given the piece of chalk and the blackboard, the larger muscle sets of his arm, legs, and torso need very little practice—nor even any particularly concentrated effort—in order to reproduce the same characteristic shapes several times larger.

We can say neither that his fingers were "stupid" to have taken so long to have developed the pattern, nor that his arm, legs, and torso were "smarter" for having picked it up so quickly. None of Mr. X.'s muscles are either "smart" or "stupid." Each motor unit can do nothing but respond to the pattern of stimulation it receives, and the organization of that stimulation obviously takes place elsewhere. A muscle cell may be thicker or thinner, stronger or weaker, but not more or less clever than any other muscle cell.

The Site of Patterning

So the muscle or limb which actually performs any skill is not the site of the learning of that skill. Since we know that the spinal cord and the lower brain are largely made up of the genetically established internuncial circuits of reflex responses and the stereotyped building blocks of movements, it would seem logical to press our search for overall organization of acquired skills up into the more conscious cortical regions of the brain. And of all the areas of the cortex, none appears to be a more logical candidate than the motor cortex, where the terminal cell bodies of the alpha motor neurons are clustered, reaching their axons and end plates out into the motor units all over the body.

But experimentation with the motor cortex of laboratory monkeys has yielded a rather puzzling result:

> Removal of small portions of the motor cortex that control the muscles normally used for the skilled activity does not prevent the monkey from performing the activity. Instead, he automatically uses other muscles in the place of the paralyzed ones to perform the same skilled activity.[2]

The organization and the memory of the skill, in other words, do not take place in the part of the cortex which we would logically assume most directly commands the motor system. The alpha cell bodies on the map of the motor cortex which normally activate the specific motor units involved in the performance of the skill may be destroyed, rendering those motor units useless, and yet the pattern of performance remains intact, easily transferable to new motor units and a corresponding new cluster of cell bodies on the surface of the motor cortex.

And this newly established cortical area may be destroyed in turn, with the same transference repeated again and again until all the possible alternative motor cell bodies have been killed and all their corresponding motor units paralyzed, before the ability to perform the skill is terminated. Even then we cannot say that we have extinguished the skill, but only its means of exercising itself.

If, on the other hand, we do not tamper with the motor cortex, but instead destroy cell bodies in the *sensory* cortex which correspond to the areas of muscles, skin, and joints involved in the performance of the original skill, we find something even more arresting: The monkey loses all ability to repeat the skill. And furthermore, relearning takes almost as long as the original training period.

The conclusion to be drawn from these experiments is as inescapable as it is surprising—the entire motor side of our nervous systems appears to have little to do with the *organization* of acquired motor behavior and the perpetuation of learned skills. Both alpha and gamma neurons form complex lines of communication from the brain to the muscles, but they are no more the source of conscious decisions and developed sequences than telephone wires are the source of the voices they carry.

> It is not the motor cortex itself that controls the pattern of activity to be accomplished. Instead, the pattern is located in the sensory part of the brain, and the motor system merely "follows" the pattern.[3]

But perhaps this decisive role of the sensory cortex in motor control is not as surprising as it may at first appear. The alpha muscles themselves generate no feedback sensations in connection with their own activity, so it is only our proprioceptive organs—principally the anulospiral elements of the spindles, the tendon organs, and the joint and skin receptors—which give us a "feel" for what our muscles are up to. It is therefore primarily in the sensory cortex that we consciously experience the effects of our motor movements. The data which define for us a particular effort of our motor movements, the data which refine repetitions, are sensory impulses, and the memories which preserve an established skill are sensory memories.

The Engram

Each discrete sensory record of a particular gesture or series of gestures is called a *sensory engram,* and once the feeling of it is firmly established as a clear, recallable memory, this "engram" works something like the templates which produce different stitching patterns in a sewing machine. When a person wishes to accomplish some act, he first recalls the sensory engram associated with past repetitions of that act. That is, he begins by remembering how it felt to do it. The motor systems are then set into motion to *reproduce the remembered sequence of sensations* laid down in the engram.

The translation of an engram into motion involves the entire collection of sensory and motor apparata that we have examined up to this point—the skin and joint receptors, the deep tissue pressure receptors, the Golgi tendon organs, the stretch receptors of the spindles, the reflex internuncial circuitry of the spinal cord, the stereotyped sequences of reflex responses organized in the brainstem, the gamma motor system, the alpha motor system, the cerebellum, and the successively more conscious centers of choice and voluntary sequential integration—the thalamus, the hypothalamus, and the cortical regions. The sensory cortex has memorized the feel of a gesture or a series, and each time it is recalled for the purpose of repeating the action the proprioceptive feedback of all the body parts is compared against that memory for each step of the intended repetition, and cerebellar corrections are made automatically and unconsciously.

And of course, this translation of desire into movement must all take place within the context of the overall levels of tonus, the characteristic postures, the chronic limitations, and prevailing state of mind of the actor. Any template can only function within the range of possibilities inherent in the sewing machine as it presently stands.

Sensory Memory

We do not know just what an engram is physiologically, nor where it is located anatomically. The sensory cortex appears to be crucial to their organization, but not exclusively so. Some theorists hold that each memory "bit" is specific to a single cell or reverberating loop of cells. Others maintain that memory is stored in a more holographic manner, such that all of the cells of the sensory system participate in some way in all sensory memories.

Whatever their actual form of storage, their existence and their control of motor behavior cannot be doubted. The "learning" of a new motor skill is the process of establishing a new series of sensory engrams, and the ability to repeat the performance of that skill depends absolutely upon the preservation of the intact sensory engram. Motor elements merely supply the motion; it is the sensory side of the nervous system which establishes the control of that motion.

The impulses carrying the information contained in sensory engrams are fed out to the muscles through various routes contained within the *direct corticospinal path* and the *multineuronal path* in the spinal cord. These pathways are themselves organizational features of the nervous system, each coloring in a specific way the translation of sensory memory into muscular activity.

Some of them make direct uninterrupted connections from the cortex to the skeletal motor units (the direct corticospinal path); these have the advantage of producing almost instantaneous effects upon the muscle cells from cortical activities. Some synapse onto many internuncial reflex circuits on their way down the spinal cord (the multineuronal path), and these have the advantage of utilizing fixed, inherited reflex movements in order to build up more complicated sequences. Still others pass through the brainstem and all the basal ganglia, and hence can use the reptilian brain's entire vocabulary of stereotyped gestures and fixations in order to carry out still more complex commands.

Current Sensations

But the impulses generated by an engram, the memory of a sensation, is not all the information that is necessary to successfully direct a movement. Constant account must be taken of *current* sensations concerning my body's orientation and my surroundings as well. I can take up a pen and sign my name; fine. But am I sitting or standing? Am I stabilized or swaying back and forth in a car? Am I using a thick pencil or a delicate quill? Is the paper smooth or rough, tough or fragile? All these factors make large differences in the specific *muscularity* used to reproduce my signature, even though the *artifact* looks much the same under any of these variable circumstances.

For this reason, all of my various *descending motor* paths carrying engramatic impulses synapse richly with *ascending sensory* paths, which continually modify my efforts with regard to local sensory events. And in turn, any part of this continually incoming sensory background can be either suppressed or heightened as it is more or less relevant to the task at hand.

Engrams and Acquired Reflexes

The sensory engram is in some functional and anatomical respects at the opposite end of the neurological and behavioral spectrum from our primitive reflexes and the stereotyped responses triggered by the spinal cord and the brain stem. These older patterns have been established over millions of years of species development, and are inherited as anatomical structures that are faithfully recreated during the genetic unfolding of every normal foetus.

Engrams, on the other hand, are built up largely from the life experiences of every individual, and are in many ways unique to that individual. They cannot be localized into any anatomical unit of fixed connections, as can the primitive reflexes associated wtih specific internuncial groups in the spinal cord. Engrams are a means of arranging into meaningful sequences the firings of these primitive reflexes units; they are an *organizing factor* that cannot be materially pinpointed. They are among the ghosts in the machine, the direct observation of which has so far been denied us, but whose practical effects are everywhere evident.

This functional and anatomical antithesis of engram and reflex, however, does not long remain a clear one in the developing individual. As the gesture which builds up a specific sensory memory is repeated over and over, the engram—wherever it is formed—takes on a more and more stable nature, follows increasingly predictable neuronal paths, and requires less and less focused attention, until many of our laboriously learned skills become almost as automatic as any primitive reflex.

We all experienced this gradual maturing of an engram as we learned to walk, or ride a bicycle. In the same way, a musician, or an expert swordsman, a typist, or an assembly line worker establish many learned repetitions that eventually require little conscious directing. A sensory memory, after all, must use a series of motor reflexes in order to translate itself into movement, and a specific series of motor reflexes can itself become a "megareflex," a collective "all or nothing" response which, once begun, has a powerful tendency to run through to completion, very like the complex swallowing reflex present in the infant at birth, which utilizes some twenty muscles and a distinct series of contractions lasting four or five seconds that is always completed once it is begun.

In this fashion many engrams become developed to the degree that volition and ongoing sensory feedback play only minor roles in their function. This happens to virtually any skill or posture that is repeated often enough long enough, and it is the reason why the carpenter, the artist, the race-car driver, or the seamstress can perform many of the repetitious motor aspects of their activities unconsciously, while focusing their attention upon higher level problems and decisions, or even while they are day-dreaming.

A Sliding Scale

Thus the engram is not really a fixed element in our control of muscular activity. It exists upon a sliding scale, from the initial attempts at a skill, which rely heavily upon consciously directed muscular contractions and consciously assessed current sensory feedback to guide our efforts, to habitual repetitions which function in a semi-conscious manner, to intensively trained reactions built upon highly specific sensations, which resemble a "knee-jerk" as much as they do a consciously controlled response.

This sliding scale of stability in the establishment of behavioral patterns is one of the major triumphs of the process of encephalization and the expansion of the cerebral cortex. Spinal reflexes and the species-specific behavioral patterns that are encoded in the lower brain centers have taken millions of years of trial and error and normal usage to build up. The foundations of these patterns are anatomical, and therefore they are highly reliable and relatively implastic. The sliding scale of sensory engrams represents a stunning advancement over this older, more conservative, and infinitely slower method of establishing modes of behavior. By relying more and more on higher and newer and increasingly flexible brain centers to organize and execute movements, encephalization has tremendously accelerated the evolutionary development and modification of motor behavior. Periods of trial and error may now take years, or hours, or even minutes for an individual to establish a new stable pattern, rather than generations.

Furthermore, this new organizing principle is not anatomical and implastic. It may be endlessly refined and modified throughout a lifetime. New series of reflexes may be built up to adapt to new situations, to accommodate new desires. Old series may decay when their usage is no longer appropriate, or they may be preserved for years to be called forth at the right moment. Above all, the cortex and the engram make possible the unfolding of behavioral patterns that are not merely species-specific, but are utterly unique to each individual. The mind has developed a way to compress eons of evolutionary differentiation into a single lifetime. This is what gives man his vast alternatives of behavior, his endless capacity for adaptation, his ability to consciously direct his own futher development. With regard to our motor behavior, the cerebral cortex and the sensory engram have done nothing short of placing within the sensory and cognitive processes of the individual the same kind of potential—only greatly accelerated—that exists in the evolutionary and genetic processes of the species.

Counter-Productive Engrams

Such a heady power and freedom is not without its attendant dangers. The conservatism of the older nervous system did not provide these novel possibilities but it did insure the fact that no pattern was genetically established that did not have some demonstrable survival value for the species. Encephalization has advanced to such a degree in man that he can over-ride many of these tried and true responses; he is free to side-step their limitation, but at the same time he is free to develop styles of behavior that are positively destructive to his person, even to the species and the environment.

These newer, more consciously developed patterns (which represent our pre-ferred adaptations to our unique experiences) may be at such variance with our older, less conscious patterns (which represent the accumulative wisdom and usage of our predecessors) that our minds and our physical responses may seem to be absolute enemies to one another. Vicious circles and degenerating condi-tions typically are initiated and sustained by such conflicts, with the cortex ignoring the unconscious but inexorable demands of our physiological pro-cesses, and the body in its turn producing symptom after symptom in an effort to arrest an irresponsible cortex.

The Dangers of Editing

The burgeoning of the cortex has provided our species with a means of interpreting and using sensory information in completely new ways, and of creating new kinds of behavior. But no less awesome than this new creative power is the ability of the human mind to *edit* sensory information in ways that are new as well.

We have noted that any particular bit of sensation may be either heightened or suppressed, as it is more or less relevant to our present activity or interest. The need for such editing powers is obvious; without it we have no way to focus our attention upon any particular thing. But the very strength of this editor opens the door to significant problems—it operates with equal force under the influence of neurotic compulsions as it does under the influence of intelligent decisions.

For instance, sensory information that is irrelevant to my present actions may be amplified to the degree that it completely disrupts my ability to focus upon what I am doing, as is the case when distracting noises disturb my typing or my piano playing. This disturbing influence can reach agonizing neuraesthenic pro-portions, until the slightest distraction can thoroughly undermine my normal competence.

On the other hand, a sensory engram and its motor responses can operate so compulsively that they can override almost any amount of current sensory input. This can be a life-saving advantage when it allows us in a state of emergency or injury to focus upon crucial defensive actions rather than succumbing to pain, fear, and shock. But it also offers us the destructive option of pig-headedly ignoring significantly painful symptoms while sustaining our efforts in the same old manner, thus creating more and more actual damage to ourselves.

In other words, encephalization and the engram not only open the door to unique individual skills, but also to a host of malfunctions and neuroses. A new system is exercising its power over an older, formerly autonomous one, and the take-over is not proceeding without its serious skirmishes. Such conflicts may in fact be *inherent* in a situation of enormously expanded choices, and these conflicts are not always resolved to the advantage of the organisms in which they occur. What we have is the continual renewal of choice; what we have not is any assurance that we will choose best.

Traumatic Engrams

One of the factors that greatly contributes to these conflicts arising through encephalization is the inevitable inclusion of many traumatic physical and emotional events in the long course of an individual's development. To the degree that we rely less and less upon genetic programming of the lower brain centers, and more and more upon the development of associations in the cortex by means of life experiences and training, the negative aspects of our experiences take on a more important role, and may even become foreground in our responses.

Some of these influential traumas may be physical. For example, a boy may be exposed to an abusive parent, sibling, or instructor whose habit is to register their annoyance by slapping the back of his head. Eventually, the repeated withdrawal away from the anticipated slaps becomes a chronic ducking of his head and stooping of his shoulders, until these become a characteristic feature of the child's postural structure, outlook, and general response.

Or alternatively, the trauma may be of a more emotional nature. A young girl develops prominent breasts well ahead of her peers. The attention they attract causes her embarrassment and confusion, so she seeks to conceal them as best she can by lowering her head, stooping her shoulders, and withdrawing her thorax, with the result that she develops a postural attitude very like that of the physically abused boy, and will eventually suffer the same kinds of postural discomforts and limitations.

The Ambiguity of "Feelings"

The postural results stemming from these two different kinds of trauma are a further indication of the great ambiguity which exists between "feelings" which reflect sensory stimulations and "feelings" which reflect mental states. It is the memory and association of things which are *felt* that develop and direct the mind's expression through the musculature, and in this constant feedback between gestures and the feelings that accompany them neither tactile nor emotional feelings can be said to dominate the other. Either physical slaps or emotional embarassments can, eventually, produce the same kind of postural stoop. And my posture and my actions may appear to be happy or depressed, aggressive or fearful, just as clearly as they appear to be practical or clumsy, efficient or wasteful.

It has even been theorized that all of my emotions are rooted in physical sensations of various kinds. But it does not really seem necessary to reduce our emotional lives to such a simple mechanical or chemical principle in order to understand that they have just as practical and observable effects upon our physical states and our behavioral competence as do our ability to feel objects or feel ourselves in motion. My emotions regarding a particular activity, or regarding activity in general, *function much like sensory engrams,* influencing my degree of skill and coloring my style just as surely as do a greater or smaller number of repetitions.

We have even seen that in fact my inner emotional states may amplify or suppress to a large degree all incoming sensations and all outgoing motor impulses, thus playing a major role in the transmission to and from the cortex of the very bits of information and response that would constitute the development of any engram whatever. And even further, a prominent emotion such as fear or elation clearly has the ability to color all motor behavior, whether that emotion was present during the original learning process or not.

In this way, not only all sensorimotor skills learned in the past, but all present mental states are equally reflected in the activities of the muscles. We observe engramatic templates of all kinds, emotional and sensory, shuffling apart and merging together in the normal flow of human behavior.

Generalized Engrams

This process of shuffling and merging of engrams illustrates a most interesting duality in the nature of these building blocks of voluntary behavior. On the one hand, every firmly established engram is a memory record of specific sensations associated with specific movements. It is brought into focus by the trial and error phase of repetition of the specific movement, and serves primarily to provide a reliably constant sensory blue-print, purged of errors, for all future repetitions of that specific movement.

On the other hand, there are definitely elements in any highly developed engram which are not limited to the specific local muscular activities with which it was originally associated. A strong engram can somehow transfer the skill it gives to one set of muscles over to another set quite readily, or can even diffuse itself throughout the entire musculature. This was clearly the case in the example of the two signatures of Mr X. A skill that had been honed in the manipulation of a pencil with his fingers was quickly transferred to the manipulation of a piece of chalk by a far larger collection of muscles in his arm, legs, and torso.

Some more generalized law of association seems to be operative here, so that if I learn first to walk with ease, roller skating will be far easier to learn, and if I have learned to handle snow skis, then surfing will come more readily, or if I have first developed several other skills involving strength, grace, rhythm, and balance, I will learn to dance much more rapidly. And the emotional elements of my feeling states tend to reinforce one another as well, profoundly influencing not only the manner in which I perform tasks already a part of my repertoire, but also my ability to learn new skills.

Thus both the sensory and emotional contents of any movement may be projected far beyond the specific activity with which it was originally associated. Any given engram, then, may not merely encode a particular movement, but also a sense of "style" which can permeate all movements. It is possible—it is even common—that we may, late in our development, master certain skills which in one way or another alter the style of nearly everything we do. It is not that we have wiped out any previous experience, but rather that a new piece of behavior has redefined old relationships and associations in a sweeping fashion, just as a new paradigm changes the course of a science by looking in a new way at facts which had been well established before.

This aspect of engrams gives them tremendous potential, both as negative, destructive forces and as positive, constructive ones. They can be not only bits of organized behavior to be repeated, but also a source of a whole sense of style about ourselves. The strongly developed engram is itself a general organizing factor, whose powers can be exerted on many levels.

The Birth Engram

The experiences associated with birth and early infancy offer a particularly striking example of this diffuse nature of engram encoding. The act of being born is for most of us the most intense physical and emotional stimulation we will ever receive, and it brings with it a powerful cluster of sensations and feelings that we could call the "birth engram." Some psychologists maintain that this birth engram operates throughout our lives as one of the major organizing principles of our personality, the context in which all other engrams are built up.[4]

The stages of birth—the blissful suspension in the undisturbed womb, the inescapable squeezing of the contracting uterus, the crushing and distorting pressures of the birth canal, and the final release into the open air—provide the first massive wake-up of our proprioceptive system and our conscious mind. The sensations and associated emotions of these progressing stages establish very deep relationships between sensory, cortical, and motor elements in each individual, and to a very large degree establish the manner in which all future sensory information is processed and all future motor responses are elicited.

This impact of birth trauma is especially acute in man because of the relatively undeveloped state in which he enters the world, and because of his reliance upon sensory experience and memory to build up the developed behavioral norms with which other animals, such as the colt, are born. What is waked up in man by the sensory barrage of labor and delivery is not a more or less complete genetic program for his various behavior patterns, but rather the *apparatus for learning these things*. This is why each individual's unique experience of birth and the period of infancy immediately following condition in very specific and far-reaching ways everything else that is to follow.

Sensory feelings, emotional feelings, and muscular responses are never more solidly fused together than during birth, and from the intensity of this fusion come some of the most long-acting and generally influential behavior patterns that we will ever develop.

Archetypal Engrams

It is fascinating to consider with regard to this generalization of powerful engrams another sort of cluster of feelings and sensations which direct our behavior, the psychological archetype. Possession of our thoughts and actions by an archetype may be thought of as the sudden appearance of a "spontaneous engram" of a full-blown and highly complex nature. It is almost as though the *experience of another* has entered our neural precincts, dramatically changing the hue of our perceptions and our reactions, often pushing us into feats of stength, or agility, or sensitivity that are quite alien to us in our normal state. And again we can see that this potent kind of "engram" is not a fixed anatomical entity in our nervous system. They usually come and go with their own autonomy, and they seem to have little relation to "learning" in the normal sense of the word.

Sensory and emotional clusters of this potency are never singular in their use or significance; they are like words in our behavioral vocabularies, and like words they may be arranged in any order, uttered in any tone of voice, used literally, or as metaphors, as symbols, as elements of style whose possibilities of expression reach far beyond their individual definitions, as pieces of line and bits of color whose meaning and application may vary widely depending upon the context and manner in which they are used.

Bodywork as Engram

The engram is the cortex's means of learning new skills and behavioral patterns, and of imposing them upon the primitive levels of our motor organization. What is most forcefully implied by the way in which they are built up and the manner in which they guide our behavior is the overwhelming importance of sensory information to our neuromuscular organization. We are at all times literally "feeling our way" through life.

Hence it follows that to the degree which we are without an acute awareness and appreciation of our tactile senses, both exteroceptive and proprioceptive, we are cut off not only from the world around us but also from the chief organizing principle of our own organism. In the absense or suppression of specific peripheral sensory input, a great deal of specific mental and physiological function is lost. It is the simple truth of the matter that our minds and thoughts exist in our skin, our joints, and our muscles just as surely as they do in our craniums. It is in this sense that Blake spoke of "the improvement of sensual enjoyment" as the key to man's mental and spiritual growth, and that Yeats spoke of "singing in a marrow bone."

It is precisely here, in the sensory programming of behavior and attitude, that bodywork is able to exert its most dramatic, most beneficial effects. A bodywork session is essentially a carefully controlled tactile environment—a highly developed and generalized engram—which the client is unable to effectively build up for himself, given his present habits or injuries, and is therefore unable to incorporate into his feeling states and his behavior.

We have remarked how some sensory associations may have a powerful, diffuse effect, changing the manner of many or even most of a person's existing patterns to one degree or another. Bodywork can be a sensory influence of this kind—a sort of reverse trauma. By creating through artful manipulation a sustained series of sensory impressions suggestive of pleasure, of softness, of length, of relaxation, the reduced muscle tone *normally associated with these feelings* can be evoked. And with either local or general reductions in muscle tone comes a veritable parade of sensory effects: Pressure is relieved on proprioceptive receptors throughout the body's tissues; circulation and metabolic activity are accelerated in formerly constricted and immobile areas, so that pleasant and encouraging sensations creep into these areas to displace pain or numbness; body positions that formerly produced discomfort are now adopted with ease, and so on. All of this constitutes the restoring of the complete and coherent flow of sensory information with which we organize our bodies and our minds.

This "engramatic" quality of bodywork is why the *manner* of working is ultimately so much more important than are particular procedures or techniques, more important than merely pushing tissues and structures back into the "right" place. Even though structural changes are to be expected, the crux of the therapy is not *material,* but has to do with the sensory evocation of *feeling states.* It is art, not science in the strict sense of the term, and it is aimed towards shifts in mental response, not mere physical adjustments. This is why it must always be subjective and intuitive, in addition to being well-informed.

These changes make the client *feel good,* both sensorily and emotionally, and in a very deep and diffuse way. There may be specific areas of relieved tension and improved ranges of motion, and these of course will have specific effects upon future movements. But far more important than the local effects of a loosened muscle is the general feeling that accompanies the accumulation of many small releases—the feeling within the client that his comfort and his competence are not at the mercy of blind neuromuscular reflexes, the feeling that on the contrary his reflexes are under the control of his mental states, and that—most important of all—his mental state in turn responds powerfully to enjoyable sensations.

This experience can constitute a break in a vicious circle of discomfort, withdrawal, and subsequent disability that may have been perpetuated for years. To experience pleasant peripheral sensations changing inner feelings, those inner feelings changing habitual attitudes, and those changing attitudes changing the tone and functionability of the muscles, which in turn produce more pleasant sensations—this is the reverse rotation of the vicious circle, a *new organizing engram* at work. And this new engram does not merely add a new activity to the client's present repertoire; it adds a new dimension into *all* activities, and a new insight that can influence the development of all future activites.

What is ultimately to be gained from an effective bodywork session is not just "relaxed muscles." If this were the only goal, then there are available pharmaceuticals which can accomplish the job quite nicely, albeit temporarily. What bodywork intends to produce is a conscious reaffirmation of the kinds of thought forms and feeling states that are themselves the effective mechanisms of relaxation, the reassertion in the conscious awareness of the client of his tactile sources of information, body image, and self control. The physical body and the activities of the mind are restored to one another through the interface of touch, and each finds in the other the very element without which it languishes and suffers in isolation.

10

Movements Toward Disease and Movements Toward Health

Our bodies are our gardens, to the which our wills are gardeners; so that if we will plant nettles or sow lettuce, set hyssop and weed up thyme, supply it with one gender of herbs or distract it with many, either to have it sterile with idleness or manured with industry, why, the power and corrigible authority of this lies in our wills.
—William Shakespeare, *Othello*

Making an Effort

One of the main practical features which distinguishes the function of the human central nervous system is our ability to produce an extremely wide—indeed, a seemingly infinite—range of voluntary movements and patterns of behavior. It has been the mushrooming of the cortex with its dense interconnections that has given us this vast voluntary repertoire, lifting us out of the relatively limited varieties of behavior provided by the older spinal cord and lower brain centers, and giving us the means of imagining and executing all kinds of activities that are new to the species, unique to the individual. Humans, more than any other creatures, have choices. In fact, we have many more choices than we are often willing to admit with regard to our daily routines and responses. With a considerable degree of latitude, we can decide what sort of individuals we will be.

Further, voluntary actions provide us with much more than simply the ability to do this or do that as we choose. It is the voluntary quality of my conscious muscular activities that provides the point around which my sensations of movement are organized. It is only when I decide to move an object that I gain both a clear sensory evaluation of its mass and of my own strength and coordination in relation to it. The voluntariness of my action is absolutely crucial to my ability to arrive at a clear sense of the precise relationships between my own mass and movement and the mass and movement of the object:

> From a theoretical point of view the importance of body movement and particularly self-produced movement derives from the fact that only an organism that can take account of the output signals to its own musculature is in a position to detect and factor out the decorrelating effects of both moving objects and externally imposed body movements.[1]

In other words, my "sense of effort," upon which rests so much of my concrete perception of my self and of objects, tells me nothing *unless I make a conscious effort.*

This situation is in marked contrast to that of my other special senses. Light "enters" my eye, sound "falls upon" my ear, I feel an object bump up against my skin whether I move or not. In these cases I am far more passive—I *receive* their sensations. It is true that I can choose to see or hear or feel touch with greater or lesser intensity; I can suppress or facilitate these sensations. But if a bright light or a loud horn are turned on close to me, I am aware of them without having to "decide" to be.

It is altogether different with my "sense of effort." Effort is not something that happens to me; it is something that I must do, and without the element of conscious doing, I not only have no gauge with which to measure, I have nothing to gauge. "Hence we know which direction our eyes are pointing *when we move them voluntarily,* and this must be because we make an unconscious estimate of the effort put into moving them."[2] No matter how unconscious this estimate is, it absolutely requires voluntary activity. My sense of effort does not function apart from the exertion of my will to move an object, whether that object is myself or something else.

Voluntary Behavior

Now the importance of voluntary activity to our sense of effort brings us to an interesting problem concerning the actions of my muscles: It is not at all clear to what degree I actually have full conscious control over their operations. What is more clear is the fact that I am always doing more than I am consciously directing, or even consciously aware of. I am continually responding with my musculature to everything that happens to me, both externally and internally:

> Almost all sensory and even abstract experiences of the mind are eventually expressed in some type of motor activity, such as actual muscular movements

of a directed nature, tenseness of the muscles, total relaxation of the muscles, attainment of certain postures, tapping of the fingers, grimaces of the face, or speech.[3]

Needless to say, many of these continual muscular reactions occur without my consciously commanding them, and often without my noticing them at all. They even extend into the smooth muscles of my organs—the expanding and contracting of my arteries, the tightening and loosening of my sphincters, the churning of my stomach, the peristaltic contractions of my intestines, the activities of my sweat glands, the coming and going of goose bumps, and so on—responses mediated by my autonomic nervous system which have traditionally been regarded as being categorically beyond my voluntary control.

Picking a Flower

Even when I isolate a gesture that I am certain I am making voluntarily, such as leaning over to pick a flower, my conscious directing of that gesture becomes more and more ambiguous the more closely I examine it. I am certainly not, after all, aware of each of the millions of individual muscle cells whose contractions are coordinated to make the necessary movements of my limbs and torso. I am aware that I want the flower, that I stoop, and that my hand reaches out, stops at the correct place, grasps and plucks the flower, and then raises it to my nose. But can I then say that I have "consciously directed" the countless internuncial circuits in my spinal cord, the constant refinement of my movements by my cerebellum, and the complex participation of my gamma motor system, all of which are necessary to make this simple voluntary gesture possible?

Obviously, a great deal of the neuromuscular events involved in my voluntary picking of the flower are not directed by my conscious will. No matter how "voluntary" a gesture may be, in the sense that I choose to do it and could easily do otherwise, its actual execution depends upon long chains of intercellular events that are for the most part submerged below my level of conscious awareness, events whose speed, number, and complexity would simply overwhelm me if I were to try to direct them one by one. "A large part of every voluntary movement is both involuntary and outside consciousness."[4]

Nothing is clearer to me than my ability to do or not do any number of things as I choose. In fact, as we have noted, some of my most sophisticated sensory information is not generated in the absence of this volition. Yet there is no anatomical structure or neurological process that we know of which pinpoints the source of my will to do this or do that. And further, the execution of that will relies upon the unconscious operation of millions of spinal circuits and reflex arcs which orchestrate innumerable contractions and lengthenings into smooth and accurate gestures. I know with complete certainty when I am doing something that I want to do, but I have no precise idea how it is that I came to want to do it, nor exactly how it is that I am doing it.

The only way that an impulse towards movement can reach my muscles is through my spinal cord, and everything that we know about the circuits of the cord suggests that they are among the most fixed features of the nervous system. Whenever a particular nerve cell in a spinal circuit is given a specific stimulation, the impulse is obliged to follow a specific path towards one or more specific motor neurons, which transmit that message to the muscle cells of their motor units, obliging them in turn to twitch in a specific manner.

And so far as we know this is always the case, whether the movement in question is a knee jerk in response to the tap of a rubber mallet or a voluntary gesture in response to an internal command.

> The cord patterns of response are integrated into the overall control of muscular activity even when this control is initiated in the cerebral cortex.... Impulses from above ordinarily control only the sequence in which these patterns of contraction occur rather than the contraction of individual muscles.[5]

"Voluntary" does not appear to be the opposite of "reflex" at all, but rather a special case of the use of reflexes, implying a high degree of variable choice in the selection and ordering of fixed patterns available in the spinal circuits.

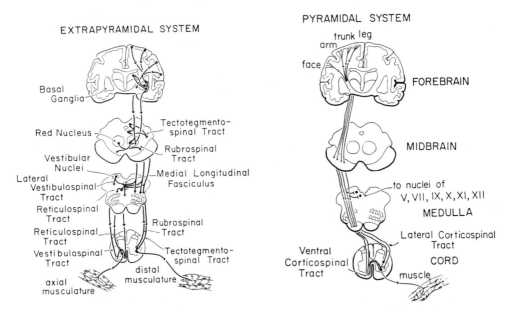

10-1: The pyramidal system of motor nerves consists of pathways running straight from the motor cortex to motor nerves in the spinal cord which stimulate alpha muscle cells. These pathways are the most direct routes available for voluntary command impulses to reach the appropriate muscles. We rely upon them heavily to execute movements that are not familiar, and to guide us in the early stages of learning a skill. They are also necessary to fine tune familiar movements to fit the immediate circumstances.

The extrapyramidal system is not nearly so direct. It channels voluntary commands through many internuncial circuits which organize the desired movements into sequences of stereotyped responses that are preserved in the brainstem and the spinal cord. The more familiar we are with a gesture or a skill, the more we rely upon this unconscious sequencing to guide our "voluntary" efforts.

A Continuum

In the language of movements, then, the neurons of the spinal cord and their specific interconnections are the words and the grammatical structures. I am not free to alter the patterns of the spinal circuits any more than I am free to alter the conventions of the language I am speaking if I am to make a coherent statement. The execution of a voluntary gesture, like the uttering of an original sentence, can only be accomplished by the appropriate arrangement of basically inflexible sub-units. These circuits and reflex arcs, like the definitions of words and the rules of grammar, cannot ordinarily be changed, and yet there is no limit to the number of original and meaningful statements that can be built up with them. The cortex can initiate literally any sequence that it can distinctly imagine, as long as it is consistent with anatomical limitations.

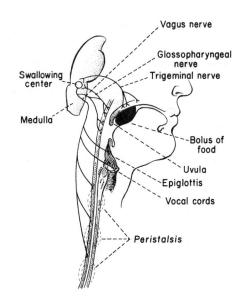

10-2: The swallowing mechanism, an example of an "all or nothing" sequence that utilizes fixed sequences of internuncial circuits to guide to completion a movement once it is initiated by the cortex. The only voluntary portion of a swallow is the pushing back of the bolus of food by the tongue; from that point on, reflex circuits control the muscles of the floor of the jaw, the throat, epiglottis, larynx, and esophagus to propel the food down to the stomach.

There is also another sense in which my voluntary actions appear in a ambiguous light. I am born with the ability to swallow, a rather complex coordination of twenty or so muscles that lasts about five seconds, beginning with contractions in my jaw and tongue and descending in a rippling wave down my throat and throughout the length of my esophagus. I am usually free to swallow or not to swallow, but once I have started, I am not free to interrupt the sequence until it has completed itself; once begun, it will run to the end.

Now many of my practiced skills and ingrained habits attain some measure of this all-or-nothing automaticity: If the appropriate cue is presented to me, I will produce my typical response, without pausing to reflect or consciously command it.

> Most muscle behavior is neither purely voluntary nor purely involuntary, but falls at some point on a spectrum between these two extremes. But even this statement is of little help because patterned muscle movements shift along the spectrum according to the frequency with which they are performed.[6]

All learned movements are initially both voluntary and crude. The more times they are practiced, the more refined and automatic they become. The sensory engram which encodes the movement becomes more and more precise, and the motor response to the engram becomes faster and more uniform, to the point that almost any learned movement can be trained to function very like a reflex.

This, of course, is why it is so difficult to drop long-standing habits, so difficult to make fundamental changes in the ways in which I do things that I have repeated as often as I have the way I sit and stand and walk, or the way I sign my name. Nor is this effect limited to isolated gestures or skills. Engrams encode not only the things I do repeatedly, they also encode the manner in which I do them, my style of doing. And even further, the things that I do the most often, and the manner in which I do them, begin to influence the style with which I do everything else.

Thus my most highly developed mannerisms become influences upon all of my movements and preconditions for all new skills. So it can not only be difficult to break an old habit, it can be difficult to alter the way in which these mannerisms influence the general style with which I do *everything*. My style itself becomes a habit. Normally I cannot just decide to stop doing things the way I have "always done them." And this habit of style is constantly reinforced by the very things that I do the most often and the most automatically.

The Riddle of Volition

What then has become of my concept of "voluntary behavior"? The more carefully I examine the objectively detectable events in my nervous system, the less "voluntary" the ultimate movement of my muscles seems to be, and the more mechanistically determined the chains of causes and effects which underlie my patterns of behavior appear. Is it merely self-delusion to think that I can choose to do things or not to do them, or even that I can change in any meaningful way the way in which I have "always" done them? If we limit our physical investigations to the lower brain and spinal cord, then our logic is definitely pushed in this direction. And if we direct them towards the cortex, we encounter another sort of cul de sac : The number of cell bodies involved is so large and their interconnections so legion that the number of objectively verifiable statements that we can make about the precise origins of our behavior within their mazes diminishes to very nearly zero.

No matter in which direction we look, we cannot even find a solid clue as to where "volition" might be located, or a specific series of impulses that we can identify as being pivotal in the process of "choosing this" or "choosing that."

> Someplace in the brain, and it is not known where, immediate afferent information must be associated with data acquired from past experiences and with the neural activity of present feelings (I want to, I feel like doing, etc.). It is the output of this synthesis which ultimately commands the motor cortex.[7]

Indeed, the empirical problems connected with the discovery of concrete manifestations of these kinds of associations and syntheses have led many neurological and psychological theorists to conclude that we are in fact interesting and puzzling sorts of robots, and to make a logically tidy "-ism" out of human "behavior" as a whole.

But perhaps when we are dealing with volition we have to do with a phenomenon which by its very nature cannot be located in a specific place in the brain or identified with a specific burst of impulses. Perhaps the mechanisms of choosing are far more diverse than that, and perhaps they are not "mechanical" at all.

In any case, the facts that we do know concerning the kind of volition that we exercise in our everyday behavior suggest that it is not meaningful to think of it in terms of my conscious control of all the motor units and the complex cooperation between them that is necessary to produce a movement. However it does work, the nervous system does not seem to work in this way. I do not normally orchestrate my actions in terms of individual motor units, but rather in terms of whole groups of units and muscles; I do not consciously organize my movements in terms specific impulses in specific circuits, but in terms of general shapes and directions and feelings.

Whether or not it allows us to locate volition, it squares more with the observations we do have to say that somewhere, somehow in the mind is formed the *idea* or the *image* or the *feel* of a movement, possibly made up of the appropriate snippets from a catalogue of sensory memories. This idea, or image, or feeling memory then acts like a sensory engram, guiding the musculature as a whole by means of the appropriate sequencing of the circuits in the lower brain and spinal cord. Upon its first use, this "imaginative engram" may produce a halting and awkward performance, but as the action is repeated current sensory feedback further clarifies the finer details of the desired movement, with the result that it becomes smoother and more faithful to the original intent.

The more substantial and similar to the desired movement the sensory memories are, and the clearer and more precise the image formed of that movement is, the more rapidly it will be mastered. This "visualization" approach to motor training has been used with great success in a wide range of applications. A look at some of the ways in which it has been utilized may serve to bring us a little closer to an understanding of the nature of our volition.

Visualizing the Goal

If a marksman is observed closely as he takes aim, it can be seen that many parts of his body are moving back and forth, oscillating in a manner that he is powerless to stop; but the gun remains quite still. It is true that the grosser degrees of this oscillation can be refined out with practice, but there is a limit to this. It is not possible to bring the musculature to a locked halt. What many marksmen actually report is that rather than thinking about individual muscles and their oscillations (which more often than not only makes the situation worse), they cultivate the ability to fix the concept of an immobile gun and its stable relationship to the target so firmly in their mind that their extraneous motions adjust in such a way as to cancel themselves out and produce the desired effect.

> The marksman ends by thinking only of the exact position of the goal, the singer only of the perfect sound, the balancer only of the point on the pole whose oscillations he must counteract.[8]

This concept has become a major cornerstone of teaching technique for many dance instructors. Martha Meyers makes the following observations while discussing the proper execution of the plies, a movement that is notoriously difficult and potentially dangerous to the knees:

> After several years of training, students have responded in shock: you mean those are the muscles I should have been using? I can't even feel them! One of the keys to relieving this frustration is, I have found, to help students change their 'initiation' and 'spatial intent,' both concepts from Bartenieff Fundamentals. My interpretation of the concepts is that *spatial intent begins in the mind as an image guiding the movement.* The initiation and the point in space to which the movement is directed alter the bony alignment; if the bones are not in good alignment, the muscle groups needed to achieve the action are at a mechanical disadvantage. *Precision of intent is crucial, for once a movement has begun, any effort to correct it leads to a distortion of the alignment of the pattern intended.*[9]

The use of this method to more efficiently direct the movements of dance students is by no means limited to Ms. Meyers or Mrs. Bartenieff.

Swinging and Swimming

Clearly in dance the movements are too complicated and follow one another too quickly to admit the possibility of consciously controlling them muscle by muscle. The students who are stuck in this approach are the ones that appear the most awkward, because the complexity of the task is far beyond the simplistic crudity of such means.

The same is true for music. Saxophonist Bob Wilbur has this to say about attempting to think ones way through a jazz improvisation:

> If the musician thinks about improvisation for long, he won't succeed. The less he thinks, the more successful he is going to be. It's like swimming, which is an extraordinary combination of muscle and timing. If you think about each breath, each stroke, your arms and legs get out of sync, your breathing falters, and you sink. Swinging is swimming well, and it has a distinctive rhythmic feel. It's the contrast between a steady pulse and the syncopated figures against it. It's the excitement of simultaneously going with the beat and battling it. The rhythmic element is what releases the intuitive powers in great players. Intense swing sharpens their reflexes and carries them away. They might be thinking melodic lines and chords, but then the intuitive powers take over and it's like going on automatic pilot. The notes are ripping, and they are totally free.[10]

The Experience of Dr. Sacks

Dr. Oliver Sacks is a Professor of Neurology at the Albert Einstein College of Medicine. As a hobbyist he is a very athletic individual, and as a professional he has spent his lifetime studying the nervous system, following in the footsteps of both his parents. He had already spent many years studying patients who had severe problems with motor control, but it was an accident of his own that most clarified for him the differences between "thinking" on one hand and "swimming" or "swinging" on the other when we seek to consciously control our musculature.

One afternoon while hiking in Norway, he had a bad fall which tore loose the quadriceps of one of his legs. The nerve which empowered his thigh was severed, and had to be surgically reconnected, along with the damaged tendons. Because of his background and the acuity of his mind, he was able to make some startling observations during the period of time it took him to heal and to learn to use his leg once again. The experiences which he reports bear so directly upon my discussion at this point that I will describe his circumstances and quote his responses at some length.

Finding Volition

As soon as Dr. Sacks had recovered sufficiently from the surgery on his leg, his surgeon and the supporting physical therapists tried repeatedly to get him to flex the muscles that had been repaired and begin to recover their use. Repeated attempts only produced repeated failure, to the consternation of everyone involved. The surgeon assured Dr. Sacks that the tendon had been reconnected and that there was "nothing to worry about." These assurances had no positive effect, nor did the coaxings, the commandings, nor finally the scoldings of the physical therapists. The leg would not be moved, in spite of the most fervent desire of its owner to do so.

Then early one morning while Dr. Sacks was alone in his room, something happened that began to change the situation.

> When I awoke I had an odd impulse flex my left leg, and in that self-same moment immediately did so! Here was a movement previously impossible, one which involved active contraction of the whole quad—a movement hitherto impossible and unthinkable. And yet in a trice, I had thought it, and done it. There was no cogitation, no preparation, no deliberation, whatever; there was no "trying"; I had the impulse flash-like—and flash-like I acted. The idea, the impulse, the action, were all one—I could not say which came first, they all came together. I suddenly "recollected" how to use the leg, and in the same instant of recollection I actually did it. The knowing-what-to-do had no theoretical quality whatever—it was entirely practical, immediate—and compelling. It came to me suddenly and spontaneously—out of the blue.[11]

Throughout the day Dr. Sacks attempted this spontaneous willing and flexing over and over, sometimes with success, sometimes not. During rest periods, he combed his mind for a clue as to why his quadricep responded to his wishes so erratically.

> What I kept thinking of, in particular was a remarkable chapter in A.R. Luria's *The Man With a Shattered World*—a chapter entitled "The Turning Point." This was in essence, for the patient, the recovery of 'music'. [A.R. Luria is a prominent twentieth century Russian neurologist; the patient referred to was a man suffering from a kind of aphasia—in this case, the loss of the faculty of reading and writing words.] At first writing was as difficult as reading, and perhaps more so. The patient had forgotten how to hold a pen or to form letters. He was completely helpless.... But a discovery he made one day proved to be the turning point: writing could be very simple. At first he had proceeded just as little children do when they first learn to write—he had tried to visualize each letter in order to form it. Yet he had been writing for almost twenty years and as such did not need to employ the same methods as a child, to think about each letter and consider what strokes to use. For adults, writing is an automatic skill, a series of built-in movements which I call 'kinetic melodies'. Hence, why shouldn't he try to use what skills he still had?...In this way he started to write. He no longer had to agonize over each letter, trying to remember how it was formed. He could write spontaneously, without thinking. [12]

Not without thinking altogether, perhaps, but certainly without thinking in the "normal" fashion. What Luria's patient discovered was the same thing that William James' marksman, Martha Meyers' dance students, and Bob Wilbur learned—that for the purposes of muscular performance thinking about *how* exactly a thing should be done is not nearly as effective as thinking clearly about *what* is to be done. In getting the words on the page, it is of very limited help to "consider what strokes to use"; a far better approach is to imagine clearly what one wants to say and then abandon oneself to the sensory memory, the "feel," the "kinetic melody" of writing.

Bringing the Will to Bear

The strength of this clue from Luria was dramatically demonstrated to Dr. Sacks the next day when, encouraged by his new flexing abilities, the physical therapists decided to help him to take his first steps after his injury. They stood him up, supported him, and coaxed him, but Dr. Sacks once again found himself unable even to twitch his quadricep on conscious command.

> She [the physical therapist] gave me a long look and then, seeing the useless-ness of words, wordlessly moved my left leg with her leg, pushing it to a new position, so that it made, or was made to make, a sort of step. Once this was done, I saw how to do it. I could not be told, but could instantly be shown— and she showed me what such a movement was like, so that, having been shown, I could bring my will to bear, and do it actively for myself. Once the

first step was made, even though it was an artificial, not a spontaneous, "step," I saw how to do it—how I might flex the hip in such a way that the leg moved forwards a reasonable distance.[13]

During the subsequent attempts to take steps, Dr. Sacks quickly discovered the enormous difficulties inherent in trying to think analytically about each movement of his leg, as opposed to simply recalling a forgotten "kinetic melody":

> In order to judge what was a "reasonable distance" in a "reasonable direction," I found myself entirely dependent on external, or visual, landmarks—marks on the floor, or marks triangulated with reference to the furniture and walls. I had to work out each step fully, and in advance, and then advance the leg, cautiously, empirically, until it had reached the point I had calculated and designated as secure.
>
> Why did I "walk" in this ludicrous fashion? Because, I found, I had no choice. For if I didn't look down, and let the leg "move by itself," it was liable to move four inches or four feet, and also to move in the wrong direction—for example, sideways, or, most commonly, at randomly slanting angles. On several occasions, indeed, before I realized that I would have to "program" its movements in advance and monitor them constantly, it "got lost," and almost tripped me up, by somehow getting stuck behind, or otherwise entangled with, my normal right leg.[14]

Remembering the Kinetic Melody

After taking a series of halting and exhausting steps in this manner, Dr. Sacks paused to rest and reflect:

> "This is walking?" I said to myself, and then, with a qualm of terror: "Is this what I will have to put up with for the rest of my life? Will I never again know a walking which is natural, spontaneous, and free? Will I be forced, from now on, to think out each move? Must everything be so complex—can't it be simple?"
>
> And suddenly—into the silence, the silent twittering of motionless frozen images—came music, glorious music, Mendelssohn, *fortissimo.* Joy, life, intoxicating movement! And, as suddenly, without thinking, without intending whatever, I found myself walking, easily-joyfully, *with* the music. And, as suddenly, in the moment that this inner music started, the Mendelssohn which had been summoned and hallucinated by my soul, and in the very moment that my 'motor' music, my kinetic melody, my walking, came back—in this self same moment "the leg came back." Suddenly, with no warning, no transition whatever, the leg felt alive, and real, and mine, its moment of actualization precisely consonant with the spontaneous quickening, walking and music.... I said to the physiotherapists: "Something extraordinary has just occurred. I can walk now. Let me go—but you had better stand by!"
>
> And walk I did—despite weakness, despite the cast, despite crutches, despite everything—easily, automatically, spontaneously, melodiously, with

a return of my own personal melody, which was somehow elicited by, and attuned to, the Mendelssohn melody.

I walked with a style—a style which was inimitably my own. Those who saw this echoed my own feelings. They said: "You walked mechanically, like a robot before—now you walk like a person—like yourself, in fact."

It was as if I suddenly remembered how to walk—indeed, not "as if." *I remembered how to walk.* All of a sudden I remembered walking's natural, unconscious rhythm and melody; it came to me suddenly, like remembering a once-familiar but long-forgotten tune, and it came hand-in-hand with the Mendelssohn rhythm and tune. There was an abrupt and absolute leap at this moment—not a process, not a transition, but a transilience,—from the awkward, artificial, mechanical walking, of which every step had to be consciously counted, projected, and undertaken—to an unconscious, natural-graceful, musical movement.[15]

This moment, coming after decades of neurological study and clinical experience, and at the apex of a personal crisis, was decisive for Dr. Sacks' grasp of his motor control and for his understanding of the peculiar—even paradoxical—blending of reflex response and voluntary effort.

Again I thought instantly of Zazetsky [the patient] in *The Man With a Shattered World,* and *his* "turning point," as recounted by Luria—the sudden discovery he made one day, that writing, previously desperately difficult, as he agonized over each letter and stroke, could become perfectly simple if he let himself go, if he gave himself, unconsciously and unreservedly, to its natural flow, melody, spontaneity. And then I thought of countless, though less spectacular, experiences of my own—times when I had set out to run or swim, first counting and calculating each step or stroke consciously, and then, quite suddenly, discovering that I had "got into it," that I had, mysteriously, without in the least trying, got "into the hang," "into the rhythm," "into the feel," of the activity, and that now I was doing it perfectly and easily, with no conscious counting or calculation whatever, but simply giving myself to the activity's own tempo, pulsion, and rhythm. The experience was so common that I had hardly given it a thought, but now, I suddenly realized, the experience was fundamental.[16]

A Fundamental Experience

We have all shared this common and fundamental experience of Dr. Sacks countless times—so often, in fact, that we only rarely consciously note the spontaneous ease which most characterizes it. We are far more apt to be aware of the absence of this spontaneous ease when we are fumbling self-consciously than to be specifically aware of its presence when we are doing something smoothly and successfully.

The experience of ease occurs to us exactly when we are so immersed in the doing of a thing that we do not have time for reflecting upon what we are doing, when our entire neuromuscular system is so completely engaged in a movement that there is simply no room for linear, analytical thinking about the separate elements of that movement. Even in retrospect it is difficult to analyze exactly how we were doing a thing, and we are most likely to fall back upon simile and metaphor in order to describe it; it is the sort of experience when "swinging is like swimming well," when good swimming is letting go to a "natural tempo and rhythm," when hitting a high note is like good marksmanship, when walking has its own "kinetic melody."

The fundamental thing revealed by this common sort of experience is that the normal voluntary use of the muscles is not directed by the conscious mind in a step-by-step analytical manner, with each discreet portion of a gesture actively willed and executed in sequence. This piece-meal approach is, in fact, typical either of severe pathology—such a Dr. Sacks' laborious first attempts to guide his newly repaired leg through its initial steps—or of the earliest stages of motor learning—such as a child arduously attempting to reproduce the forms of the alphabet with a pencil. This sequentially analytical way of going about things can never produce a normal smooth unbroken gesture of any complexity, simply because very quickly there are far too many elements involved, all going much too rapidly, to command and monitor each of them individually.

Rather, learned gestures—and new movements that are made up by selecting and rearranging learned gestures—are initiated and controlled by engrams, memories of how specific muscular actions have *felt*. These sensory memories function much more like blueprints, or templates, than they do like a linear sequence of commands. That is, each quantum of engrammatic memory contains the whole of a particular movement—whether short and simple or longer and more complex—stored as an image or an outline, instead of a fixed and precise sequence of commands directed at constant and specific motor neurons and muscle cells.

What I evoke when I call up an engram to initiate a voluntary movement is not a set of step-by step instructions that tells me how to do a thing, but a full and complete sense of how it has felt to do it in the past. It is a memory of a specific *accomplishment*, not a memory of exactly how the thing was accomplished. Indeed, if this were not the case, think of what enormous problems even minor differences in context would present: The adjustment of a piano stool up or down a few inches would render me incapable of performing a piece that I had learned in one particular position; I would have to learn to drive all over again each time I got into a different kind of car; it would take me as long to learn to write my name on a blackboard with chalk as it did to write it with a pencil on a piece of paper. What happens instead, of course, is that clear memory of an accomplishment, *in toto,* can select any number of actual reflex circuits and motor units to reproduce the sensations of successful *doing,* automatically taking into account and compensating for all sorts of variations in my actual physical relationship to the keyboard, the steering wheel and foot pedals, or the writing surface.

Appropriate Thought Forms

Thus the proper job of my thought process in directing successful voluntary behavior is not to consciously order sequences of muscular contractions, but to fix clearly in my imagination the feelings associated with accomplishing a particular thing, and then to surrender my muscular actions to that feeling. This is the way that I get the hang of a thing, get the feel for it, get into the rhythm, into the swing, into the swim.

It is certainly an intelligent thought process, but it requires for its success a very different manner of thinking than the analytical one, the one we normally associate with "planning," or "thinking a thing through." "If the musician thinks about improvisation for long, he won't succeed"—and almost all of our behavior is improvisation, that is, the application of learned skills to constantly changing situations. In order to be graceful and appropriate in our movements we must evoke a sensory memory, not analytically attack the situation, surrender to a flow, not attempt to control every part of the current, fix our imaginations clearly upon a goal, not try to marshal all the incremental steps that will get us there. Feeling states, the clear sensory memories of *doing* are the things that accurately guide our muscular actions, not the analysis and conscious sequencing of steps.

Purpose and Muscular Imagination

> The most succinct statement [about voluntary behavior] I know is one put forward by the Swedish neuro-physiologist Ragnar Granit in his recent book, *The Purposive Brain:* "What is volitional in voluntary movement is its purpose." From this viewpoint the volitional features of a motor act should be considered in terms of the goal of the action. Meanwhile, the actual events that underlie the achievement of the goal are built up from a variety of reflex responses[17]

It seems impossible to overestimate the importance of this concept: An action or an overall style of behavior is stored in the form of sensory memory, and without the power to recall that specific feeling we may not reproduce that specific action of that specific style. Man, in a way that was not possible before the evolution of the cerebral cortex, is both freed and limited *by what he imagines* he can do. He has at his disposal hundreds of thousands of possible patterns of movement, built up by the millions of years of development of the spinal cord and the brainstem; he has in addition the ability to arrange these reflexes into as many new series as he wants. Yet he is powerless to organize these bits of motion into any purposeful sequence without *first imagining clearly* that sequence. An individual's relative strength of will and sense of coordination are in fact nothing other than his relative ability to imagine clearly and in detail those acts he wishes to accomplish.

This is the crux of both the limitless possibilities and the grave dangers inherent in the addition of the cortex to the older behavioral centers. All reflex responses come under the control of the thought forms of the cortex, and any new sequence which can be clearly imagined can be learned; but on the other hand, this advancement of encephalization is not reversible, so that for us none but the most primitive orderly and purposeful sequences are possible without first imagining them clearly. And even more crucial to our habitual movements and styles is this principle's corollary—whatever elements are operative in our imaginations and our feeling states will find themselves inevitably expressed in our motor behavior, either as specific actions, as aberrations of those actions, or as the underlying style of actions in general.

We are, every cell of us, directed by our minds, materially conditioned for the better or for worse by our thoughts. And yet the kind of thought that we as a culture most readily associate with mental activity—analytical thought, conscious arrangement of sequences—is by itself quite inadequate to significantly change our material conditions and the behavior that they evoke. Other kinds of thought forms—feeling states, sensory memories, kinetic melodies, focussed imaginings—are in many concrete ways much more to the point when we address the problems involved in our motor behavior.

Learning to evoke and to surrender to a flow of sensory information are in the end far more effective means of learning or altering behavior patterns than are analyzing and commanding particular muscle twitches. With regard to controlling our posture, our movements and our style, it is not a question of pushing our thoughts to a more exact understanding of the empirical elements involved in our motor behavior, but rather a question of cultivating the sort of thought forms which by their nature effortlessly manipulate those empirical elements into successful coordination and control, whether or not we ever understand them fully.

"Voluntary" and "Autonomic"

This ambiguity between voluntary and involuntary, conscious and unconscious control, runs through the entire range of our muscular activities, both skeletal and visceral, creating something of a paradoxical situation both in functions which we usually regard as "obviously" voluntary—such as spooning exactly the right amount of sugar into one's cereal—and in functions traditionally held to be wholly involuntary—such as the digestion and internal distribution of that same cereal.

These latter examples of digestion and distribution are accomplished by the action of the muscles of the viscera, from throat to stomach to intestines to anus, and of the circulatory system, from heart to capillary to vein. These "smooth muscles" and "cardiac muscles," as distinct from the striated skeletal muscles, are directly linked to the autonomic nervous system, whose very name expresses the independence from voluntary intervention that it once was thought to have.

Because we normally do not consciously command the contractions and expansions of these visceral muscles, it was assumed that their healthy activities and their pathologies were beyond our conscious control.

The Autonomic Nervous System

All of the autonomic cell bodies and ganglia are located in the body cavity, outside the spinal cord and the anatomical boundaries of the central nervous system, and their interconnections appear to be a loose, wandering mesh-work, in contrast to the highly specific segmental arrangement of the spinal cord and the spinal nerve trunks. These structural features of the autonomic system strongly reinforced the idea of its distinctly different and independent mode of operation: Its anatomical circuitry is physically separate from the central nervous system, and it appears to be designed to broadly disperse any stimulation throughout its web-like network, creating a general response, quite unlike the highly specific pathways, targets, and responses which characterize the spinal cord and its nerve trunks.

Directly related to these structural features and functional theories was the conviction that the autonomic system was in basic ways less complex, more diffuse, less precise and acutely controlled than its spinal counterpart. It was held that the viscera were altogether refractory to most of the kinds of voluntary training and adaptation possible in the skeletal muscles. Certain organs and glands could be affected by the simple techniques of classical, Pavlovian conditioning, such as the prompting of the salivary glands by a bell that has been previously associated with food. But they were not susceptible to the more complicated learning procedures of instrumental (operant) conditioning, which involve conscious choices, trial and error, and voluntary cooperative attempts on the part of the animal being trained.

Learning Autonomic Functions

Happily for those many people who suffer from a wide variety of visceral malfunctions, both ancient meditational disciplines and more recent scientific observations have conclusively demonstrated that the autonomic system is not by any means as autonomous as all that. We are not completely at the mercy of some internal, genetically programmed (or misprogrammed) engine whose operations we are powerless to aid or modify.

> The autonomic nervous system is not self-governing at all. Its functions are integrated with voluntary movements no less than with motivations and affects. In short, its roots are in the brain: one's experiences from moment to moment dictate not only the contractions of one's skeletal muscles but also large functional shifts in the body's internal organs.... After all, affect and motivation find observable expression in visceral and endocrine changes.[19]

This exertion of voluntary control over autonomic functions has been successfully pushed to astonishing degrees of specificity.

> Recent experiments have indicated that visceral and glandular responses can be learned. For example, to avoid an electrical shock, a rat can learn to selectively increase or decrease its heart rate, and a rabbit can learn to constrict the vessels in one ear while dilating those in the other.[19]

Various animals have been taught to increase or decrease their heart rates, increase or decrease their blood pressure, increase or decrease the peristaltic contractions in their intestines. A trained animal can alter its rate of urine formation, or control the diameter of its blood vessels; rabbits have even been taught to dilate the vessels in one ear while simultaneously contracting those in the other ear. Even the voltage of brain waves can be voluntarily effected, and animals have further learned to isolate and amplify single brain wave forms—alpha or theta waves, for example. And all of this laboratory training of autonomic responses was done with operant—not classical—conditioning. The viscera are responsive to exactly the same training techniques and voluntary control as are the skeletal muscles. And as the author of the previous quote concludes, "The implications of such voluntary control of autonomic functions in human medicine are enormous."[20]

Meditation and Autonomic Response

Indeed, it can be shown that the human autonomic system may be influenced by a wide variety of factors that are within our conscious control. One of the main goals of the ancient arts of meditation and yoga has been the achievement of control over the healthy functioning of internal organs and glands by means of cultivating certain states of mind and physical postures. Again, these disciplines do not have to do with the analysis of particular synaptic connections, or with the conscious dictating of specific commands to specific muscle cells, but with the evocation of particular feeling states which in turn produce verifiable effects upon the visceral muscles, and even upon the blood chemistry.

If the physiological processes of a person in deep meditation are carefully monitored, we may observe distinct and consistent shifts in many internal activities. The heart rate slows, as does the respiratory rate, with the direct result that the body's use of oxygen and production of carbon dioxide drops, indicating an overall reduction in the rate of metabolism. And these changes occur to a markedly greater degree during meditation than during normal rest or even sleep.

At the same time, an increase in blood flow to the muscles can be observed, stepping up their oxygenation even though the breath and the heartbeat have slowed down; this has the effect of reducing the level of lactate in the blood, because it lessens the degree of anaerobic glycolysis taking place in the muscle tissues, a chemical process which produces a toxic waste product—lactate; injections of lactate into a normal individual typically causes local pain and acute mental anxiety. There is also a measurable increase in the electrical resistance of the skin, an index used in the lie-detector test to indicate a low-anxiety state. There is an increase in the acidity of the arterial blood. And the subject's electroencephalogram shows a rise in the brain's production of alpha waves, indicative of a deep relaxation bordering on sleep.

Nor are these physiological changes that have been observed in meditating subjects difficult to achieve. They do not require exhaustive knowledge of the mechanisms involved; they require only the adoption of a mental attitude, a feeling state, which can be readily taught to almost any willing subject.

> These physiological modifications, in people who were practicing the easily learned technique of transcendental meditation, were very similar to those that have been observed in highly trained experts in yoga and in Zen monks who have had fifteen to twenty years of experience in meditation.[21]

These physical and related chemical changes in the body during deep meditation have rather obvious implications for our mental and physical health. Referred to as a "wakeful, hypometabolic state,"[22] researchers have observed it producing a range of symptoms which are complementary opposites of the body's responses to pain or stress, lowering its metabolic activity and its muscle tone (and hence its expenditure of energy) while thought processes remain at a conscious and alert level. These are certainly among the necessary conditions for vital and relaxed health.

Bodywork and Autonomic Response

Now these changes observable in meditation are very like those that can be induced by effective bodywork—a slower and deeper respiratory rhythm, a slower heart rate, a diminishing of muscular tension both in chronically contracted areas and in overall muscle tone, an increase in blood flow through the visceral and skeletal muscles, and a more efficient use of available energy. There seems to be no reason to doubt that the intensified *alpha* activity of the brain and the beneficial chemical changes associated with the "wakeful, hypometabolic state" also accompany similar physical symptoms induced by bodywork.

The circularity of our internal feedback/response system is such that it does not matter whether we begin with the cultivation of an inner mental calm and allow its influence to project out into the muscles, or whether on the other hand we manipulate the sensory-motor reflexes in such a way as to decrease their normal tone and thus induce a calmer inner state. The bridge goes from Minneapolis to St. Paul as well as from St. Paul to Minneapolis, and the same beneficial shifts in the electrical activity of the brain and the chemistry of the blood can be expected to accompany the same physiological changes, regardless of where specifically in the circular sensory-motor loop these changes are initiated.

For bodywork just as for meditation, the desired end result is an individual who is both relaxed and alert, in control. Each of the two disciplines begins at opposite ends of a continuum to achieve their purpose, but due to the completeness of the integration of the nervous system their net results are very similar. And both are capable of exerting positive effects upon internal functions that have long been regarded as being beyond our control, "autonomic."

So we can see that the term "autonomic" does not describe a wholly separate type of neuromuscular behavior any more than does the term "voluntary." The autonomic system

> is in no sense functionally separate from the central nervous system, but receiving axons from cells within that system, forms one of the routes by which the central nervous system controls the tissues of the body. The significant difference between this route (visceral efferent) and that which supplies the muscles of the body wall and limbs (somatic efferent) is that the cells of the visceral efferent route which actually innervate the tissues lie outside the central nervous system, while those of the somatic efferent system lie inside the central nervous system.[23]

In other words, we here encounter once again the dangers of allowing the properties of our abstract visual and verbal models to condition our reasoning about natural processes, rather than keeping our focus firmly fixed upon the actual functions which our words and models seek to describe. Our methods of dissection and observation, and our subsequent labels of "autonomic" or "voluntary," suggest a separation and a difference of functional modes which we do not in fact find reflected in the fully coordinated activities of the these two systems of trunks and ganglia. In the end we find our terms and descriptions reduced to near-tautologies in order to preserve these presumed distinctions. "Autonomic," after all, simply means "self-governing." *Which* self is the question at hand.

Learning to be Sick

The degree of control over internal conditions which an individual can learn to exert by first adopting and maintaining a calm and alert state of mind, such as is taught to students of meditation, is indeed remarkable. What is even more remarkable is the degree to which psychological and medical research have ignored, even actively resisted, the neurological implications and therapeutic possibilities of this kind of training. After all, it has long been known that *negative* states of mind can *adversely* affect specific organs in specific ways. At this point, no one can doubt that sustaining high levels of anxiety for extended

periods of time is likely to produce stomach ulcers; researchers can even effectively manipulate the size and number of gastric lesions in laboratory animals,
simply by controlling the intensity of the anxiety-provoking agent and the length
of exposure. Other studies have indicated similar direct correlations of various
organic damages to sustained states of anger, grief, hatred, and apathy. Why
then, one cannot help but wonder, has medical research as a whole not been
keener on following up hints that specific positive states of mind may in fact have
restorative effects on specific organs as well?

Given the success that many researchers have repeated in training both human
beings and laboratory animals to exercise voluntary control over many of their
autonomic processes, there seems to be no justification whatever for regarding
the actions of our organs or our blood vessels as separate from or inaccessible to
our conscious choices. With reference to a trained animal's ability to constrict or
dilate its arteries, Leo DiCara, a prominent researcher in this field, concludes the
following:

> These striking results suggest that vasomotor responses, which are mediated
> by the sympathetic division of the autonomic nervous system, are capable of
> much greater specificity than was believed possible. This specificity is compat-
> ible with an increasing body of evidence that various visceral responses have
> specific representation at the cerebral cortex, that is, that they have neural
> connections of some kind to higher brain centers.[24]

In other words, there may be no organ, perhaps not even a cell of the body that is
"autonomic" in the sense that it is not affected by the sensations, the feeling
states, the attitudes the opinions, the fantasies, and the voluntary choices of the
conscious mind. And there is no association in the cortex that fails to find
concrete expression in a muscle, a gland, a chemical reaction.

Now if a laboratory animal or a human being can learn to selectively dilate
arteries, or alter the rate of urine production, or isolate and amplify a single form
of brain waves, what on earth may the conscious mind *not* do to the body, for
both its well and its woe? As a matter of fact, doesn't this kind of evidence suggest
that perhaps all of our internal conditions, either healthy or pathological, are
directly affected by our habitual states of mind, at least as much as they are by our
habitual diets, our occupations, our exposure to viruses or bacteria, our genetic
predispositions? Might we not grow up teaching our organs patterns of functioning just like we teach our skeletal muscles how to stand and walk and gesture?
And might not organ function suffer from slipshod training just as much as does
muscular coordination? It may be the case that, whether we were noticing it or
not, we have in many ways learned to be sick or be well, be depressed or be
robust, in exactly the same fashion that we have learned to speak a language or to
drive a car.

> The evidence for instrumental learning of visceral responses suggests that
> psychosomatic symptoms may be learned. John I. Lacy of the Fels Research
> Institute has shown that there is a tendency for each individual to respond to

stress with his own consistent sequence of such visceral responses as headache, queasy stomach, palpitation, or faintness. Instrumental learning might produce such a hierarchy. It is theoretically possible that such learning could be carried far enough to create an actual psychosomatic symptom.[25]

And what a list of "learned" symptoms might include! In addition to headaches (by far the most frequent medical complaint in the United States), there are queasiness, faintness, palpitations, and high blood pressure; there are ulcers and a host of other digestive, circulatory, and metabolic disorders; there are secondary responses to trauma; many skin conditions and allergies; neuropathies of various kinds; any number of imbalances of glandular secretions, including the most powerful hormones of the body; very probably even lapses in the body's immune system, and the legion of foreign invasions that can then take possession of us. Even some types of cancer have been tentatively linked to the failure to successfully cope with powerful, chronic, negative emotions such as anger, hate, or despair.

The significance of effective bodywork in this kind of a learning process could be enormous. If we can learn to respond viscerally one way, then we can learn to respond in another. And manipulating sensory input is potentially an extremely potent means of evoking new responses, quite simply because sensations are one of the primary sources of information which the mind uses to establish motor patterns.

Pleasant, calming physical sensations can be instrumental in learning to cultivate calm states of mind; we use this principle instinctively whenever we stroke an excited animal or caress a frightened child. Sensory stimulation also creates a flood of information that supplies the brain with the data it requires to make intelligent decisions and adjustments. Pleasant sensations and the positive feelings that are associated with them can themselves become a "psychosomatic symptom," triggering productive circles in the same way that pain and anxiety trigger vicious ones.

> If visceral responses can be modified by instrumental learning, it may be possible in effect to 'train' people with certain disorders to get well. Such therapeutic learning should be worth trying on any symptom that is under neural control.[26]

The various forms of effective bodywork are direct sensory approaches to such a therapeutic learning.

Psyche and Soma

Of course if we view all illnesses as being the results of genetic deficiencies, physical traumas, chemical toxins, or the invasions of micro-organisms, then "learning" to be sick or well does not make any sense. The word "psychosomatic" itself usually suggests a disturbance that is not after all a "real" sickness. Psychosomatic disorders are "in the head," and have little to do with the actual function or dysfunction of our nerves and organs.

And yet over and over, in many different situations, we find demonstrated that this strict separation of what is "in the head" from what is "in the tissues" is not an accurate representation of reality. Whatever is happening in the brain will inevitably find its way into the tissues, and through these avenues depression, anxiety, anger, and the like are as capable of damaging the organism as are accidents, diphtheria, or cirrhosis.

The relationships between our experiences, our feelings, and our body chemistry are undoubtedly far more intricate than we can presently imagine. We have seen how specific mental states effect specific glandular secretions, circulatory patterns, organ functions. If we now remind ourselves that every nerve cell is itself a type of gland, a gland whose chemical secretions are the mechanisms for carrying action potentials from cell to cell, we can appreciate the fact that there is probably no limit to the influencing of function and behavior by feelings and attitudes.

What we are given by genetics is the schematic layout for this system of neural glands, a layout that replicates itself in astonishing detail in individual after individual. But the number of impulses, the patterns of the impulses, and the material effects of those impulses by this genetically standardized layout can fluctuate so widely from individual to individual, and even from time to time in the same individual, that our functional and behavioral differences are equally as striking as are our genetic constants.

One of the things that is becoming increasingly clear in neurological research is that mere *anatomical* constants in the structure of neural circuits does not necessarily imply *functional* constants in the actual activities in those circuits. Nowhere in the body do we find experience, attitude, and chemistry more reciprocally interwoven than in the performance of the neural cell itself. Both *habituation* and *sensitization*—by far the most common modes of processing sensory information to establish selective awareness, memory, habit, associations, and so on—appear to operate by virtue of variations in the chemical secretions of the presynaptic cell membrane, and in the fluctuations of those secretions lies one of the principle mechanisms for the organization of our thoughts, our actions, our postures, our mental outlook.

Habituation

Habituation is the gradual decaying of a nerve cell's response that occurs when an initially novel stimulus is repeated over and over. I habituate a sensation when I cease to hear background conversation while I am reading intently, or when I cease to consciously feel a shirt that I have put on, even though it continues to rub my skin. Although its mechanism is very simple, habituation is probably the most prevalent of all forms of learning. Without this screening device, we could establish no orderly background/foreground relationship of stimuli in our consciousness, and every sensory message would register itself just as forcefully as all the others and demand an equal response—a hopeless cacophony of sensations and twitches.

The presynaptic membrane of each nerve cell is the site of this dampening of repeated stimuli. Less and less transmitting substance—acetylcholine or one of the other neurotransmitters—tends to be released from the presynaptic membrane of a cell when it is stimulated over and over in the same manner. If it is established over a relatively brief period of repetitions, this decrease in chemical secretion gives rise to *short term memory*—even after the repetition is stopped, the cell remains indifferent to the renewed onset of an identical stimulus for a short period of time. And if the repetitions continue for long enough, the amount of neurotransmitter released remains diminished for long periods of time, perhaps even permanently in sóme cases. When this occurs, a datum of *long term memory* is established, an enduring neurological shift, a chemical storage of a bit of our experience.

Notice here the remarkable plasticity of the nervous system, even at the level of individual cells: Even though the physical circuitry remains unchanged, the actual nature of every synaptic transmission may either fluctuate rapidly or be set more or less permanently, as these bits of memory come and go or accumulate and reinforce one another. And even though the outside world has not changed, my awareness of and response to a bit of it has been diminished. It is easy to see why this dampening effect is absolutely necessary in order to focus my attention, but it is also easy to see how it could become dangerous as well: My attention is shifted away from certain stimulations, but in some cases those stimulations— numbingly repetitive or not—may in fact be very significant to this or that function over long periods of time.

Sensitization

Sensitization is opposite in its effect, and is a little more complex a form of learning. It is the long-lasting *amplification* of an individual's response to a specific stimulus, as a result of the association of a second conditioning stimulus. In the laboratory, sensitization seems to take place most rapidly if the second, conditioning stimulus is a painful one. For example, if a small red light is set to blinking in a rat's cage, the rat will soon grow accustomed to it and go back to behaving as though it were not there. But if an electric shock is then delivered soon after the light is turned on, and this sequence is repeated a few times, the rat will begin to notice the blinking light very much indeed. It will actually evoke the same anxiety and avoidance behavior as the shock itself, even though the light is nothing like the shock.

Sensitization is a completely different function than habituation, yet its site of operation is also the presynaptic membrane. In the case of sensitization, more transmission substance is released from this membrane, because the sensations from the second, conditioning stimulus facilitates the response of the cell being sensitized. In this way, a single sensory input may be acutely pinpointed and brought into prominent foreground, usually because it is associated with some kind of threat which requires immediate attention.

Once again, the survival value of such a mechanism is clear, but once again it is equally clear that it has its dangers as well. Habituation can cause us to ignore persistent messages that may be important in the long run. Sensitization can so heighten our response to a relatively neutral stimulus that we may begin to structure much of our behavior around an avoidance that may or may not always be appropriate. And needless to say, those things that are habituate—that is, put out of conscious awareness—and those things that are sensitized—forced unavoidably into the foreground of our conscious awareness—vary widely from person to person. And the degree of their appropriateness—and hence their survival value—varies widely from situation to situation.

Bodywork and Synapse Performance

We are here concerned with two of the nervous system's most common methods of sorting out sensory input and of structuring responses to them. And both of them, opposite in their effects, are initiated by specific qualities and associations of *sensory inputs themselves*. We would be lost without both of them, and yet each can be pushed to an extreme that is positively dangerous to the organism. Once we have thoroughly habituated a symptom, it is out of our awareness and therefore beyond our ability to deal with it voluntarily; once we have sharply sensitized a symptom, it becomes a fixed point in our awareness around which much of the rest of our experience must be arranged.

Bodywork can be very helpful in defusing these extremes because it is itself a specific form of sensory input, input that can both alter previously established synaptic tendencies and condition newer, more appropriate ones. Parts of the body that have been deadened to our awareness can be given new sensory life, can once again begin feeding the brain with the information that is necessary to the tissue's normal, healthy function. "The synapses that were functionally inactivated (and would have remained so for weeks) were restored within an hour by a sensitizing stimulus...."[27]

Likewise, acute sensitizations can be quieted by flooding an affected area with pleasant sensations until the sensitizing fear of the expected pain abates, or by building up small, painless motions until larger, more problematic ones are disentangled from inhibiting associations. This sort of process is far from merely rubbing where it hurts, indulging in a bit of distracting touchy-feely; nor need it be merely temporary in its effects. It is, after all, a kind of sensory education that uses the same principles with which the nervous system structures all experience. E.R. Kandel, Director of the Center for Neurobiology and Behavior at the College of Physicians and Surgeons of Columbia University, states that

> Hence, there are synaptic pathways in the brain that are determined by developmental processes but that, being predisposed to learning, can be func-

tionally inactivated and reactivated by experience! In fact, at these modifiable synapses a rather modest amount of training is necessary to produce profound changes.[28]

Such shifts in synaptic secretions as a result of sensory experience would seem to imply, Kandel concludes,

> that even during simple social experiences, as when two people speak with each other, the action in the neural machinery in one person's brain is capable of having a direct and perhaps a long-lasting effect on the modifiable synaptic connections in the brain of the other.[29]

And if this is true of social conversation, what may be done with the far more sensorily direct means of physical touch? Painful stimuli are not by any means the only ways to achieve useful sensitization; a piece of information may also be thrust into the foreground because it is sensorily exquisite or emotionally moving, or because it has real practical value. Nor are we stuck with painful, anxiety-provoking associations until we either outgrow them or are successful in talking ourselves out of them. A much more direct route is available—namely, physical manipulation of the same sensory/chemical processes that established the counterproductive associations in the first place.

Stress

Learning to be sick involves my own physiological and emotional responses to certain stimuli I do not like, responses which either reinforce or add their own flavor to the potential negative consequences of noxious experiences. These reinforcements can be relatively mild and temporary, such as making a bad situation worse by developing a headache, or they can be chronic and devastating, such as the long-term withering of mental and physical vitality due to constant and unresolved stress.

As a matter of fact, the highly individualized manner in which each of us deals with stresses of all descriptions reflects the way that we have *learned* to react to potential problems. Have I learned to minimize my losses, recuperate efficiently, and move on? Have I learned to ignore unpleasant symptoms until the damage is so extensive that I am forced to succumb? Or have I learned to react to symptoms with additional diverting symptoms, discomforts of my own making, which may effectively mask the original problem, but which can join it in a vicious circle that can make its consequences far more severe?

As an example of how these vicious circles can be generated by our own conditioned responses, let us examine our system's reactions to the threat of bodily harm or severe stress—the so-called "flight or fight" responses—which are designed to prepare us for emergency situations. This stereotyped series of internal reactions is one of the best documented examples we have of specific sensations evoking a particular feeling state (defensiveness), which precipitates both neurological reflexes and chemical shifts, which in turn trigger observable changes in tissues and in behavior. What we know about these events will help us begin to understand how our perceptions, based on sensations, can profoundly effect our actual material conditions.

Stress and the Productive Life

The effects of stress on our mental and physical health is receiving a great deal of attention these days. We have stress specialists who write books and articles about stress syndromes, stress-prone personalities, stressful environments; we have stress seminars, stress clinics, and stress institutes, all working hard at teaching us how to recognize signs of stress, how to avoid it, or at least reduce it, how to recover from it, cope with it, live with it.

In all of these studies and the pathologies with which they concern themselves, a genuine bind consistently emerges: If we are unwilling or unable to tolerate certain levels of stress—occasionally even high levels—we are probably not going to get very much accomplished in our lives. Life and work are often inherently stressful. But on the other hand, if we overload ourselves with stress in an unrelenting effort towards more and more achievements, we can literally die trying, and quite prematurely at that.

It is clear then that neither the extreme of altogether avoiding stress nor the extreme of altogether ignoring it are really very helpful to the individual who wants to live a long and productive life. The real point is to learn how to *handle* stress, how to sustain our efforts without burning out, how to keep the nose to the grindstone without grinding it off.

A professional dancer, for instance, must continually push her body to the limits of its performance and endurance in order to reach and maintain her peak during her prime years. The pressure of these demands often creates painful symptoms in her body, weaknesses in muscle, tendon, or technique that are repeatedly revealed by the efforts of making the best possible dancer out of the existing materials. If she habitually ignores these symptoms, they have every chance of becoming worse, and her art will suffer; but if she avoids the movements which cause problems, her art will suffer just as surely.

The dancer, if she is to be a successful professional, has no choice but to confront each symptom and discover the way to work through it, not by ignoring nor avoiding its causes, but rather by learning to form a working relationship with the stresses of her activity. She trains herself to produce strength at every point where a weakness develops.

The Potentials of Stress

Notice what a double potential stress has here. On one hand, it is the factor which produces painful symptoms, causing limitations which can end a career.

On the other hand, it is the signal of a weak point, which, if managed correctly, can guide the dancer to greater strength, more endurance, fewer limitations, finer art. Clearly it is not stress per se that is either the good or the evil, but what we learn to do with it.

The following experimental format has been reproduced over and over, with young animals of various kinds, with mature animals, and with humans, and always with the same results: A population of individuals is selected so that it is as homogeneous as possible with regard to hereditary characteristics and signs of biological health. This population is carefully segregated into three supervised groups—one whose environment is kept as neutral as possible, one whose environment is conditioned with significant stresses, but is provided with some means of escaping or modifying these stresses, and one whose environment presents the same stresses but no way to avoid or modify them.

Invariably, it is the group which is given both stress and some means of coping which will continue to exhibit the soundest physical health and the most alert, well-adjusted responses to psychological testing. In fact, if the whole population is very young animals, *both* of the other groups will demonstrate signs of "deprivation dwarfism" in their growth patterns! The strictly neutral environment and the no-escape environment both produce various retardations, while the stress/adaptation environment produces growth patterns which are characterized by stronger bones, more muscle tissue, bigger brains, earlier sexual maturity, more aggressive behavior, and more outgoing curiosity when presented with novel situations.

> Just as the studies of young animals showed, contrary to expectations, that some degree of stress in infancy is necessary for the development of normal, adaptive behavior, so the information we now have on the operations of the pituitary-adrenal system indicates that in many situations effective behavior in adult life may depend on exposure to some optimal level of stress.[30]

This should not be too difficult to understand. After all, every gardener knows that the greenhouse does not produce the most resilient strains, and every athlete knows that without forcing himself past the limits of his physical comfort, no new limits of strength or endurance may be reached. The nerves and muscles and glands of our body are genetic givens; the optimal strength and efficiency of their operations are not. Just as the hand requires constant sensory feedback in order to develop its potential power and delicacy, so do our internal organs and glands require adequate levels of stimulation—which may include occasionally high levels of stress—in order to develop and maintain their functions.

The Pituitary-Adrenal Axis

We are all familiar with the palpable effects of adrenaline, the principle chemical secreted by the adrenal glands situated on top of kidneys. It is released into our bloodstream whenever we encounter high levels of stress or the threat of immediate danger, producing a rush of physical and mental energy which gives us an extra boost to meet the difficult situation. It usually gives us an exhilarating feeling, sometimes even euphoric, heightening the acuity of our senses, the speed of our reflexes, and the strength of our muscles. Often after the emergency is past, it leaves us with a chemical surplus in our blood which makes us feel queasy, shaky, abnormally excited for a brief period until all of the adrenaline that has been released is reabsorbed.

We associate adrenaline primarily with these subjective feelings, and we identify it primarily with the adrenal glands that secrete it. But the adrenal glands do not in any sense function in isolation; they are members of a large complex of elements designed to mobilize the body and the mind against threat, and for their switching on and switching off they rely directly upon chemical and neural messages from the pituitary—the master gland of the body. The interrelated secretions of the pituitary and the adrenals then circulate freely and alter the functions of a large number of our internal systems.

Many of these changes can be seen as the complementary opposites of the changes initiated by the calm state of meditation. In particular, the activities of the *parasympathetic* branch of the autonomic nervous system (which are the most pronounced during periods of rest and recovery) are markedly dampened by the secretions of the adrenal glands, while those of the *sympathetic* branch (whose activity is largely stimulatory, precluding states of rest and recovery) are markedly heightened. In the autonomic system, these conditions are the reverse of the "alert, hypometabolic state." By tipping the balance of parasympathetic/sympathetic activities in this opposite way, the pituitary and the adrenals affect a whole array of internal processes, such as circulation, respiration, digestion, muscle tone, and the secretions of various other glands, all of which become part of the body's overall response to stress or danger.

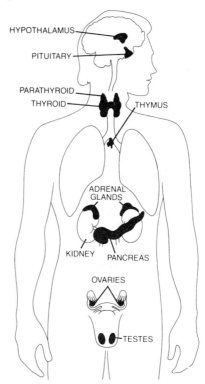

HYPOTHALAMUS

PITUITARY

PARATHYROID
THYROID

THYMUS

ADRENAL
GLANDS

KIDNEY   PANCREAS

OVARIES

TESTES

10-3: The pituitary and the adrenal glands work together as a team to orchestrate most of the other glands of the body.

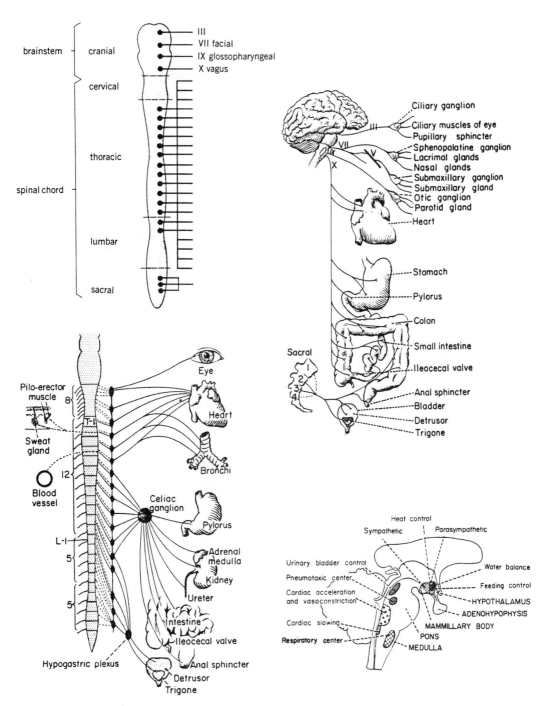

10-4: The sympathetic and parasympathetic branches of the autonomic nervous system. The parasympathetic branch originates in the cranial and the sacral nerves; the sympathetic branch originates in the cervical, thoracic, and lumbar nerves.

Note that they both contact a number of organs, where they have the opposite effects: The sympathetic branch stimulates the organ into a higher state of activity, while the parasympathetic branch lowers these levels of activity for periods of rest and repair. The sympathetic branch is very active during the "flight or fight" response, while the parasympathetic activity is more prominent during meditation.

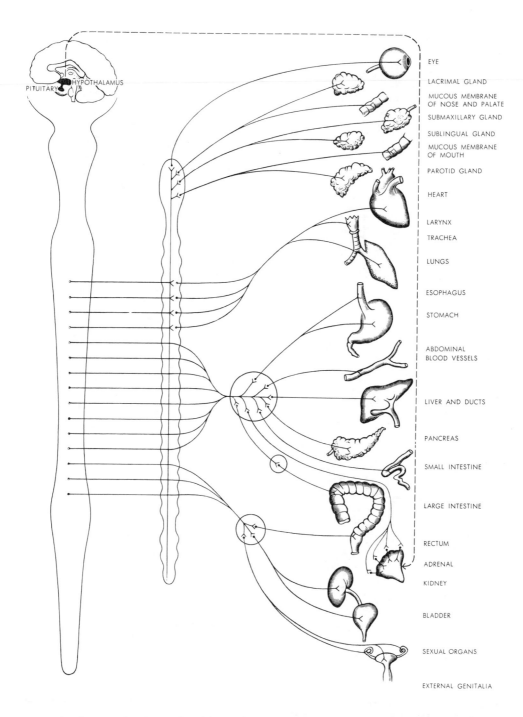

EYE

LACRIMAL GLAND

MUCOUS MEMBRANE
OF NOSE AND PALATE

SUBMAXILLARY GLAND

SUBLINGUAL GLAND

MUCOUS MEMBRANE
OF MOUTH

PAROTID GLAND

HEART

LARYNX

TRACHEA

LUNGS

ESOPHAGUS

STOMACH

ABDOMINAL
BLOOD VESSELS

LIVER AND DUCTS

PANCREAS

SMALL INTESTINE

LARGE INTESTINE

RECTUM

ADRENAL

KIDNEY

BLADDER

SEXUAL ORGANS

EXTERNAL GENITALIA

PITUITARY   HYPOTHALAMUS

10-5: The pituitary/adrenal axis (dotted line), which directly activates the autonomic nervous system. The
reactions to stress and the "flight or fight" response are partly controlled by the illustrated pathways.
The column to the left is the spinal cord; the column in the center is the chain of autonomic ganglia lying outside
the spine; the circles are autonomic plexi, the largest one pictured being the celiac plexus.

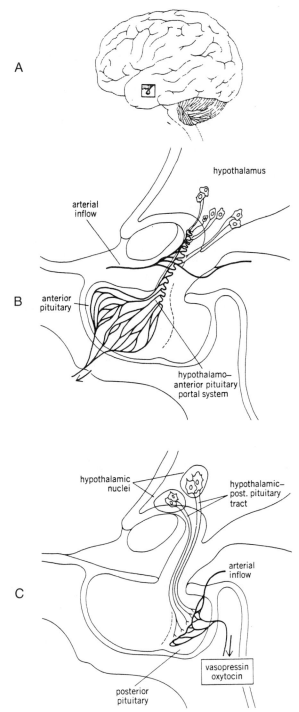

10-6: The pituitary gland is intimately associated with the hypothalamus and the higher brain. Capillaries from the hypothalamus branch out into the anterior lobe of the pituitary, and nerve axons from the hypothalamus branch out into the posterior lobe. Thus chemical messages in the blood and neural impulses from both the brainstem and the cortex converge upon this master gland.

Now "stress" and "danger" are in themselves very vague terms for general conditions which usually have many components. They may be purely physical, or purely psychological, or (as is most often the case) some combination of the two. What constitutes extreme duress for one individual may not elicit any particular response at all in another. A tremendous number of factors—sensations from all over the body, memories, associations, expectations, emotions, intellectual analyses—are included in the assessment of stress or danger.

It is in the brain's hypothalamus where all these factors are collated and assessed with regard to their urgency, and where the preparations for a state of emergency are initiated. The hypothalamus is, among other things, a dense crossroads of interconnections between other major portions of the brain. It is, as it were, the ceiling of the spinal cord and the basal ganglia, and the floor of the cerebral cortex, and serves as a neurological bridge between the older and the newer parts of the central nervous system. It also has a rich two-way connection with the cerebellum, and many links with the various segments of the autonomic system, both sympathetic and parasympathetic.

In addition to these neural interconnections, it is also situated in such a way as to receive the most freshly oxygenated and most vigorous thrust of arterial blood leaving the aorta; thus it is one of the most sensitive analysts of our internal chemistry as well as of our neural activity. And one of its principal appendages, with which it is connected by both neurons and capillaries, is the pituitary gland.

This hypothalamus/pituitary partnership is one of the most potent influences in our entire organism, because in response to these various chemical and neural inputs they in turn either secrete or trigger the secretion of the most powerful hormones in the body, and mediate the responses of the entire autonomic nervous system.

And in all matters relating to stress, they are joined by a third partner at the other end of a vertical axis, the *adrenals*—small but extraordinarily important accessories of both the neural and hormonal systems.

> The two adrenals of an average human adult weigh together not more than ten grams—about one 7000th of the total body weight. But they are extremely important organs; if they are destroyed, as in Addison's disease, death is inevitable. With the exception of the brain centers for breathing and vascular tonus, there is no other equally small part of the body whose destruction or removal results in so quick a death. You can remove both legs, two thirds of the liver or a whole kidney and life will not be endangered; but if you remove an animal's adrenals, it loses its resistance to the slightest damage and dies within a few days. Obviously, then, the adrenals hold a key position as regulators of vital functions.[31]

The Chemistry of Stress

Once a stress or a danger has been positively identified in the mass of messages converging on the hypothalamus (and we must remember that in this initial interpretive phase of the overall alarm reaction there is an extremely wide latitude of association for "stresses" and "dangers"), the hypothalamus secretes into the bloodstream a chemical known as *corticotropin-releasing factor*. This corticotropin-releasing factor (CRF) then quickly circulates into the pituitary gland, where it triggers the release of a second chemical that is manufactured and stored there, the *adrenocorticotropic hormone* (ACTH).

As soon as the blood carrying this ACTH reaches the adrenal glands, they in turn are stimulated to release their chemicals into the bloodstream. There are at least twenty known adrenal hormones, the principal and most understood of which are *adrenaline* and *cortisol*. These freely circulating substances, adrenaline and cortisol, then react with many target tissues throughout the body to trigger an entire range of physiological responses in the anticipation of an emergency. Normally, as soon as these responses are initiated the excess levels of adrenaline and cortisol are noted by the hypothalamus, and the original CRF catalyst is turned off, effectively stopping further pituitary and adrenal secretion until a new alarm is sounded.[32]

Cortisol has three general effects on the body's tissues: 1) It promotes the breakdown of proteins in the body to form glucose, the fuel burned by the muscles; 2) it acts as an anti-inflammatory agent, primarily by decreasing the accumulation of white blood cells and fibroblasts at injured areas; 3) it exerts a powerful anti-allergenic action. It was these latter two effects which made synthesized cortisol—cortisone—appear to be a "wonder drug" at one time; it can miraculously reduce inflammation and eliminate allergic symptoms—but without, unfortunately, having any effect whatever on the actual causative agents. This purely symptomatic action, coupled with the severe consequences of the sustained protein breakdown generated by sustained dosages of cortisone, have proved to be very dangerous side-effects, and have cooled considerably the enthusiasm for this drug.

Adrenaline is the same chemical compound as epinephrine, the neurotransmitter secreted by the ends of sympathetic nerve fibers. It therefore has the general effect of stimulating the entire sympathetic branch of the autonomic system, and of phasing out the rest and restorative actions of the parasympathetic branch.

All together, CRF, ACTH, adrenaline, and cortisol precipitate a large number of specific changes in neural activities, tissues, and behavior, all designed to aid the organism in an emergency—the so-called "flight or fight" responses, which quickly prepare us for either physical conflict or rapid escape. First of all, there is an immediate arousal of the central nervous system, stimulating it to a state of increased alertness. This hyper-arousal sets the stage for other responses. One of its immediate results is an increase in muscle tone and a facilitation of

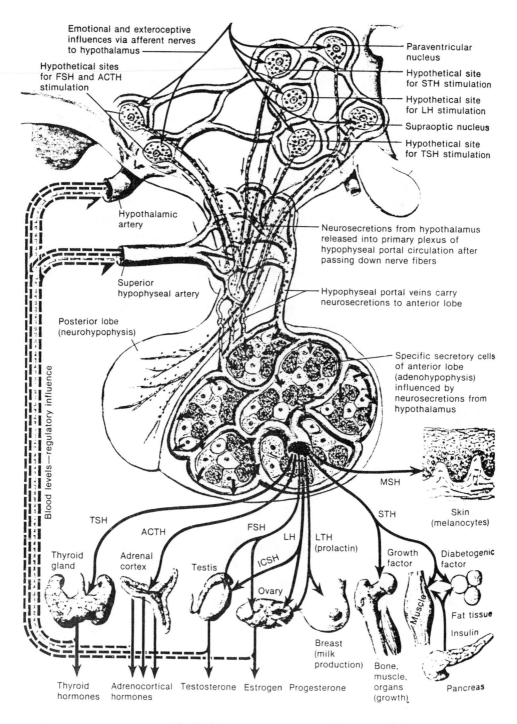

10-7: A detail of the pituitary gland—its inputs and its targets.

sensory reflexes and muscular contractions, poising the entire musculature in extreme readiness to react.

As a necessary sequel to this charging up of the nervous system and the muscle tissues, respiration deepens and accelerates, the heart rate increases, and blood pressure rises. In addition, the arteriols of the circulatory system adjust their relative diameters so that this increase in blood flow is directed away from the stomach and intestines, towards the skeletal muscles and the nervous system. And the spleen dumps an extra supply of red corpuscles into the bloodstream to carry an increased amount of oxygen to and an increased amount of carbon dioxide away from the charged muscles.

Digestion stops. Or, more accurately, we stop digesting the food in our stomachs and intestines and begin digesting ourselves. Cells in the liver and other protein-rich tissues are broken down to create glucose, instant fuel for the muscles. Fatty tissues are catabolized as well, providing even more materials for the rapid production of glucose, and increasing the supply of fatty acids required for oxygenation. Glucagon and growth hormone further mobilize internally available nutrients.

Besides neural arousal and muscular supercharging, potential physical injury is also anticipated in the "flight

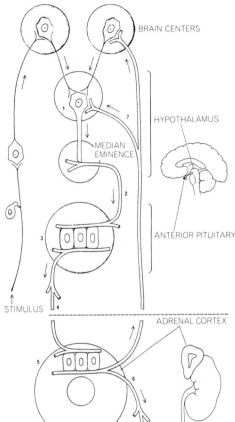

10-8: The pituitary/adrenal axis involves nerve cells and hormones in a feedback loop. A stress stimulus reaching neurosecretory cells of the hypothalamus in the base of the brain (1) stimulates them to release corticotropin-releasing factor (CRF), which moves through short blood vessels (2) to the anterior lobe of the pituitary gland (3). Pituitary cells thereupon release adrenocorticotrophic hormone (ACTH) into the circulation (4). The ACTH stimulates cells of the adrenal cortex (5) to secrete glucocorticoid hormones (primarily hydrocortisone in man) into the circulation (6). When glucocorticoids reach neurosecretory cells or other brain cells (it is not clear which), they modulate CRF production (7).

or fight" responses. Chemicals are added to the blood which increase its coagulability in case of bleeding. Aldosterone and antidiuretic hormones are released to cause water retention by the kidneys—an important factor in the event of either heavy sweating or blood loss.

In addition to triggering all of these bodily changes, there is also evidence that CRF, ACTH, and adrenaline have specific effects upon the brain beyond its general arousal, and may play a role in heightening learning, sensory selection processes, sense of time-lapse, and other behaviors that would be helpful in dealing with danger. Nor is this list of changes by any means complete; the activity of every hormone and every target tissue in the body is probably affected in some way by the chemistry of the "flight or fight" responses.[33]

Adrenaline/Cortisol Poisoning

The long list of responses which we have so far examined are the *biochemical* effects of the secretions of the hypothalamus, the pituitary gland, and the adrenals. Each of these effects plays its part in the preparation of the nervous system, the muscles, and the organs for dealing with an injury, a threat, an emergency; all together they supercharge the body and the mind so that we can vigorously take flight or fight, or take whatever other steps are necessary for our survival.

But if the dosage level of these hormones is sustained for longer periods of time in the bloodstream, either by artificial injection or by repeated episodes of stress, very different changes begin to occur in the nervous system and other target tissues. These are referred to as the *pharmacological* effects of those hormones.

Initially, the most obvious of these pharmacological effects is the inhibition of inflammation around injuries or infections. This is, indeed, why large doses of cortisol—or its synthetic, cortisone—are administered in dangerously acute cases of infection, allergy, arthritis, and similar inflammatory conditions.

But inflammation is a natural response to injury that is more often than not of great value. It is the body's method of flooding the area with oxygen, nutrients, antibodies to fight invading organisms, fibroblasts to carry on the wound-healing process, and so on. It is not hard to see, then, how sustained levels of cortisol in the bloodstream, far from shielding us from threat, may actually interfere with the natural healing process and steadily decrease the body's resistance to infections, swellings, and tumors of all kinds.

Adrenaline also has significant pharmacological effects, as opposed to its primary biochemical ones. Recall that it is the same compound that is secreted by the nerves ends of the sympathetic branch of the autonomic nervous system, the compound which stimulates its target organs and prevents them from lapsing into a phase of rest and recovery for as long as it contacts them. Eventually this will have the same effects upon the effected organs as chronic loss of sleep has for an individual—exhaustion and dysfunction.

In addition to the suppression of healthy inflammation and the overstimulation of the autonomic nervous system, sustained levels of adrenaline and cortisol have other disturbing effects. These include high blood pressure, gastric ulceration, atherosclerosis, suppression of the immune system (which would lower one's resistance to diseases and infections of all kinds), sterility, and significant personality changes.

> The correspondence of this list to the major causes of mortality in industrialized countries makes it tantalizing to speculate about the possible role of psychological stress, acting by means of increased cortisol [and adrenaline], in all these diseases.[34]

And of course, we must not forget that when these "flight or fight" hormones are in our bloodstreams, we are constantly digesting our body's own proteins, just as we do when we fast, when we are seriously ill, or following severe injury or surgery. This is the mechanism by which over-stressed children are retarded in physical growth and mental development, and by which over-stressed adults wither away.

This seems to be something of a paradoxical situation. The very chemicals released by the body in order to safeguard itself in the case of an emergency are the same ones whose long-term effects are demonstrably destructive, even fatal. Cortisol, a drug which can be used to promote healing by reducing severe inflammation, can weaken the entire system by suppressing the inflammatory response altogether. Adrenaline (epinephrine) is normal and healthy in the parasympathetic nerve ends, but can be terminally exhausting in the bloodstream.

But this is not really so paradoxical. We know that many drugs have "side effects," often as potent as the one desirable effect for which they were administered. As is the case with these drugs, massive or prolonged levels of stress hormones in the blood begin to produce increasingly noticeable side-effects, which eventually overshadow their normal healthy functions. Rather than a genuine contradiction, what we have here is simply too much of a good thing. Chronic stress creates a sort of adrenaline/cortisol poisoning, in a manner similar to the way in which repeated doses of cocaine or heroine eventually obliterate with physical and psychological horrors their original salutary pain-relieving effects.

This chemistry and physiology of stress is one of the best documented examples of how we learn to be sick, how we cultivate—sometimes even lust after— the very reactions which set into motion vicious circles that systematically destroy us. The factors that converge upon the hypothalamus to produce an alarm are so complex, and the things that are regarded as "threatening" vary so widely from individual to individual, that in all but the most clear-cut cases of actual danger we really must say that we learn to overdose ourselves with adrenaline and cortisol, and by sustaining this overdose we force ourselves to learn to adapt our bodies to the conditions imposed by chronic stress. And as this learning takes place, profound changes occur in the development of the pituitary/adrenal axis, changes which both impose worsened conditions and limitations on our adaptability.

Development of the Pituitary/Adrenal Axis

We have observed that a total absence of stressful stimulation is not really better for our healthy development and adaptive behavior than too much, that "in many situations effective behavior in adult life may depend on exposure to some optimal level of stress."[35] Some "optimal level" of adrenal secretion evidently must take place in order for the physical and mental faculties to fully develop and assert themselves. The proper functioning of the hypothalamus/

pituitary/adrenal system—which includes an occasional demonstration of full-scale alarm—is an absolutely necessary element for the healthy growth of many of the body's tissues. And it turns out that *adequate tactile stimulation* is absolutely crucial for the normal growth and function of these precious nerves and glands.

Let us consider again an experiment discussed in the previous "Skin" chapter in which young laboratory animals were divided into three groups: One regularly fondled during infancy and development, one regularly shocked with a painful electric jolt, and one stimulated as little as is clinically possible. This latter non-stimulated group, remember, failed to match up to either the fondled or the shocked group in a whole range of physical and psychological comparisons.

If we look more closely at the internal processes of these three groups as they mature, it becomes obvious that the non-stimulated animals specifically do not develop a vigorous "flight or fight" response, and further, that this sluggish activity of the pituitary/adrenal axis begins to impose adverse conditions on many other tissues and responses in the growing animal.

The differences in pituitary/adrenal reactions among the three groups may be assessed as follows: When they are examined in a neutral situation, all members of all three groups are found to have more or less identical and normal levels of stress hormones in their bloodstreams. If a painful threat is then presented to all three groups, distinct internal differences arise.

Both groups that had been stimulated regularly during their infancy and growth—regardless of whether that stimulation was fondling or shocking—elicit a healthy pituitary/adrenal response. That is, they very quickly achieve high blood levels of stress hormones, and when the threat disappears these levels drop again just as quickly. On the other hand, the non-stimulated animals achieve these same blood levels of stress hormones much more slowly, and these high levels are maintained for a much longer period of time after the threat has been removed.

> This observation acquires its full significance when it is considered in the light of the biological function of the stress response. The speed and short duration of the response in the stimulated animals obviously serve the useful purpose of mobilizing the resources of the organism at the moment when it is under stress. The delay in the response of the non-stimulated animals is thus, by contrast, maladaptive. Moreover, the prolongation of the stress response, as observed in these animals, can have severely damaging consequences: stomach ulcers, increased susceptibility to infection, and eventually death due to adrenal exhaustion.[36]

There is also another extremely important difference between stimulated and non-stimulated laboratory animals with regard to their pituitary/adrenal responses. Not only do the non-stimulated subjects evidence more sluggishly initiated and more sustained secretions, giving them higher levels of adrenaline in their blood for longer periods of time than their stimulated counterparts, but also the non-stimulated animals will show adrenal reactions to many more kinds of situations than will the stimulated ones.

> It is a striking fact that the system's activity can be evoked by all kinds of stresses, not only severe somatic stresses such as diseases, burns, bone fractures, temperature extremes, surgery and drugs, but also by a wide range of psychological conditions: fear, apprehension, anxiety, a loud noise, crowding, even mere exposure to a novel environment.[37]

The non-stimulated animal will not only learn to respond in a manner ill-suited for coping with actual danger, it will also learn to respond far more readily to lesser physical threats and to more innocuous psychological stresses.

> The maladaptive nature of the stress response in the non-stimulated animal is further manifested in the fact that it may be elicited in such a neutral situation as the open field test. The animal that has been manipulated in infancy shows no physiological stress response in this situation, although it exhibits vigorous and immediate endocrine response when challenged by the pain and threat of an electrical shock."[38]

In the psychologically complicated life of modern man, this conditioned shifting of the threshold of pituitary/adrenal response has considerable significance. In addition to situations which threaten actual physical injury, people exhibit adrenal responses for a wide variety of reasons. Flying in airplanes, awaiting surgery, anticipating final exams, competitive athletics (either playing or watching), divorce, death of a loved one, driving in rush-hour traffic, many job experiences, many motion pictures, any circumstance so novel that we do not know what to expect, anxiety of any kind—all these things have been clinically demonstrated to cause adrenal reactions in human beings,[39] and they tend to provoke a much more sustained response in those whose pituitary/adrenal system shows the same sluggish sort of behavior as in the non-stimulated laboratory animals. Thus the pituitary/adrenal axis tends to become at the same time more sensitive, more easily triggered, yet more sluggish and more sustained in its secretions. This tendency may even be built up into a whole personality profile, such as aggressive, ambitious, time-urgent, anxious, paranoid, and so on.

Now these psychological sensitivities are primarily a case of an essentially *appropriate response* elicited at an *inappropriate time*. So the hormonal machinery which is set into motion cases the body to feed on itself, weaken itself, rather than successfully met an external challenge.

> During man's early history the defense-alarm reaction may have had high survival value and thus have become strongly established in his genetic make-up. It continues to be aroused in all its visceral aspects when the individual feels threatened. Yet in the environment of our time the reaction is often an anachronism.... There is good reason to believe the changing environment's incessant stimulations of the sympathetic nervous system are largely responsible for the large incidence of hypertension and similar serious diseases that are prevalent in our society.[40]

And with regard to bodywork, the important thing to remember in this context is the crucial role of adequate tactile stimulation in establishing and maintaining the healthy glandular responses which underlie the appropriate onset and the appropriate withdrawal of the stress hormones.

Interrupting Vicious Circles

Ulcers are responsible for approximately 10,000 deaths per year in the United States alone, and they will affect one out of every twenty in our population at some time in their lives.[41] Gastric ulceration is probably the pathology that has been the most empirically pin-pointed in its relationship to physical stress and anxiety. And it has been clearly demonstrated that in many instances it is *psychological factors* which are related to the stressful situation—and not necessarily the actual stress factor itself—that are without a doubt the main causes of this organic disorder.

The factors which appear to be the critical ones are in some ways rather surprising. Animals who ulcerate the most under experimental stress conditions are those that are made helpless or confused in conjunction with painful stimuli. But when the experimental conditions around the painful stimuli are changed in such a way as to lessen the helplessness and the confusion, these animals ulcerate far less in spite of the fact that they still receive physical abuse.

It does not seem to be the case, then, that it is necessary to protect the animals from painful shocks in order to eliminate ulceration and other signs of physical stress. Several other factors emerge as being more significant as protectors. If the animals are systematically given a warning before the painful stimulus, if they are provided with some means of controlling or escaping the stimulus, or—most helpful of all—if they are provided with some positive feed-back device to signal successful avoidances, they prove to be far more resilient to strain than their helpless or confused counterparts.

Inducing Ulcers

Again, let us consider three more groups of experimental animals. One group is periodically given an unpleasant electric shock, preceded each time by a warning beep; a second group receive exactly the same shocks at the same times, but they are given no warning as to when the shocks would occur; the third group received neither warnings nor shocks. Naturally enough, the group which received no shocks ulcerated the least in this experimental situation. But, perhaps not so expectedly, the animal who received warnings preceding their shocks ulcerated only slightly more, while those deprived of warnings consistently produced much more numerous and massive ulcerations.

In other words, it was not the shock itself, but rather its unpredictability which produced the most debilitating symptoms of anxiety. The receiving of the shocks was not really very detrimental as long as the animals could accurately anticipate them.

> In short, the results demonstrated clearly that the psychological variable of predictability, rather than the shock itself, was the main determinant of ulcer severity.[42]

Also beneficial to the animals under these stress conditions is to be given some means of controlling the source of the shock. Those who could escape the shocks by jumping up on a platform likewise ulcerated much less severely than their helpless counterparts, even if their warning beeps were withheld.

But what proved to the most strikingly beneficial of all was the addition of some form of *positive feedback reinforcement* relating to successful avoidances. If a tone was sounded each time the correct avoidance behavior was executed, this tone then became associated not with the negative stressor, but with the positive act of successfully avoiding it. The mere presence of this positive feedback was enough to reduce ulceration nearly to the levels of the group which received no shocks at all.

> The effectiveness of coping behavior in preventing ulceration depends upon the relevant feedback that coping responses produce; simply to have control over the stressor is in and of itself not [as] beneficial.[43]

If this positive feedback is tampered with after it has once been established, thereby creating inconsistency and confusion, then the catastrophe is markedly increased. When animals that have been trained to perform an avoidance response are then given a brief pulse of shock whenever they give the formerly correct reaction (producing, incidentally, a laboratory imitation of the painful and cruelly arbitrary trials of the biblical Job), they develop the most severe ulcerations of all, even more than the completely helpless animals receiving the same shock.

> Hence by manipulating the feedback consequences of responding, rats could be made to develop extensive gastric ulceration in an otherwise [relatively] non-ulcerogenic condition, or could be protected almost completely from developing ulcers in a condition that was normally quite ulcerogenic.[44]

Appropriate Feedback

These hints at the helpful influences of accurate anticipation, effective control, and positive feedback to coping responses are pregnant with suggestion concerning the possible role of effective body work in teaching an individual to deal with stress and discomfort. If establishing the habit of encountering stress with helplessness, confusion, or other counter-productive reactions is one of the main ways that we learn to court illness, then surely establishing some means of early warning, self-control, and positive feedback for our efforts to cope can take us a long way towards learning to be well.

Tactile stimulation, in and of itself, offers a potent method of redressing a sluggish and imbalanced pituitary/adrenal axis, insofar as impaired response of these glands can be directly related to the lack of tactile stimulation during early development. Further than the general efficacy of almost any kind of touch and manipulation in this context, skilled body work can be used specifically to enhance the sensory awareness of our bodies as a whole and of problematical areas in particular. This enhanced awareness can create exactly the same sort of early-warning system for the onset of symptoms that protected one group of laboratory animals from the more severe ulcerations suffered by their counterparts that had no such warnings.

Moreover, if the releasing of local tensions and their attendant discomforts can help to induce a calmer state of mind, or if soothing sensory input can calm the mind and help induce the release of local tensions, then an important mechanism of *control* can be established. If an individual can learn to unwind the exacerbating reactions to anxiety or pain, then he is not helpless in the face of them. He is not only forewarned, but forearmed as well; he has acquired a technique or two for successful avoidance, and learned to identify a number of positive feedback messages that can reinforce the effectiveness of these techniques. Once initiated and identified, these positive feedback loops can then guide the individual more and more clearly towards greater and greater control of his internal symptoms, including those that erupt in organs related to the autonomic system.

In short, he can learn to protect himself, not necessarily from stress, but from his own ineffective and counter-productive reactions to it. He can learn to cultivate *habits and states of mind* which help him to avoid some stresses altogether and to take the sting out of many others. What can be particularly avoided, and what certainly *must* be avoided if the individual is to learn to get better, is the irritated and even desperate thrashing about from one ineffective reaction to another, creating more tension, more anxiety, and more of the physically and emotionally debilitating results of fear, confusion, and failure. If it is arrived at consciously, and through the means of concrete sensory feedback, a moment of relief need not be temporary or merely symptomatic, but may instead be a bit of positive learning, a clue that can be built upon to create more self-awareness, more self-control, more productive responses to stress.

Stress Addiction

Gastric ulcerations, atherosclerosis, and other specific tissue changes are not by any means the only negative results of the failure to cope with stress effectively. Entire personality profiles can be extensively disturbed as well, and these give us clues as to how learning to accurately sense and control bodily reactions can have profound effects upon our psychological as well as our physical health.

The changes brought about by the secretion of stress hormones into the bloodstream are not all strictly physiological ones; some of the chemicals involved in the alarm state directly effect the nervous system, our process of learning, our moods, and our modes of perception. As mentioned earlier, there is evidence that learning processes in the brain may be temporarily heightened, both by the boost in neural activity of the hypothalamus and by the chemistry of the various secretions.[45]

This heightened state, necessary in emergencies, can be experienced as pleasantly exhilarating, and is probably behind many people's desire to cultivate dangerous sports, turbulent personal affairs, or hectic work schedules. In this respect, adrenaline is quite like any other psychoactive drug and can—quite as effectively as amphetamine or cocaine—produce a dependence upon the heightened energetic state, and upon a lifestyle centered around its repetition.

Stress and Sensory Perception

The qualities of sensory perception itself seem to be significantly effected by the relative levels of some of the stress hormones in the blood. It has been shown that abnormally low levels correspond with hypersensitivity to sensory input.[46] Increased sensitivity may in fact be a plus in some ways, but it can be overburdening, confusing, maddening without any means to select or suppress; unmodifiable sensitivity is the basis of many miserably distracting nervous conditions. This seems to suggest that those who go to great lengths to avoid external stresses may well be setting themselves up for less escapable internal ones: Without occasional infusions of stress hormones in the blood few relevant details must be clearly differentiated from a barrage of inputs, and when pain from injuries must not distract us from doing whatever must be done to survive the confrontation. But chronic dulling is not really very helpful in most of our day-to-day observations and interactions. There is a hint here that some of the less attractive social qualities of the "macho" stance may in fact be reinforced by the chemistry involved in an aggressive, confrontational lifestyle. And there is the hint as well that to the degree that these hormone levels are due to the activities of the pituitary/adrenal axis, and to the degree that these activities can be adjusted by experiencing appropriate sensory inputs and by learning to cultivate appropriate states of mind, body work can have a potentially important role to play in alleviating the discomforts and limitations inherent in the extremes of both of these personality types.

Norepinephrine and Helplessness

Both the alarm reaction and the subsequent success or failure in dealing with the threat trigger a great number of neural and chemical events whose influences are felt in various ways throughout the nervous system. The *catecholamines* are a whole group of chemicals produced in the central nervous system itself which are currently exciting much neurological research because they seem to have marked effects upon mental processes. *Norepinephrine* is one of this family, and its activity appears to be linked to the successful avoidance of threat. It is, in fact, closely related to, if not identical with, noradrenaline, one of the secretions of the adrenals. Norepinephrine is thought specifically to have a major role in mediating active, assertive responses, and there is evidence that its depletion in the brain brings about chronic depression in humans. Jay M. Weiss concluded from his research that

> animals able to avoid and escape shock showed an increase in the level of brain norepinephrine, whereas helpless animals, which received the same shocks, showed a decrease in norepinephrine.... It may well be that the causal sequence leading from "helplessness" to behavioral depression depends upon biochemical changes in brain norepinephrine. This would indicate that depressed behavior often can be perpetuated in a vicious circle: the inability to cope alters neural biochemistry, which further accentuates depression, increasing the inability to cope, which further alters neural chemistry, and so on.[47]

It is this sense of "helplessness" exactly which traps us into beliefs concerning the inevitable outcomes of our genetics, into the exclusive reliance upon surgical and pharmacological crisis-intervention models of health care, into relying upon "experts," into wilting before the advance of noxious influences. Here is the chemistry of self-fulfilling prophecies of doom. It can be the role of effective body work to provide this helpless mental state with pleasant and relevant sensory feedback, feedback which can be an invaluable aid in re-establishing an individual's sense of control and ability to cope.

Lactic Acid and Anxiety

The possibilities within our vast network of neruo-chemical processes for these kinds of imbalances which generate vicious circles are legion. An emotion or attitude initiates a neural or metabolic change, this change deepens or complicates that attitude, which results in increased chemical changes, which even further exacerbates the attitude, until major mental or physical symptoms appear. Another example is the rise of lactate, or lactic acid, levels in the blood during periods of anxiety: "Patients with anxiety neurosis show a large rise in blood lactate when they are placed in a position of stress."[48]

Lactate is the principal waste product of *anaerobic glycolysis,* the synthesis of of ATP in the muscle cells without the use of oxygen. Its rise is related to anxiety in the following manner: One of the common effects of anxiety is an increase in muscle tone; this can be either because of the general charging up of the musculature during an actual "flight or fight" response, or because of local areas of tension that are generated by one or more of the constrictions or postures characteristic of chronic states of worry and anticipation. As muscular tension increases, local metabolism increases also, in order to support the higher level of tonus; but due to pressure exerted by the constricted muscles on the local capillaries, local circulation *decreases*. When more ATP is demanded in a muscle cell than the available oxygen can help to synthesize, the anaerobic process takes over, which produces the waste lactic acid. So anxiety, through the mechanism of increased muscle tension, produces lactic acid.

But if we inject lactic acid into the bloodstream of a calm subject, we find a curious thing: The lactate will produce anxiety attacks in a neurotic subject, and even produce all the symptoms of anxiety in a healthy one. So once again we find a mental state and a chemical reaction reinforcing one another, reciprocally increasing until serious local or system-wide imbalances—even damages— occur.

As was the case with the norepinephrine levels in the brain and stress hormone levels in the blood, this production of lactic acid in the muscles becomes much more significant to the health of an individual whose pituitary/adrenal response has been trained to be sluggish and prolonged. Both anxiety and depression, and the array of symptoms that follow in their train, can be sustained and deepened over much longer periods of time, and can be irritated by more and more situations.

The Two Sides of the Circle

It is of extreme significance that each of these vicious circles requires a *feeling* or an *attitude* just as much as it needs a chemical change to perpetuate itself. Indeed, in the absence of these negative emotions, many destructive processes are unable to continue. No depression, no drop in brain norepinephrine; no anxiety, no rising lactate levels in the blood; no helplessness, no gastric ulcers. And conversely, if in the laboratory we can produce a specific negative emotion in a rat or a monkey or a man, the animal will then produce specific chemicals and specific physical symptoms.

This highly correlated relationship between certain feeling states and certain physiological functions may seem like a riddle or even like nonsense, if we do not recognize that we are to a large degree *learning* to cultivate our various states of health and disease, just like we learned to walk or to read. And the efficacy of body work may seem like more nonsense unless we recognize that it also precipitates learning, the learning of self awareness and control reinforced by positive sensory feedback.

Touch, Stress, and Modern Man

These studies of laboratory animals who have been separated into groups and subjected to various levels of stimulation, deprivation, and stress strongly suggest that there are distinct features of our physical and emotional development which are wholly dependent upon sensory nutrition, and as such are not in any way an inevitable unfolding of a genetic code. If we are either deprived of tactile stimulation or over-overburdened with stress, the very mechanisms in the body which are designed to protect us turn and begin to devour us instead. P.C. Constantinides and Niall Carey summed up this situation in 1949, while they were two young members of the research team working for Dr. Hans Selye, one of the most prominent investigators of the stress syndrome:

> We seem to see the merest outlines of a great biological chain reaction which can be set off by almost any stress and which may frequently lead to the suicide of the organism. Some of the links in this chain are still missing, but its essential structure has been amply confirmed.[49]

As we have observed in experimental results, adequate tactile stimulation in general and appropriate sensory feedback in particular are absolutely crucial for the avoidance of this self-destructiveness, and so are periodic episodes of successfully resolved stress. Furthermore, it is the two former conditions (adequate stimulation and appropriate feedback) which provide the learning and the neurological bases for any successful resolutions. The unfortunate thing is—and for some individuals it is disastrous—that adequate stimulation and appropriate feedback can be difficult to comeby in our culture. Our two researchers continue:

> It would appear that the most frequent and fatal diseases of today are due to the "wear and tear" of modern life. One might question whether stress is peculiarly characteristic of our sheltered civilization, with all its comforts and amenities. Yet these very protections—modern labor-saving devices, clothing, heating—have rendered us all the more vulnerable and sensitive to the slightest stress. What was a mild stress to our forebears now frequently represents a minor crisis. Moreover, the frustrations and repressions arising from emotional conflicts in the modern world, economic and political insecurity, the drudgery associated with many modern occupations—all these represent stresses as formidable as the most severe physical injury. We live under a constant strain; we are losing our ability to relax; we seek fresh forms of physical and mental stimulation.[50]

Deprivation dwarfism, ulceration, depressed immune response, high blood pressure, and suicide by glandular secretion are merely the most extreme distortions of the normal development of essentially sound genetic materials. All degrees of deprivation or unresolved stress produce their corresponding degrees of aberration and maladaption. And these aberrations often develop just as if they were themselves genetic flaws, because they have no discernible cause—only a *lack* of a subtle but decisive cause for optimum health. Many of these conditions *seem* to be congenital only because this lack does its work quietly and slowly, and because we are never given the opportunity to compare our present condition with how we would have been had we not suffered that lack, of which we may not even be aware.

This not knowing what we do not know about the course of our own learning and physical development has the result of surrounding these issues and symptoms with a great deal of ambiguity. Because our behaviors are not mere assemblies of fixed reflexes, because our voluntariness has graduated into the vast arena of choices presented by the cortex, the principal developmental problem for human beings is decidedly a psychological one. We have the power to invent obscure causes and blank out obvious ones to an awesome and dangerous degree.

The following experiment has been repeated many times, and its results have shown a high degree of reliability: Two groups of adult humans are given identical electrical shocks; one group is falsely informed that by pressing a button they can lessen the shocks, while the other group is promised no such escape mechanism. The group that is offered no possible avoidance will consistently produce more physical stress responses than will the group who believe they were given an escape mechanism, even though it has no real effect upon the strength or duration of the shocks! This once again emphasizes the enormous significance of training, attitude, and the individual interpretation of relevant feedback in human beings.

> The people who thought they had control over the shock perceived their responses as producing relevant feedback. In contrast, the people who thought they were helpless necessarily perceived their responses as producing no relevant feedback. Thus for humans, as for rats, the same variables seem important in describing the effects of behavior in stress situations. On the other hand, the experiments with humans alert us to how important higher cognitive processes are in people, showing that verbal instructions and self-evaluation can determine feedback from behavior, which will subsequently affect bodily stress reactions.[51]

These ambiguities are a very large part of the problem for us as we experience and attempt to resolve stress in our lives. Because the cortex has such enormous associational powers which ultimately extend into all tissues and all behaviors, we are very capable of turning bumps in the road into full-blown crises, and are particularly apt to cling to patterns of response that we have learned previously, regardless of their effectiveness in the present instance.

"Thus it would not be surprising to find," conclude our two researchers working with Dr. Selye,

> that much of our organic disease derives from psychological trauma, with the general adaptation syndrome as the bridge that links one to the other. If this is true, medicine may eventually find a cure for the consequences of stress.[52]

Constantinides and Carey make it clear that for them that

> in searching for ways to combat the diseases produced by too much hormone production, one of the most obvious targets would be to try to neutralize the hormonal excess, in other words, to find a chemical antidote.[53]

"But the prevention of the basic causes," they conclude, "will remain a task that lies beyond [medicine's] reach."[54]

But are these last conclusions necessarily true? Are we genuinely stuck with waiting for a "chemical antidote" to the natural glandular and neurological processes taking place in our own bodies? Might not research and modern health care have available to it, in the form of bodywork—or more precisely, in the increased self-awareness, relevant sensory feedback, and sense of self-control which bodywork can foster—an effective means of beginning to redress some of these imbalances? To the extent that tactile stimulation in general and relevant sensory feedback specifically appear to play a great role in mediating our psychological identification of stress, our physiological reactions to it, and the success of our coping behaviors, might not intelligently and sensitively administered sensory input be expected to be helpful? It may or may not be immediately clear exactly *why* positive sensory input is often helpful, but this input can in fact provide relief to many, even though we do not have at the present time an exhaustive empirical understanding of all its mechanisms. And this relief is no small favor to those who are presently suffering.

After all, the central problem in all of these inquiries is a very practical one. In their minds, where feeling states, habits, and the tissue changes related to them are born, people are not suffering from "hormone imbalances," and they are not yearning for "chemical antidotes"; they are suffering from discomfort and pain, and they yearn for relief. Of what value to them is a "chemical antidote" that will come along for them "someday," after research can elucidate every material nuance of their suffering? Of how much more value to them would be a readily available means of gaining some degree of control over themselves and their symptoms, of creating a little slack in the vicious circles that entrap them, and of alleviating some of their discomforts and limitations, so that they can get on with the productive parts of their lives?

I am not suggesting that the ravages of serious disease or the aftermath of extensive trauma can be simply rubbed away. Our bodies are not, alas, quite like magical lamps that will give us our wishes just for the rubbing. But what is evidently true is that many uncomfortable and painful conditions, including some of the most widespread and some of the more lethal ones in our modern culture, are in part either manufactured or facilitated by the neural and glandular interconnections within our own systems. And if those conditions can be generated by our own mental and chemical processes, it does not seem unreasonable to expect that relief and regeneration might possibly be precipitated by those same internal systems. And if the amount of sensory stimulation and the specific qualities of sensory stimulation apparently have so much to do with the healthy development and optimal functioning of so many neural and glandular systems, might it not also be reasonable to expect that sensory input could be intelligently used in such a way as to redress some of the underdevelopments and dysfunctions we find associated with various pathological conditions?

If feeling states, attitudes, behavioral and physiological habits can start large circles of inter-related processes turning in vicious directions, might not different states, attitudes, and habits start them turning in healthy ones? And have we not somehow *learned* the attitudes and habits that are crippling us? And can we not learn new ones? Are not voluntary attempts, monitored by relevant sensory feedback, our most universal means of learning *anything*?

We are losing the ability to relax, and the loss is maiming us. And wouldn't it be prudent, even if the mechanisms are not thoroughly understood, to seek out and use whatever we can use to teach ourselves this important skill again? What could theoretical understanding have to offer that would be half so valuable to an individual slipping into one of these vicious circles as some practical steps that can be taken to regain some measure of this relaxation and self-control? And to what practical steps do our instincts immediately guide us when we would soothe a frightened child, or calm an over-excited animal?

Perhaps if discomfort and limitation were themselves clear-cut matters, then some of these issues would have long since been a little clearer as well. But the catch-22 for many individuals is the fact that often the very attitudes and habits that are doing them the most real damage are the very ones that are for them the most cherished and ingrained. To make the necessary shifts can seem like more trouble than tolerating the present condition. Well-worn routines, and even world views, may well be involved in such shifts. Giving up the pain may appear to be too painful.

It is an axiom of popular wisdom that we are usually better off living with the pain we are familiar with than trading it for a pain we are not. Or, to choose an alternate adage, "no pain, no gain" can be elevated from an athletic rule of thumb to a theoretical underpinning for a whole lifestyle that borders on out and out masochism, even suicide.

And by the time that the discomfort we are suffering becomes greater than the discomfort of changing our ways, the physical damage has already been done. It is this extreme psychological ambiguity of discomfort and pain themselves that diverts us from the real causes of our limitations and from the practical steps we could be taking to move beyond them.

Pain

"Pain is the least understood of all the human senses."[55] Its primary function is to alert the central nervous system to damage or destruction in the body's tissues. The experience of pain nearly always constitutes a physical and a psychological threat, and it is accordingly accompanied by system-wide responses which are on the whole identical with those of a "flight or fight" arousal: over-all increase of mental alertness, a rise in muscle tone, faster heart rate, higher blood pressure, increased blood sugar, facilitated clotting factor, shunting of blood from the viscera to the skeletal muscles, dilated pupils, sweating.[56] In other words, pain is very commonly an activator of the pituitary/adrenal axis and all of its neural and glandular responses.

We have noted the practical value of these responses to an endangered organism. And we have also seen that the "flight or fight" response can itself be a threat to the individual if it is triggered too easily, too often, and sustained too long. Now the chronic anxiety and alarm that accompany this sluggish and debilitating state are more often than not simply anticipations of painful events.

Our experiences of and our associations with pain are therefore very closely linked to the functioning of our pituitary/adrenal alarm system, and it may in fact be the case that our attitude towards pain is the most decisive single factor in regulating our body's release of stress hormones. And from the stoic endurance required by many excruciating tribal initiation ceremonies to the pitiful whimpering of a neurasthenic with a hang-nail, there is no single sensation that presents such a vast array of kinds and degrees and provokes as many varied reactions as does pain. Our relationship to pain is consequently of major importance to the health and resilience of our organisms.

Ambiguities of Pain

Pain, along with touch, pressure, hot, and cold, is normally regarded as one of our primary sensory modalities, and like each of the other modalities it has a system of cell-bodies, axons, and nerve endings all its own. But even here at the cellular foundation of the pain experience we encounter ambiguity. It is often the case that "too much" sensation from any of the other modalities is experienced as pain. And "too much" does not necessarily mean "to the point of tissue damage"; it may just mean "more than we are used to." So pain from its very inception is often not a clear-cut sensation from a clear-cut source: Is it an announcement from the pain-specific nerve endings (the *nocioceptors,* from Latin *nocere,* to harm), or is it the brain announcing an overload from one of the other modalities?

All pain-specific pathways terminate in free nerve endings which branch out to contact a local area of tissue cells. These endings are densely supplied in the skin, in the periosteum encasing the bones, the arterial walls, joint surfaces, and the connective tissue partitions within the cranium. In these densely supplied areas, a relatively small injury is enough to cause acute sensation. Pain endings are found in fewer numbers in organs and other deep tissues, where a fairly widespread disturbance is necessary for neural summation to produce an ache or a burning sensation.

Just what initiates a pain signal at the nerve end is not known with certainty. Damaged cells release two substances (along with their other spilling contents), bradykinin and histamine. When these chemicals are artificially injected underneath the skin, they cause intense pain. Perhaps it is the bathing of the nerve endings in these substances which triggers the reaction of the nociceptors.[57]

Once initiated, the pain sensations do not follow a single pathway to the spinal cord and the brain. The sensations travel on two different classes of fibers, one larger and faster, one smaller and slower. The faster fibers carry a "pricking" pain which reaches the brain first, where it signals the original fact of damage. This original pain prick is generally not felt after the damage is done, but only while it is occurring. Slower fibers then transmit a longer lasting burning or aching which apprises the brain of the *extent* of the damage.

These two modes of pain can be easily distinguished by quickly pinching the web of skin between the fingers: The initial sharp stab is the pricking sensation carried by the fast fibers, and the ensuing ache is the second message carried by the slow ones. It is this secondary response that

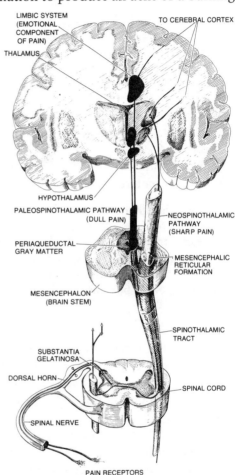

10-9: Pain pathways. Sharp pricking pains travel on larger, faster axons, announcing to the cortex the instant of damage. Duller summations of pain stimuli travel on slower axons, and inform the brain somewhat later of the full extent of the damage. These latter tracts contact the emotional centers of the brain on their way to the cortex, giving pain its "suffering" quality; this component can be enormously heightened, making the slightest pain an occasion for full alarm or extended debilitation.

tends to become more intense over a period of time, as all the endings in the affected area sum up their impulses, and as the central nervous system either dampens or facilitates the results of this summation. This is the sensation which brings with it the lingering and more intolerable feelings we associate with severe pain.

Surprising as it may seem, with regard to the actual initiating of pain sensations due to tissue damage humans are all nearly identical in their response.

> It is almost never true that some persons are unusually sensitive or insensitive to pain. Indeed, measurements in people as widely different as Eskimos, Indians, and whites have shown no significant differences in their threshold for pain.[58]

That is to say that the original "pricking" signal of damage is initiated in all of us in the same fashion. With the application of slowly increasing heat, for instance, almost all subjects experience pain sensations at 45 degrees centigrade; this is the lowest temperature which begins to actually damage cells.

But this initial signal is the briefest and in some respects the least significant phase of the pain response.

> Even though the threshold for the recognition of pain remains approximately equal from one person to another, the degree to which each one reacts to pain varies tremendously.... Conditioning impulses entering the sensory areas of the central nervous system from various portions of the central and peripheral nervous systems can determine whether incoming sensory impulses will be transmitted extensively or weakly to other areas of the brain. It is probably some such mechanism as this that determines how much one reacts to pain.[59]

The possible ranges which these dampening and heightening effects may have are truly remarkable. They are capable of producing responses so widely varied that we are often simply incredulous that others around us are not sharing our experiences of discomfort: "Aren't you too hot?" "Doesn't that hurt?" "Aren't you tired?"

In itself, there is nothing unique to pain in this wide range of possible responses. Alternations of inhibition and facilitation are, after all, constantly influencing almost every cell of our nervous systems, providing the mechanisms for the sorting out process without which we would be powerless to isolate a single sensory experience or a single motor reaction. It is this selectivity and this selectivity alone which produces the fundamental relationships of foreground and background upon which our concepts of reality and behavior rest.

But because of the specifically threatening nature of pain, and because of the havoc that can be caused in our systems by excessive preoccupation with threat, certain features of this general selection process leap into bold relief when it is associated with pain.

> Descending pathways capable of altering the transmission of information in the afferent neurons, spinal pathways, or brain centers are known to exist in

most sensory systems, but they are particularly important in pain. They are thought to be one means by which emotion, past experiences, state of attention, etc., can alter sensitivity to pain. When the descending pathways reduce the activity in the pain pathways, the unpleasant emotions and response behavior, as well as the specific pain perception, are diminished.[60]

A Sensory Experience Plus a Psychological Response

Needless to say, these descending pathways are equally capable of magnifying pain sensations as well as diminishing them, to the point that the mere anticipation of a specific pain may cause all the dread, the adrenaline and cortisol, the sweating and trembling, and the emotional suffering that the excitation of the pain receptors themselves could produce. This anxiety and anticipation may become so closely associated with a painful experience that the distinctions between the fear, the actual tissue damage, the glandular responses, and the behavioral reactions can be utterly lost to the sufferer.

> The experience of pain includes an emotional component of fear, anxiety, and sense of unpleasantness, as well as information about the stimulus' location, intensity, and duration. And probably more than any other type of sensation, the experience of pain can be altered by past experiences, suggestions, emotions (particularly anxiety), and the simultaneous activation of other sensory modalities. This complex nature of pain can be accounted for by saying that the stimuli which give rise to pain result in a sensory experience plus a reaction to it, the reaction including the emotional response (anxiety, fear) and behavioral response (withdrawal or other defensive behavior). Both the sensation and the reaction to the sensation must be present for tissue-damaging stimuli to cause suffering.[61]

Indeed, this *reaction* to the sensation can be so powerful, and so influential to our experience of the situation that some of our most effective pain-relieving drugs—morphine, for instance—accomplish their effects by acting upon the emotional, reactive centers of the brain alone, and not upon the actual local sensations of tissue damage.

> The sensation of pain can be disassociated from the emotional and behavioral reactive component by drugs, e.g. morphine. When the reactive component is no longer associated with the sensation, pain is felt, but it is not necessarily disagreeable; the patient does not mind as much. Thus, satisfactory pain relief can be obtained even though the perception of painful stimuli is not reduced.[62]

The Pain/Spasm Cycle

The kinds of mechanisms which combine to create this acute heightening of pain and tissue involvement may be illustrated by an examination of the genesis and some of the consequences of a familiar type of pain—that which is associated with muscle spasm. This pain often has an innocuous enough primary source, commonly a minor strain or bruise, a source small enough and peripheral enough to be unproblematic for the system as a whole, and yet it can be potentially excruciating and thoroughly debilitating.

One of the reflex reactions, even to a limited amount of tissue damage, is the direct transference of the sensory pain signal to the motor neurons in the spinal cord that are associated with the muscle cells in the injured area. This reflex spinal relay stimulates the muscle tissues surrounding the injury to contract in order to support and protect the injured tissue. This is well and good, and it is what the pain/contraction reflex was designed for.

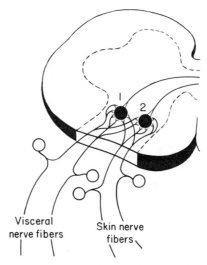

Visceral nerve fibers   Skin nerve fibers

10-10: The mechanism of "referred" pain. Pain from internal organs may seem to come from areas of skin or muscles, because the signals may be transferred in the internuncial circuits of the spinal cord. This referral mechanism can sometimes make the actual source of the pain very ambiguous. Indeed, referrals can create muscle spasms which themselves begin to generate pains in addition to the original visceral one. This can become a very vicious circle.

But, just as in the case of the "flight or fight" response, it is not difficult to press this contraction too far and sustain it too long, engaging reactions that were never meant to be part of the self-protective mechanism, and which in the end work far more to the detriment of the organism than did the original injury.

In the first place, if my reaction to the pain of the minor injury is extreme, the supporting muscle contraction may be too strong, squeezing the injured area and thereby causing more pain; or the contraction may become too widespread, involving much more muscle tissue than would be necessary to brace the actual injury, limiting far more movement than would otherwise be the case, and wasting much more metabolic energy.

Or a more complex situation may begin to develop, one which not only increases the pain coming from the original injury, but which also creates a new source of pain to add to it. Remember that a sustained contraction of muscles tissue reduces the blood flow through the capillaries in the constricted area; this restricted flow and the resulting lack of oxygen delivery to the tissue is called *ischemia,* and if it persists for even a short time the cells in the area become very painful, whether they suffered any of the original damage or not.

The higher the rate of local metabolism—that is, the more forceful and widespread the contraction—the more quickly the available oxygen is burned, the more quickly this pain appears, and the more intense it becomes. The actual cause of ischemic pain is not known for certain; it may stem from a build-up of lactic acid wastes in the constricted tissue (created by anaerobic glycolysis in the muscle tissue due to the lack of oxygen), or from a release of histamine or bradykinin from distressed muscle cells. In any case, it is evidently directly related to the lack of oxygen in the area, because it is immediately relieved by supplying oxygen to the affected cells.

So we now have to do with *two* sources of pain in the area—the original injury, which causes muscular contractions, and the secondary ischemia which is started up and accelerated by increased and maintained contractions. It is not hard to see how these two sources of pain can form a circular relationship, and a vicious one at that:

> Any local irritating factor or metabolic abnormality of a muscle…can elicit pain or other types of sensory impulses that are transmitted from the muscle to the spinal cord, thus causing reflex muscle contraction. The contraction in turn stimulates the same sensory receptors still more, which causes the spinal cord to increase the intensity of contraction still further. Thus a positive feedback mechanism occurs so that a small amount of initial irritation causes more and more contraction until a full-blown muscle cramp ensues.[63]

Needless to say, the erupting of this vicious circle can be either effectively suppressed or powerfully amplified by the higher brain, as its descending tracts either inhibit or facilitate the spinal sensory-motor reflex. If the original pain is situated in a critical area—deep in the back and close to the spine, say, where almost every movement or gesture will irritate it, or in the ribs or diaphragm where every breath will cause a stab of pain—bracing reactions and emotional anxiety may be aroused to such a degree that all motion is frozen, or breathing becomes labored and acutely painful. Or if an individual's tolerance for pain is low and their reactions exaggerated, a full-blown cramp, achiness, and restricted movement may result no matter how minor the injury or how innocuous its location.

In all sorts of injuries, big and small, these secondary reflex and metabolic reactions frequently complicate both the pain symptoms and the healing process. It is not at all uncommon for them to actually cause more discomfort and difficulty than the original injury, and to last long after its damage has been mended.

It is also very much worth noting with regard to this vicious circle that it need not originate with an actual injury at all. The discomfort of local ischemia alone is enough to begin a contractile reaction that can perpetuate and accelerate itself with no other prompting, engaging the full spectrum of oxygen depletion, metabolic complications, mounting pain, spreading spasm, and heightened anxiety. And ischemia can arise by means of virtually any of the internal circumstances which result in a sustained contraction in a specific area—a habitual posture or gesture, long spells of sitting, habitual emotional states that are expressed in chronic contractions, conditioned reactions that are often stimulated and repeated, and so on. All of these things can function exactly like an injury as far as the pain/spasm cycle is concerned.

And if the cycle is kept up long enough, then an even more serious situation may evolve: If a local area is deprived of adequate oxygenation for an extended period of time, impaired metabolism and waste removal can begin to build up considerable amounts of toxic chemicals, enough to poison and weaken all of the cells in the area. When this happens, their resistance to all sorts of pathological developments is lowered, and they become targets for any number of diseases and breakdown processes. They may give viruses a foothold, or spawn serious infections, or constitute the first step in the over-all weakening of the entire system.

Toxic wastes may even actually kill individual cells outright, and when this happens their spaces are filled in by scar tissue which accumulates to form fibrosis, connective tissue deposits that can permanently limit stretch and movement in local areas. These toxins can even build until they begin periodically dumping sizeable doses into the bloodstream, ultimately affecting our nervous system as well as local tissues. Recall the fact that lactic acid, one of the principle metabolic wastes of contracting muscle, causes not only acute pain but also acute anxiety when it is injected into healthy subjects. All of these catastrophes can follow in the train behind symptoms that may be absurdly small—small, that is, until we allow their self-perpetuating cycles to get out of our control.

Breaking the Cycle

For most people in our culture, the most frequently used means of breaking this pain/contraction cycle are drugs of various kinds—pain killers, tranquilizers, muscle relaxers, and the like. Some of them are narcotic and addictive, and many others have serious side-effects, some of which are not very well understood. But these drugs do perform one valuable service; they do interrupt either the physical pain component or the emotional suffering component of the vicious circle, and allow the reinforcing elements of the spasm to subside so that relief is possible. This relief, although drug-induced, may not be merely temporary; if the drug is successful in quelling the spasm, then improved circulation, decreased metabolic complications, and promoted healing may well be the result. The only danger then is to be sure that the side-effects of the cure do not conspire to generate vicious circles of their own.

Of course, heavy-handed drugs are not the only available means for diminishing pain and breaking the cycle. This is precisely one of the points at which body work can so effectively intervene into the vicious circle of physical and psychological suffering. We are all familiar with the fact that the pleasure of being comforted with soothing strokes rapidly diffuses the shock and fear in an injured child, often turning tears to brave smiles even while the scraped knee is still bleeding. The power of sensory pleasure to overcome anxiety and pain is astonishingly potent, and we use it instinctively upon crying children, barking dogs, skittish horses, and frightened birds.

This is partly for material physiological reasons. Touch sensation is carried to the brain on thicker, faster, more numerous pathways than is pain. Thus, when the area surrounding an injury (or even another area, or the rest of the body) is pleasantly rubbed, the brain is inundated with pleasurable touch stimuli which tend to drown out the acute pain signals and remove them from the foreground.

Recent research with electronic biofeedback devices, which aid the user in increasing the *alpha* brain waves, indicating deep, systemic relaxation, has even further underscored the possibilities of achieving a drugless disassociation of painful stimuli from the emotions and secondary reactions which serve to heighten them. Just as the laboratory rabbit can be trained to dilate and constrict its blood vessels, the human subject can be taught to consciously eliminate the mental and physical factors involved in the anxiety response.

> Increasing the number of *alpha* episodes has been used for the control of some kinds of chronic pain. This is successful in some cases, possibly because 1) the *alpha* training distracts attention from the pain to the feedback signal and inner feelings, 2) the patient believes that the method works, 3) the relaxation associated with *alpha* episodes decreases the patient's anxiety, or 4) *awareness of the possibility of control* over the pain changes its meaning and, therefore, the response to it.[64]

Now these are exactly the conditions which bodywork seeks to produce—the displacing of the patient's focus from his body as a source of pain to his body as a source of pleasure and comfort, the physical relaxation which diminishes emotional anxiety, and the restoration of the *possibility of control* over the situation. The magnitude of importance of even a slim measure of this last factor cannot be overstated. We will recall that it was a sense of control alone that prevented laboratory animals from ulcerating and dying, even when they received the same abuse as their helpless partners who died—died simply from the inability to establish any sense of effective influence over the source of their pain.

So the role of bodywork in learning to control these self-perpetuated cycles can be considerable, and it does not have dangerous pharmacological side-effects. If the cycle has begun in an injured area, soothing and pleasant touch can serve to isolate that specific area from the tissues around it that have become trapped in an extended ischemic pain reflex, thus leaving the patient with only the sensations of local tissue damage, and eliminating the often very-much-larger burden of discomfort generated by the perpetual motion of reflex reaction and rising anxiety. And, of course, if the primary source of pain is the ischemia itself, triggered by a chronic contraction, it is then possible to relax the surrounding spasm, increase the flow of blood and oxygen to the area, facilitate the drainage of lactic acid and other chemical irritants, and eliminate the source of the pain altogether.

In connection with these common kinds of self-inflicted pain cycles, which are bound to drag in the full range of pituitary/adrenal reactions and all of their long-term evils, the potential importance of such a simple and effective intervention would be difficult to exaggerate. Once such a cycle has gotten a start, it is utterly in vain to point out to the victim that there is no organic cause, that his or her symptom is "psychosomatic," and that the steady increase of discomfort and limitation is a neurotic response of higher brain functions. The pain is real, its increase is real, and the reflex it cannot help but trigger is real. What is needed is another real sensation, as overwhelmingly pleasant a sensation as possible, to divert the attention of the higher brain away from the anxiety-provoking discomfort. As the anxiety is lessened, the process of facilitation slows, contraction ceases to build, relaxation can set it, and oxygenation can then do its critical part in restoring comfort and mobility.

In this process, bodywork is far more an educational influence than it is a temporary treatment of a symptom. Our delicate balance of reactions is fraught with opportunities for us to hurt ourselves, and very often the most destructive self-inflictions stem from the very habits and attitudes that we are, for one reason or another, the most loathe to surrender. Bodywork can be used as a conscious means of providing not only relief from discomfort, but a new perspective upon the causes of that discomfort.

We clearly cannot rely upon any "instinct" to avoid pain. Our protective reflex withdrawal from a hot stove is unfortunately just not as reliably operative when we are dealing with psychological factors and self-inflicted damage. Too much is too often repeated to the detriment of too many lives for us to be able to believe too seriously that people have any instinctual tendency to eschew painful experience, even when they are conscious of it and have at hand the means to avoid it. And furthermore, the task of teaching someone to *accept* pleasure can be even more laborious than trying to teach them to *avoid* pain.

Our individual interpretations of "pain" and "pleasure," and our acceptance or avoidance of either, is clearly a function of many influences in our development; and the confusions possible in our behavior towards objects of pleasure and pain may require considerable untangling—an untangling that is only made more difficult and frustrating by proceeding on the convictions that any "healthy" person would "naturally" wish to behave so as to avoid pain, or that all that is needed to neutralize our self-destructiveness is a "chemical antidote." Touch, pleasurable sensation, positive sensory feedback, can be an invaluable key in this learning to accept the body as a source of pleasure and to avoid the internal circumstances which cause and magnify so much of our pain.

11

Further Views at the Turn of the Century

A New Physiology

It has been a little over a decade that *Job's Body* has been finding its way in the world, and I have been extremely gratified by its reception. I have also been thrilled by the number of dialogues with individuals and seminars with groups that the book has stimulated for me.

What appears to be emerging from the conversations, correspondence, lectures, and workshops that the book has occasioned for me has been a very pleasant surprise. I had thought that what I would be talking about would be the ways in which a working knowledge of physiology could inform the various approaches to bodywork and raise their levels of effectiveness. What I find happening instead is a much more expansive and open-ended exploration of a whole constellation of possibilities for human health and development suggested by alternative therapies of all kinds.

I encounter a collective awareness of many of the themes I had tried to articulate in *Job's Body*, and an intense interest in a very new sort of physiology. At the heart of this idea of a "new physiology" is the inclusion of a number of assumptions that are often systematically excluded from the theories that shape clinical studies: 1) that the mind tangibly exists, and wherever it is manifested it is as much a player in the world of objects and events as is any element in the Periodic Table; 2) that discriminating intelligence is present in all forms of life; 3) that dominant patterns of thoughts and feeling states impact directly and dramatically the physical development and functioning of my body; 4) that shifts in my physiological functions directly and dramatically impact the nature of my thoughts and feeling states; 5) that there are specific things that can be done, both from the mental side of things and from the physical side, that can tremendously improve the interactive functions and resilience of the bodymind; and 6) that improvements in these interactions not only improve our personal health and development, but also greatly increase the scope and deepen the texture of the world we are able to perceive.

Bodywork and Science

So my original assumption as I began *Job's Body* was that existing clinical research would surely shed light upon the concrete reasons why bodywork is so effective for so many conditions. More recently I have begun to experience something of a sea-change in this operating assumption. I now believe that experienced and effective bodyworkers have something very interesting to offer to recent developments in the biological, physiological, psychological, sociological, and medical sciences.

The first reason for this new edition of the book is that it gives me an opportunity to air some additional observations and growing convictions of my own with regard to this new way of understanding human development. One of the primary issues concerning therapy that has become clearer to me is that popular notions of science are always lagging considerably behind the latest discoveries and insights—insights which frequently up-end the concepts previously regarded to be true. By continuing to place our faith and base our actions on models that are outmoded, we often prevent ourselves from pursuing alternatives that would make our lives better. One does not tend to try things that are believed to be impossible, and previous generations of biologists have certainly placed much of the impact of the mind upon the body outside the realm of possibility.

These outdated popular conceptions create what I call "double binds in the bodymind"—misplaced ideas and efforts that our current beliefs compel us to pursue, but which cannot improve our situation because they are not aligned with the actual needs and processes of our organisms. Among these double binds that subvert our accurate self-assessment and successful self-regulation are a number of the most widely held ideas about the nature of science and the nature of human beings.

The second reason for this new edition is the inclusion of some exciting new research and speculations from two biologists whose work is new to me, and whose ideas and findings significantly expand the possible scope of bodywork and our growing sense of a new understanding of physiology. These are scientists that are not merely challenging old beliefs. They already have left them behind and are actively thinking in some altogether new terms. Their work does not merely provide bits and pieces of data that can be taken as provisional evidence of a body/mind connection. They are whole-heartedly, and with success, pursuing the details of that connection. They are managing to combine rigorous research with a sense of intuition, to entertain clarity of thought along with an immediate awareness of their own subjective experiences. And they are asking questions that were defined as out of bounds and ignored for too long.

James Oschman is a cell biologist connected with the Marine Biological Laboratory in Woods Hole, Massachusetts. After experiencing what was for him the profound effects of a Rolfing series, he shifted his research focus to the study of connective tissue and its multi-dimensional roles in the human organism. His biological background and direct bodywork experience led him to some very novel ideas about what connective tissue is up to. And what he has to add to our appreciation of the complex nature of fascia and what it contributes to our health and healing goes far beyond the biochemical and structural properties described by Ida Rolf and others cited in the first edition.

Candace Pert initiated the discovery of the neuropeptides, which I believe is an extremely significant scientific breakthrough in our conception of the bodymind. These information-carrying molecules circulate freely about the body, and their action at their receptor sites appears to be one of the principle connections not only between our various organs and biological processes, but also between our mental and physical dimensions. Not only does an understanding of neuropeptides lead us to the biochemical and physiological basis of our emotional lives, it also reveals specific pathways and mechanisms through which our feeling states continually impact our physiology. Far beyond a general notion of the mind affecting the body, Pert has demonstrated many concrete ways in which our emotions, thoughts, glandular secretions, immune responses, and behavioral patterns all utilize neuropeptide messages to orchestrate their interactions at every level. Our appreciation of these molecules has enormous therapeutic implications. And like Oschman, Pert has personally experienced the ways in which bodywork and other alternative approaches can effectively use these interactions to change our physical and emotional lives for the better.

Taken together, the fresh perspectives and discoveries of these two scientists seem to me to indicate a definite direction towards a new understanding of physiology, one in which the power of the mind is as present and concretely accounted for as that of the liver, or insulin, or the gene. And quite beyond that, I believe that their departure from the old paradigm offers some fascinating perspectives on the ways in which we envision both the varieties of human bondage and those precious bits of wiggle-room called freedom.

Sweet Science

"Bring out number, weight and measure in a year of dearth," declared William Blake in protest against the too-narrow constraints of a strict, materialistic reductionism. He spoke for what he called "sweet science"—an inquiry into nature that includes an inquiry into the nature of the mind doing the inquiring, and a description of reality that does not eliminate the significance of our rich tapestry of consciousness and of our living, evolving relationship with all that we perceive.

Perhaps the day is actually approaching when this kind of inclusiveness, this sense of mystery and of compassion coupled with rigorous method, will be added to the fundamental principles of what we mean by science. It is once again becoming possible to talk about "mind" and "consciousness" in prestigious journals. Oliver Sacks talks about describing a "neurology of the soul." The psychologist and philosopher (note the very combination!) James Hillman writes about once again "ensouling" our cosmos. The neurologist Antonio Damasio theorizes about our "emotional I.Q."

Along with Oschman, Pert, and many others, these are refreshing new voices, coming at a time when a general recognition is beginning to emerge that the disciplines of biology have for too long used exclusively the reductionist methodology adapted from physics. The major outlines of that method are: Theory and research are limited to what can be quantified and measured; experimental design isolates and manipulates a single variable to discover its specific properties; the operations of a system is then inferred as the aggregate of the properties of the variables within it; eventually we will isolate and describe all of the variables, and hence we well be able to understand all systems.

This approach works best when we use it to address clearly circumscribed and relatively simple systems with a few prominent variables, where the effects of experimental manipulation are easier to isolate and observe. But living organisms are the most complex systems in the universe that we know about, teeming with innumerable varieties of variables. The effect of any single manipulation will always be ambiguous depending upon shifting contexts within the system as a whole. Many properties emerge from such a complex system that could never be directly inferred from the properties of any given component. And attempts to rigorously isolate variables and resolve the ambiguities usually require stopping the life process in order to examine its separate functional parts. This is precisely what William Wordsworth was referring to when he observed that "We murder to dissect."

In contrast, the observations and sensibilities of the effective bodyworker plunge directly into this complex, evolving, self-regulating chaos of the living organism, manipulating and noting responses not like a physicist studying sub-atomic particles, but rather more like a meteorologist studying large weather systems. Weather systems are in fact made up of aggregates of sub-atomic particles, but there is simply no way to infer the total activity of a weather system or predict its changes from even the most exhaustive understanding of sub-atomic particles.

These observations are in no way intended to denigrate the methodology of physics, or to question the laws of nature it has established. Nor is it a point of view that is rejected by biologists themselves. Ernst Mayer and Gregory Bateson both championed this need for a theoretical and methodological shift in biological studies decades ago. The point here is simply to recognize which kinds of theoretical questions and research methods can best address which kinds of inquiries, and what kind of ideas genuinely empower us to find out more about ourselves.

Here we need to be very clear that no method guarantees freedom from ignorance, bias, and error. There are not only differing opinions and heated debates among respected scientists themselves; there are also different kinds of endeavors within modern science, each with its own goals and means. And each one has a very different impact upon popular conceptions and beliefs.

First, there is "discovery science"—the kinds of exploratory ideas and experiments that result in genuinely new and revolutionary information and insight. Scientists devoted to this sort of research typically have a passion for challenging conventional wisdom and are continually tinkering with new approaches to tease out new kinds of facts and new points of view. Iconoclasm and controversy reach their heights at this cutting edge, and there is continual abrasion between accepted ideas and new questions and answers.

Second, there is a related but distinct arena of "demonstration science." Once a new theory is developed which leads to new discoveries, there is always a great deal of follow-up research to do to confirm the salient points of the theory and to flesh out all the details and possible connections. This is the bulk of scientific work, the careful cross-checking, verifying, and refining of new information. It is the all-important crucible where enduring principles are winnowed from misconceptions, and it requires a high degree of diligence and integrity.

This second kind of science tends to attract and reinforce a very different mind-set than that of the frontiersmanship and iconoclasm often involved in original discovery. The basic task of demonstration science is to further develop and articulate a new theory, not dismantle an established one. A big part of the integrity required in this effort is a patient intellectual loyalty to the new theory. Its ultimate demonstration will only be borne out as many details are attended to and many loose ends tidied up.

Both original discovery and demonstration rely on the virtue of stubborn persistence to achieve their ends. However, here we come to a threshold whose boundary may be blurry but whose dangers are quite distinct: When this intellectual loyalty and persistence are taken too far—when they are combined with willful blindness, personal ambition, greed, fear, bias, conscious ideology, unconscious beliefs—their service in the name of truth can go seriously awry. What happens then is that the "interested parties" throw the full weight of their resources into defending propositions that have outworn their usefulness and into maintaining points of view that are simply mistaken. What begins as necessary persistence can turn into a formulaic rationalism coupled with an overweening need to control the terms of the debate.

This turn of events, and the rigid structure of truths, half-truths, and errors that it promulgates, I call scientism, and this third kind of science is a very different critter from the first two. Science is open-minded inquiry and open-to-review conclusions. Scientism, while apparently going through the pro forma of rigorous method, is really more of a dogma, a turf.

In sorting out the differences between science and scientism, it is well to remember something that Gregory Bateson was fond of observing: Science never proves anything. It is nothing more (or less) than a procedure for asking questions, proposing answers, and arriving at as reasonable a consensus as possible based upon available techniques and the data presently at hand. There is no guarantee, and there never can be, that the next discovery will not shatter that consensus and transform our view of things. As a matter of fact, this happens regularly.

Therefore, when we are dealing with something as complicated as the human organism and with something as slippery as human consciousness, and we are flatly told that certain theories are categorically unassailable and that certain kinds of speculation are utterly inadmissible, we can be sure we are encountering scientism. One of its unequivocal hallmarks is the off-hand dismissal of "anomalies"—observations that do not fit the dogma. The dismissal of anomalies is profoundly unscientific. Breakthroughs typically occur precisely when someone sets about explaining an anomaly, and questioning current theories that are unable to account for it.

There is also a fourth kind of science and scientific thinking, one in which truths, half-truths, and misconceptions can be even more slippery and ambiguous than debate or dogma—the public domain of popularly believed facts and ideas. Cutting edge discoveries are almost never included in this arena, since the long demonstration phase has not yet cleared them for general publication. And in addition to the collection of outmoded notions, popular opinion is often heavily influenced by dogma and scientism, since the general public is not usually in a position to be able to judge the relevant information, and is easily swayed by apparent authority.

So public opinion about scientific matters is informed by a hodge-podge of ideas, some correct, some outdated, and some just plain wrong. And having been taught that any sort of science is better than ignorance, most people will try to apply any and all these notions to the problems that confront them. Or, disillusioned by the muddle, many will throw it all aside and pursue one or another mode of "magical" thinking, hoping to transcend the confusion by doing an end-run around natural law.

It is with this popular muddle and the dogma of scientism that so opportunistically feeds it, that I have my real quarrel. It is here that the double binds are bred that prevent us from seeking better solutions for ourselves, where old notions and misconceptions get crystallized into enduring beliefs that continue to undermine us, even when the necessary alternatives are at hand. This muddle is a bewildering combination of enough sound ideas to command attention, enough authority to command respect, enough success to powerfully reinforce public expectation, and enough denial to effectively paper over the failures. It is utterly different from good science, and worse in some respects than no science at all, since it trades on the promise but cannot deliver.

Yet at the same time I also want to be clear about some existential limitations with even the best of science. The knowledge of enduring principles and laws takes a long time, more than any one of us has got. This cannot be helped, because, while we are patiently waiting for discovery and demonstration to do their work, our lives are rushing by and our conditions are deteriorating. The danger here is that we can become seduced into deferring action for ourselves until we complete our scientific knowledge; but often by that time it is too late for us to effectively use it. Besides rigorous scientific method and patience, we also need some interim ways of discovering practical means that can help us now before it is too late, whether or not we understand how and why they are working. And we need to recognize that there are ways of doing this that are honest, that can discern between a barefoot cure and an old wives' tale, and that do not degenerate into magicalism.

For instance, many biologists are currently convinced that the eventual deciphering of our genetic code—the mapping out of the entire human genome—will lead us to a complete understanding of how each individual has developed as they have, and to the knowledge of how to manipulate our genes to correct any developmental problems. No doubt this project will advance our knowledge and result in many useful therapeutic applications. But this will take a long time. And the real point here is that if I am in a holding pattern awaiting the outcome of genetic technology, I am less likely to search for practical actions that can address my ills now. And, in the long run even more pernicious, while I am in that holding pattern I am not looking for anything besides genes that have made me the person I am.

I fall into the danger of coming to believe that my fate is expressed entirely by my genes, and I then will miss the opportunities to discover the legion of elements that conspire with my genes to create my body and my life, elements that in fact may be much more easily addressed than my DNA code. And—here is the crux of it—I fail to even seriously entertain many of the questions that are most pressing in my actual daily life.

What and Where is the Body's Blueprint?

For instance, to me one of the most interesting—and pressing—puzzles about human development has to do with the way that the body successfully replaces and deploys new cells and molecules as it undergoes continual change, while at the same time it preserves its singular integrity.

It was not so long ago that physiologists believed that the adult body was quite stable in its structure and substance, keeping mostly the same cells and molecules while deriving metabolic energy from diet. We now know that this is far from the truth. The molecular components of our nutrition are taken up rapidly by all of the body's tissues, which use them to continually replace themselves. Different tissues turn over their component parts at different rates—intestinal lining every few days, skin cells every 28 days, and so on. And even stable tissues that do not produce new cells, like nerves and muscles, replace worn-out molecules within the continuing cells at various rates. The net result of these turn-overs of materials is that about every seven years you have completely replaced every molecule that makes up your entire body.

Now here is the question, or rather questions: Where, and in what form, is the overall pattern stored that directs the fabulous number of details involved in this ongoing process of elimination and replacement? And how is the pattern changed gradually to reflect my advancing years and accumulating experience, while at the same time maintaining an overall continuity in its myriad levels of operation?

Clearly the pattern has the sort of constancy I would infer from a genetic source—I remain recognizably Deane after every seven year cycle. However, in every cycle much has also changed: Sizes and shapes changed as I grew; and general distributions and local qualities of fat, muscle mass, and connective tissue changed—sometimes rapidly, sometimes slowly—responding to changes in diet and exercise; injuries and illnesses interrupted and ultimately rearranged many of my original patterns; dominant emotions, frequently repeated activities, focused attention or the lack of it, beliefs about what is possible for me and what is not—all these factors also materially alter my patterns of erosions and redeposits.

What template then, or what combination of templates, directs this astonishingly detailed and continually evolving mass of molecules that manages to remain "me" and yet faithfully reflects all developmental changes "I" have been through during the entirety of each cycle? And an even more penetrating question, one much more acutely relevant to health care, is: Why is it and how is it that some individuals get *better*—stronger, more robust, clearer-minded, more skilled, emotionally more developed, with healthier tissues, and so on—during the course of each cycle, while other individuals get manifestly worse and worse? What are the events, and where are the storage banks that accumulate their somatic memories and transform them into personal tendencies that manifest themselves in the processes of "learning to be sick" and "learning to be well" described in Chapter 10?

And perhaps the most important question of all in the context of bodywork is: By what means might I be able to recognize my patterns, the developing tendencies in my templates, and to consciously alter my beliefs and behaviors in ways that lead each cycle of growth and replacement towards material improvements, rather than towards degenerations that spin more and more out of my control?

Pondering these questions seems to me to be fundamental to our understanding of our development as organisms and as personalities. I see their answers as being at the heart of effective bodywork, and I cannot imagine them being excluded from the theory and practice of any variety of health care. Speculation about these questions and their answers will be a central theme of this addition to *Job's Body*.

Hangovers and Hang-ups

The entire field of bodywork is making its muscles felt, as it were, and the increasing interest in its effects appears to destine bodywork to wider acceptance, broader applications, and much new research. The shift from the term "alternative therapies" to "complementary therapies" in much of the current literature is itself indicative of reasons for optimism. More and more people are seeking out these complementaries, and paying for them out of their own pockets—definite testimony to their belief that they are receiving concrete benefits. More and more health providers are willing to refer patients to bodyworkers when it is appropriate, and more insurance companies are willing to pay the bills.

However, there continue to be inherited general attitudes, beliefs, and muddles about the body and about the mind, and divisive opinions on these matters within the fields of biology and health care themselves. These create serious impediments to expansion. In fact, it is very likely that the more momentum in the marketplace that bodywork displays, the more resistant the force of these habits of thought and these confusions will become—resistant to changing roles, resistant to new ideas that overturn old ones, resistant to sharing prestige and economic rewards.

The following discussion of some of these muddles and of their ramifications is by no means complete; everyone will have their own pet frustrations within this context of expansion and resistance. They are simply what appears to me at this time to be some of the major opinions, beliefs, and tendencies that prevent many individuals in the sciences and in the public at large from seeing and accepting some of the ways in which things could change for the better if they could entertain some new notions and new practices. As Will Rogers observed, "It's not what you don't know that will get you—it's the things you do know that just ain't so."

Double Binds in the Bodymind

The concept of the "double bind" was postulated by Gregory Bateson as one of the key elements in his rethinking of the causes of schizophrenia. One of his classic examples of the double bind is that of the abused child: That child is in many ways defenseless and utterly dependent upon its caretaker for nearly everything that ensures its survival. In addition to this non-negotiable dependence, the child also has a natural and powerful compulsion to *love* that caretaker, looking to it for a wide range of emotional as well as physical needs. But the child repeatedly experiences abuse and neglect at the hands of its caretaker—abuse and neglect that in no way diminish those physical needs and the child's compulsion to love its caretaker.

This situation imbues the abused child's early development with unresolvable paradox and emotional conflict. There is no other well available from which the child can drink; but every time it drinks from this one, it is poisoned further.

There seems to me to be an analogy between this painful *cul-de-sac* and the situation we have found ourselves in with regard to information about our own developing bodies, and to the obedient respect we have learned to extend to authoritative opinions. Many conventional beliefs about our bodies and their relationships to our minds have little or nothing to do with the actual functional needs of the organism. In the absence of better information, we are forced repeatedly to rely upon these errors in thinking in order to try to serve our bodies' needs, always hoping for the best. And every time we try to solve a problem with the wrong tool and the wrong instructions, things get worse. After all, one definition of schizophrenia is the repetition of the same behavior with the hopeful expectation of different results.

Now we certainly have every bit as much of an existential need for a healthy body and mind as we had for nurturing parents. In this case it is ourselves that we are compelled to love, and these same misinformed and misguided selves continually betray us. In addition, we are predisposed to respect the leading authorities, but their theories are far afield from our immediate needs. Thus my reading of the downfall and dilemma of Job: Merely following the rules as we have received them will only work for a limited time, because there are things crucial for us to know that are not covered by the received rules, or are even flatly denied by the authorities of the day, Job's "miserable comforters." To negotiate his way out of his dilemma, Job must turn away from conventional wisdom and look inward; he must read his own Book of the Body, the only text that can ever be faithful to his authentic needs, and to the ultimate source of his ways and means. Only in this way can he sort out these needs and means from all the "things he knows that just ain't so."

Virtually Living in Virtual Reality

Marcel Proust observed that "having a body constitutes the principle danger that threatens the mind." And of course the corollary to this alarming observation is that having a mind constitutes the principle danger that threatens the body. When the intentions and behaviors of these two domains are not in synchrony with one another's needs, a double bind is created and a great deal of havoc is sure to follow.

Our endlessly inventive imagination is the glory of human existence. More than anything else about us, it was the explosion of the frontal cortex and the resulting expansion and sophistication of every mental faculty that made us a species apart. The capacities that this imagination opened up to us may or may not in the long run prove to have been a wise experiment, but for now we are in their midst and must make the best use of them we possibly can. They are after all the very capacities that have created the arts, the sciences, community, civil law, and all of the rest of the best of human civilizations.

But there is also a real danger peculiar to humans lurking under all of this imagination and invention. Our imaginations are so vivid, and their powers of manipulation, synthesis, and creation so potent that they can construct and respond to whole worlds of events and meanings that had no existence prior to our hatching them. This is precisely why they are useful, and so enchanting, so intoxicating. And it is also the reason why they can get us into so much trouble.

That is to say, I can use these powers in very different ways—either to understand and manage my internal and external environments so as to more fully support my needs in more and more different situations; or to build up a more and more complete *virtual reality* in my mind that has less and less to do with the authentic needs of the organism that is producing and entertaining it.

This detaching from the actual substance of our lives and letting our balloon drift into an increasingly imaginative and disembodied atmosphere operates on a variety of levels and scales. Perhaps one of the simplest is that of social conventions—the avoidance of behaviors that have come to be viewed as "impolite," and the adherence to those conventions we have come to associate with "common courtesy." For instance, have you noticed that almost every reaction that the body employs to clear itself out has become impolite, embarrassing, or downright revolting?

Yawns, sneezes, belches, farts, and sweating are all simple mechanisms that the body uses in order to release toxins, accumulated energy, and tension; and yet we have developed a sophisticated squeamishness, if not a horror, of being guilty of them in "polite society." Urination and defecation could not possibly be more universal and natural, but who would be "caught dead" doing them? The truth is that they are normal healthy reflexes, and that we run a good chance of being caught dead significantly sooner for having *not* done them in a sufficiently uninhibited fashion.

And "naturally," one of the most significant releases of all, the orgasm, has in our culture been all but hopelessly mired in the double binds of courtship and avoidance, embrace and denial, surrender and repression. And with regard to the negative impacts of all these withheld releases, it is well to remember that nitroglycerin is not particularly dangerous until enough of it has been accumulated in a stoppered jar.

These scatological kinds of social rules are only the beginning of the conventional repression of normal functions and the adoption of standards of behavior that ill serve our organisms. Our schooling system could scarcely be more efficiently designed to counter bodily needs and frustrate the ways our minds are built to learn. In the interests of standardization and control, we impose on children conditions that undermine their best impulses in the learning process.

Children need to vent the large voltage of physical energy their young bodies generate; so let us put them in rows of confining desks and demand that they not fidget. (Here we might recall that all of the important questions of Western Civilization were originally asked and debated way back in the Greek *gymnasiums*, where academics was not understood as something separate from active and communal physical development.) Association, categorization, and imitation are the active methods of learning their brains are best equipped to perform; so let us make rote memorization the primary method they are allowed to use. Since *puzzling* is one of the ways in which great minds have always happened upon great ideas, let us create penalties for confusion and rewards for the quickest and briefest regurgitation of the facts we have already decided are relevant. And, for good biological measure, since adequate intake and output of water is essential to the healthy function of both their bodies and their minds, let us control their potentially disruptive behavior by restricting the times they may leave the room for those purposes.

According to linguistic studies, approximately 20% of our verbal communication is based upon the lexical meaning of the words and the syntax of the sentences. The other 80% of the information received in any verbal communication comes through tone of voice, tempo and rhythm of delivery, body language, *double entendres* of gesture versus phrase, and so on; so let us teach language as though it existed primarily on the page, and that it always means exactly what it says.

And above all, as with toilet training, belching, sweating, and the rest of all that, let us use a *shame-based* system of rewards that dries up unwanted responses and diversions as quickly as possible, and imposes a discipline onto the proceedings that will help our children unlearn what it is they came into the world with that we don't want them to know about, and will let them settle down to those pursuits and ideas that we have determined are least troublesome to our adult ways of doing things. In this way we will have honestly done our best to prepare them for survival in a society that has been designed to more or less continue the same program.

All of this is precisely how knowledge becomes abstract, learning becomes disembodied, and the world they describe becomes virtual and imaginary, not actual and organic. And the only additional thing that is needed to make this abstract world of virtual reality hermetic and unassailable is to ensure that the opportunities for individuals to *touch* one another are reduced to a minimum and kept there. Actual contact, after all, will always threaten to disrupt this strict socialization process with messy biological imperatives, and will remind individuals of what it is they may be missing. If for pedagogical, theological, moral, scientific, or political reasons you want to prevent children from going to somebody's version of hell-in-a-hand-basket, touching must definitely be judicious, sparse, and only for ethically specified reasons. Once again, shame is extremely useful here.

But, of course, the deprivation of touch is in the long run one of the most physiologically and psychologically damaging things we can visit upon a human being. Touch is taboo only to the already disembodied.

In the ways that we train our youth and in which our society is developing, these disembodied and abstract virtual realities are becoming increasingly the norm, and increasingly a self-enforcing norm. We are rushing to get schools "on-line" before anybody has their feet on the ground. The digital network is displacing whole genres of writing and face-to-face discourse. Reliance upon calculators is atrophying the mathematical muscles of our minds. Spell- and grammar-checks are atrophying the verbal ones. Campaign ads—thirty second ones—are replacing political debate.

The technical revolution, in a word, is replacing the once-ubiquitous need for human contact, and is being avidly downloaded—*not* embraced—by generations of human beings increasingly trained to be comfortable and competent only in its virtual mindspace, where reality must be reduced to an image and a sound-byte, and can only be conjured for us by a keystroke. And that mindspace is becoming large enough and complex enough within itself to easily occupy their imaginations for a lifetime with a minimum of distractions from an increasingly distant and irrelevant body.

The collective momentum of this virtual reality, as entertaining and profitable as it is, becomes extremely dangerous when it begins to displace the concrete knowledge about ourselves that can come only from actual contact with actual bodies, and to structure in a disembodied fashion our beliefs about what kind of knowledge about ourselves is useful or even possible. For example, we are told by kinesiologists that skull sutures and sacroiliacs are "immovable joints" (an anatomical oxymoron if there ever was one) because most subjects examined in our culture have no perceptible movement there. Or, that the activities of the autonomic nervous system and the organs it animates are beyond our conscious control because the authorities—unlike meditators—have simply not investigated how we might learn to control them. Feelings are not material things, and therefore cannot have any material impact upon physiology. The brain is a computer whose operations are organized around the principles of digital encoding and mathematical logic. The body behaves robotically in response to the machinations of the brain. And thought processing, to the extent that it can be meaningfully said to exist at all, is exclusively carried out in the brain.

Now the "reality" of these propositions, and the many others that spin out of them, exists solely in the fabric of scientific literature and the shared beliefs of its authors. Like geometrical points, straight lines and perfect circles, these notions are nowhere to be found in actual nature. They would be mere curiosities, like the belief in a flat earth, if we had not been conditioned to accept those beliefs and either act upon them ourselves or to acquiesce to being treated according to their virtual principles.

Hence the debilitating nature of this double bind: When I experience trauma or develop difficulties in my body, my belief in the experts leads me to choose courses of action that very often make things worse. And until we actually confront our actual bodies, and like Job "take our flesh in our own hands," these beliefs are where we will always be forced to look for aid; and each time they will only debilitate us further.

Nature "versus" Nurture

One of the most enduring debates among developmental theories that have spun out of this abstract kind of world view has been the question of "Nature versus Nurture." In its strict "versus" form it has also been one of the most absurd.

The naturists have it that I am the evolutionarily predetermined unfolding of my genetic inheritance. All aspects of myself, not only physical characteristics but personality traits and behavioral tendencies as well, are contained in the details of my DNA records. "I" am simply their inevitable biochemical and structural expression. The extreme, but by no means rare, reduction of this point of view is that bodies are primarily vehicles which "selfish genes" have evolved in order to replicate themselves. The gene is overwhelmingly the main thing that dictates both species and individual development. That is to say that the patterns, the templates that produce my body and that regulate its renewal every seven years are irreducibly molecular ones, coiled in my cell nuclei, inflexible in their operations, and requiring nothing for these operations beyond the imperatives inherent in their chemical properties.

The nurturists, on the other hand, contend that at birth I am essentially a *tabula rasa*, a blank slate being written upon by the specific realities that impinge directly upon me. The contents of my consciousness are no more nor less than the sum total of my sensory experiences. Thoughts and feelings are particular aggregates and associations of those recorded experiences. All truths must be taught, all successful skills and behaviors must be trained into existence, and all counter tendencies must be actively pruned away. The patterns that matter most come from outside myself, and reality molds my genetic potential like clay in the hands of a potter.

In this polarized form of nature versus nurture, there is something very peculiar about the terms of this debate. *Both* sides have amassed a great deal of evolutionary, physiological, and psychological evidence that supports their view exclusively; and yet each marshals counter arguments and empirical observations that effectively dismantle the major premises of the other. Both sides clearly have some of the pieces and both are just as clearly missing others, and this is why the debate continues inconclusively. What has made it so perpetual—and so absurd—is this polarized tendency to frame it in a context of "either/or." Someone must be right, and someone must be wrong. And so the evidence and rebuttals begin another cycle of debate.

An obvious alternative to this "either/or" deadlock is to shift the debate to a "both/and" plane. If both sides have evidence in their favor, then wouldn't the intelligent question be, "How do these two apparent opposites interact? What is the developmental *dialectic* between my inheritance and my experience?" Indeed, this attempt to combine these views and to sort out the real influences and limitations of both factors has led to a much more reasonable discussion and many more insights into the dynamic interface between deterministic genes and largely unpredictable life events. "Which does what, exactly?" is a more productive approach than "Which prevails?"

But to my mind, even this "both/and" is not at all a satisfactory solution. It seems to me that even after everything that nature and everything that nurture can do in the course of my development has been explored and defined, much is still missing to explain me to myself. What about genetic potentials that were never developed? What about devastating experiences that I have successfully overcome, even ended up using to my advantage? What about the parts of me that are neither clearly inherited nor obviously imposed on me by the outside world? What about my occasional novel solutions to problems that have had no precedent? What about creativity, spontaneity?

The real flaw in this nature/nurture debate is that *both* sides leave out the same crucial piece, without which even theories about their combination do not make sense for long. Notice that from both points of departure the individual life in question is *passive* in the process of its development. Either I am simply the product of deterministic genes, or I am the product of equally deterministic experiences and responses, or else I am some combination of these determinisms. Nowhere is my own active engagement and participation factored into the elements that have made me what I am.

The fact of the matter is that not only the quality and amount of stimulation I receive, but also the activities that I personally initiate have a great deal to do with the developing structure and functioning of my body, including my nervous system; my participation with the world determines a large number of physical details about myself.

For example, I was born with many more neurons than I will actually need; I have many redundant processing loops and interconnections. During my early years of development many of these redundant and unnecessary neurons die off. It is not a genetic code that dictates the patterns of this die-off, but my most consistent experiences and activities. In addition, those channels of communication that are used most frequently increase dendrites and synaptic connections, increase neurotransmitter quantities, and facilitate a favored "line of least resistance" for the flow of information and action through the circuits that remain.

Now in this process the naturists have their incontestable point: Our genetic codes have a lot to do with the fact that in the fetal stage all of our brains are formed very much alike. And the nurturists can point out that the amount and quality of stimulation in early years adds a great deal of individual specificity to this original genetic plan. But what both miss is that my own personal activities, and especially my voluntary ones—my *will*, my *conscious active engagement* with my environment—are absolutely critical in determining which pathways will be physically and functionally reinforced and which ones will fall away permanently because they are unneeded at that time. Furthermore, this selection process—while being most active during my early years of learning—continues throughout life, constantly honing the system to do more and more efficiently those things it does most frequently.

Of course, this structural and functional efficiency also means that activities which are *not* pursued become increasingly difficult to learn, because the entire system is developing an increasing tendency to favor responses that have been and are continuing to be reinforced. This is one of the main reasons why it is hard to teach an old dog new tricks, especially if it hasn't learned any lately.

The real issue here is that I am continually *finishing myself*, and that I remain always a work in progress. What I do (or don't do) is at each stage setting me up for what I can do (or can't do) later. The habits I actively ingrain most deeply become actual physical artifacts of my wiring, and are no longer merely conscious preferences.

This missing link of conscious, active engagement is of enormous importance when it comes to strategies for therapeutic intervention. Some of the habits I am unconsciously perpetuating are reinforcing—sometimes even creating—my current difficulties. I cannot simply erase these habits; they have become a part of my physical structure (and not a part that can be addressed surgically or pharmaceutically). Often what is necessary is for me to learn new habits that will reinforce different responses that will over time generate *new* pathways of least resistance, which will elicit *new* behaviors that have this time been consciously cultivated to counter the negative effects of my old ways.

These abilities to recognize self-destructive habits and to deliberately design and cultivate new ones, become even more important when we remember that these faculties do not emerge until a lot of programming has already happened to me in my first few years of growth, after my "finishing" process is well under way, and many tendencies have been set in motion. None of this early experience and behavior will ever be "unlearned." What is necessary is that I develop some means for learning patterns anew, and so correct old negative reinforcements and tendencies.

A successful life is the continual sorting out of: 1) inborn traits, genetic characteristics, 2) the events and situations I have experienced (some of the most important of which occur very early in life), 3) the personal beliefs and behaviors that have resulted from the interactions between my genetic capacities and my experiences, and 4) the exertion of my will to directly engage in actions that can generate positive adjustments in blueprints, habits, beliefs, and behaviors. "I" am not any one of these things—nature, nurture, beliefs, behaviors, will. What is truly "I" is the internal Witness, the chairman of the internal committee, the presiding point of view in my consciousness, who observes and negotiates among all these factors to produce the patterns of response and development that are continually becoming the pattern and the substance of my life.

None of the insights needed to strengthen the presence of this Witness and to inform its decisions for action come passively. They must be *sought*, and this seeking will always require active observation and engagement. Between the desire and the thing the world lies waiting; that world's secrets, and its helpful participation in the satisfaction of my desires, do not come to me until I *act*.

And in this acting and negotiating, we must also remember that our best current working theory of things is always provisional, that our deepest beliefs must always be prepared for a change, and that behavior must always be ready to adapt and to compensate. The *meanings* of the world we encounter must continually be *made* by each of us, and this absolutely requires my active, voluntary participation in both my own being and that world. This is my only reliable means of developing 1) accurate self-assessment, and 2) successful self-regulation.

The business of therapy, as I understand it, is the awakening of this Witness, and the teaching of some tools for this active process of observation, engagement, and adaptation. These—and not the therapist nor the therapy per se—are what do the healing.

Hence the debilitating double bind of this "nature versus nurture" framework: It provides me with only a passive role in my unfolding, and suggests no way in which I can personally act to make things better for myself. At best, considerations of "nurture" prompt arguments about the need for more humane social engineering, the improvement of the nurturing environment. This in itself is a good thing, but it is simply no substitute for the awakening of personal responsibility for our own condition, the active engagement of the will, and a working knowledge of how to successfully interact with the events that are affecting us. Improving in the social environment is not the same thing as empowering the individual to cope with whatever environment is available.

Both successful genetic manipulation and the legislation of a more utopian society are likely to take a long time in arriving. And in the meantime, if all we have to look to is a model that implies our own impotence and inaction, then we are simply stuck with whatever DNA or the external world hands us. And even the beneficence of genetic perfection or Utopia will wear thin and ultimately fail us if this crucial element of personal, active engagement with ourselves and with our surroundings remains dormant. Since nothing stays the same, no state of perfection, or even acceptable compromise, can last. We will never outgrow the necessity for the alert Witness, who observes the changes within us and around us, and who continually seeks the balance that is genuinely in our best interests.

Subjectivity "versus" Objectivity

Two of the most vital functions of this personal, internal Witness *are self-assessment* and *self-regulation*. Our conscious cultivation of these functions has been severely hampered for us by a third double bind that underlies the first two, that generates and reinforces them, and which therefore strikes even more deeply at our ability to feel ourselves distinctly and to think about ourselves clearly. It is a bind that threatens to eliminate much of what is immediately and authentically human from our picture of reality.

"Objectivity" is one of the key watchwords of the scientific method, and when as a value it is opposed to "prejudice" there is every justification for adhering as strictly as one possibly can to its tenets, just as truth must always oppose falsehood. The method focuses upon data that can be concretely observed, recorded, weighed, measured, cross-checked and verified, and that are repeatable everywhere and always under the same experimental conditions and procedures. These operations separate fantasy from fact, hunch from demonstration, mistaken beliefs from actual outcomes.

This fastidious consensus regarding scientific data and interpretation is precisely why the method works so well at establishing real knowledge of the enduring principles of the natural world and harnessing them for our purposes. It was developed as a means of distinguishing productive theories from hair-brained notions, hard data from accidental artifacts, rational interpretations from speciousness and bias. And it has served humanity very well in many ways; a great deal of ignorance and suffering have been eradicated by its successes.

However, as in the case of nature versus nurture, there is another conceptual danger lurking here among these payoffs, and another opportunity to step across the line from science to scientism and muddle. It is one thing to design a system of research that insures elaborate and repetitive cross-checking, independent reproduction of results, and critical debate in order to safeguard against human error and human bias. It is quite another thing to declare that an emotionless and utterly pure objectivity, a point of view devoid of all human bias, can and must be the sole source and validation of all meaningful propositions about reality.

The former sort of objectivity is meant to serve as a correction for erroneous beliefs, misconceptions, projections, personal material interests, and so on. The latter sort, on the other hand, claims that its method of inquiry, and the individuals who use it, have actually been purged altogether of these human traits and their human unreliability. The first tries as best it can to conscientiously winnow out the artifacts of faulty judgment or bias. The second claims to be a purely material point of view in which these possibilities for error do not exist. The first is science in operation; the second is quintessentially scientism.

Having grown up in the era of modern science, we all have had to develop a sense of what in general is meant by "objective" as opposed to "subjective." Objective means concrete, material, measurable, weighable, verifiable, in keeping with the known laws of physics and chemistry. Subjective on the other hand means chimerical, non-material, not amenable to weight and measure, illusory, relevant only to the contents of the human imagination. Objectivity has to do with the true laws about matter. Subjectivity has to do with the emotional and ephemeral impressions created by the mind.

But here is the central point about which we must be very cautious: There simply does not exist a point of view that is devoid of personal human experiences, beliefs, and intentions. There is no one without a stake in the answer to any important question. Proposed theory, experimental design, observation, and interpretation all inherently import personal beliefs and frames of reference. Furthermore, a human being without emotions and biases would be like a day without weather. It just doesn't happen.

Once again, the "versus" argument forces us into a sacrificial situation, the same way it tries to force us to make a choice between nature or nurture. The truth is that subjectivity and objectivity are two dimensions of our perception, and they both play crucial roles in understanding our surroundings and adjusting ourselves to our environment's demands.

Objectivity is the directing of our attention towards the world of external objects, while subjectivity is the directing of our attention inwards upon our own processes and reactions. Surely both points of view are necessary for me to successfully define and navigate my world. Objects have their uncompromising and autonomous realities that do not hinge upon my mere opinions about them. But at the same time, as long as it is human beings that are doing the observing, and as long as perception is the only medium we have available to observe with, then attention to the subjective elements of the processes of that perception would seem to be important.

This dual perspective is particularly critical when we come to observe our own bodies with our own minds. In this instance, the objects I am observing are parts of my self, and no view of them can possibly escape my subjective sense of them. In fact, this sort of self-observation makes the whole notion of separating subjective from objective rather squirrely: The parts of my body are definitely *"me,"* yet in another sense they are "external" to my mind. That is, they are irreducibly both subjective and objective. I do not *have* them, I *am* them; they are both objects under observation and active elements of the organism that is doing the observing. And when this internal/external duality comes to contemplate the part of the body called the brain—thinking about the thing we are thinking with—things can get woolly indeed, especially if I attempt to assume that my impressions about myself are irrelevant to my own material operations.

But even these stubborn ambiguities are not the real limits of strict objectivity. Its departure from our actual organic perceptual processes is far more radical than mere ambiguity; it is the point at which we step away from ourselves and into virtual reality, where we abandon our own lived biology for the rules of logic, abandon the personal for the abstract, and where we banish our human sensibilities from the forces that shape our own natures.

For here is the real heart of the issue: *There is no perception that is not fundamentally subjective.* The only things I can know in any direct way about the world are the ways in which my body is reacting to it. All the rest is inference, based upon accumulated memories and the processes of thought working upon them. I am not in direct contact with any external object at any time; "I" am only directly attached to nerve ends that are reporting certain responses to certain qualities of objects. And even here the nerve is not feeling the *object*, but rather is feeling the effect the object is having upon my own tissue immediately surrounding those nerve ends—distorting the tissue with pressure, buffeting it with vibrations, raising or lowering its temperature, and so on.

In other words, *I am never directly feeling anything but myself.* It is the responses of my own tissues that are the basis for my definition of the world. My physical being is my antenna, and I am listening to it as it is responding to the world. Therefore, subjective states of feeling are not simply distractions interfering with a more objective grasp of things. Intimate contact with my internal physical and psychological responses is absolutely indispensable for accurate perception of any kind.

What rescues each of us from being wholly isolated in our own solipsistic web of somatic reactions and whatever personal images and meanings they may produce are the facts that 1) we can also observe *others* responding to the same world, 2) that we can communicate to one another about the responses we are having, and 3) that these responses arise from similar processes in all of us. My reactions to any particular setting or event may well be very different from yours. But because we are reacting with organisms that are very similar to one another, we can pursue our communication about those reactions until I am able to arrive at a sympathetic understanding of your responses, and of the different qualities that they create in your assessment of events and their meanings to you.

That is to say *it is possible to be objective about subjective states.* In fact, no other possibility for meaningful agreement actually exists, since all primary knowledge about the world begins for every one of us with a subjective interpretation of some shift in body state and feeling state. Now to be sure, this communication, and the mutual objectifying of the significance of certain body states and associated feeling states, can be enormously facilitated by any and all reliable methods of cross-checking and verifiability. *This* is the point at which all of the discipline of objectivity properly enters into the process, where it is a powerful tool for sorting out our differing personal impressions from enduring principles we can agree upon.

In this sense *all* observation is self-observation, and all tools of clinical investigation are simply sophisticated extensions of our own perceptual faculties. These tools can expand and refine our perceptions enormously, but they can never alter the fundamental fact that it is in every instance a human being that is having, structuring, and interpreting the perceptions. And no human being can ever step aside from his or her history of feelings, impressions, and beliefs.

Objectivity is not the absence of human emotion, but rather a *cognitively useful* emotion. It is a specific feeling state, a state which is more interested than it is defensive, which cultivates calm observation, rational analysis, suspended judgment, great patience, a comfort with physically repetitive activities, and alertness for the telling detail. It is a feeling state in which an important kind of thinking can best be done, and it has more general application apart from the sciences than is usually acknowledged. On the other hand, it is a state that is not very useful for certain other kinds of thinking that are also important.

Dispassionate scrutiny may well help us to find the sort of person we should fall in love with in order to have the best chances of being happy; but by itself it does not do much to make us fall in love. There are precise laws of physics that can describe mathematically all of the forces at play as I pedal and turn my bicycle, but it is the *feel* for the thing, and not my knowledge of mathematics, that allowed me to learn to ride it. No amount of biochemical facts is sufficient to prevent chronic stress from producing ulcers in my stomach. And no accumulation of facts about the way things are can ever dispel the mystery of why there is something rather than nothing.

Some means of negotiating the potential vagaries of human emotions and their impact upon perceptions and beliefs is all to the good. But something very queer happens when we attempt to operate on the notion that this necessary objectivity is the purely rational *absence* of feeling: We arrive at the conclusion that the only thing you can possibly know immediately and for sure—what you are feeling right now—is inadmissible as evidence in an examination of reality.

This distinction between an objective feeling state and the claim of an objective point of view absent of feeling is absolutely pivotal, and the intellectual pretense that emotionless, bias-free observation is possible has created enormous dislocations between what we know about our bodies and our minds and how we are able to use that knowledge to our benefit. Indeed, it is exactly this denial of feeling and of the ultimately personal significance of our interpretations of the world that has given rise to the reality that is virtual—slick, colorful, and entertaining, but devoid of the genuine contents and meanings of our constantly emotional human lives.

In the end, "objectivity" can never really mean anything other than "honesty." It is nothing more nor less than my best attempt to think as clearly and openly as I can, and my willingness to invite examination of my methods and conclusions. I simply cannot make out why it should be that this honesty is a good thing when applied to some kinds of invisible things, like gravity or subatomic particles, but is supposed to be categorically impossible when applied to other kinds of invisible things, like feeling states. After all, if you set about looking for it properly, the trace of an emotion's track through my body and my behavior is just as visible as the vapor trail left by an electron in a cloud chamber. And I can, if I put my honest mind to it, infer just as much about the nature of that emotion as I can about the nature of that electron.

This intellectual elimination of our invisible worlds of responses and feelings from the proper scope of scientific investigation has had a singularly unfortunate consequence. It has prevented us from learning how to apply the powerful tools of objective inquiry to one of the most complex puzzles—and for us the most important puzzle—in our universe: The continually evolving relationships between our minds and our bodies, and the consequences of these relationships for our physical and mental health. By eliminating everything that cannot be weighed, measured, and trimmed to fit the Procrustean myth of pure objectivity, we have eliminated the consciously felt substance of our actually lived lives from the catalogue of things we believe to be verifiable.

In the process of refining our intellect, we have turned away from the intimate internal realities of the creature that is producing it, and relegated our experience of those subjective realities to the status of misguided fantasies. By attending too exclusively to the abstract language of logic we have forgotten how to listen to the yeasty, wordless bubbling of the biological information constantly churning within us. And by maintaining the belief that the mind—an immaterial thing, the improbable ghost in the machine—cannot possibly have any material impact upon the physical processes of the body, we effectively prevent ourselves from recognizing the countless impacts the mind is in fact having whether we choose to be conscious of them or not.

Surely this is one of the most debilitating double binds ever devised by the human mind. Because we have learned to discount the validity of the subjective side of our experiences, we do not tend to them seriously. Because we do not tend to them seriously, the information they are generating becomes sketchy and degraded for us. Because we must then try to regulate our complicated lives with sketchy and degraded internal information, processes are always going awry within us. And because our primary means of having any control over those processes is precisely to listen to them internally and respond to their messages, we find ourselves peering into the darkening cave of our bodies with no torch to see by.

And finally, because our internal awareness has become so dim, we never quite grasp the fact that the very definition of reality that has given us tremendous control over every other domain of nature has also obscured our primary means of accurately assessing and successfully regulating the very matter that matters most to us. Once complete, this vicious circle draws in on itself, and the self-fulfilling prophecy of pure objectivity comes to pass: My mind is left with little or no means with which to actively intervene in the internal processes of my body. And in order to protect ourselves from the knowledge of our own participation in all this, we dutifully embrace this induced helplessness as yet another of the important laws of nature.

It is my conviction that this degradation of subjective experience is one of the chief catastrophes of the modern world, a tragic price we have paid for the rapid advancement of sciences and technologies. Out of it spins our dehumanized virtual reality, and much of the suffering and abuse that goes on around us. It keeps us from seeing what we do to ourselves, and from seeing what we could do to help it. It gives us a false sense of power as we confront the forces of nature, and it gives us an equally false sense of impotence when we confront our own lives.

Above all, by silencing the internal voices of all that is subjective and human within us, it puts us in the midst of a world in which our deepest needs and sensibilities are alien and irrelevant. This disconnect then leads away from ourselves and towards many forms of personal and collective self-destructiveness. It is what goads us to launch any and all innovations as quickly as we can, never asking whether or not we should. It is what allows us to ignore the havoc we wreak upon the environment in our pursuit of what we have come to call progress. It is what has seduced us into valuing material accumulations more than the quality of life itself, and into viewing any destruction we perpetrate along the way as some combination of inevitable collateral damage and a God-given right.

The world views of Western religion and Western science are opposed over many issues, but they do have one core presumption in common: Our bodies and our immediate experiences of them are to be discounted and mistrusted. Some higher authority outside of ourselves must be the final arbiter of our reality and our relationship to it. Science in the name of discovering nature's laws, and religion in the name of a higher moral purpose, suppress and degrade much of what is most human and most personal about us; and both of them demand adherence to ideals that do not encourage us to attend in any ongoing daily way to the genuine needs of our organisms and their authentic satisfaction. For religion, the body has traditionally been mere "dross;" for science our experience of it is merely "subjective." For both, our too-close-attention to it is to be purged.

All of this suppression naturally has consequences upon the evolving templates that are directing my physical development and the cycles of growth and renewal that are constantly taking place. Gardens definitely take a different course of development when no one is tending them. While they may continue to proliferate cycle after cycle of renewal, the organization suffers and functional arrangements are disturbed. Weeds find footing, opportunistic species invade space and take over resources. Servicing paths are choked off, and thickets become impenetrable. The sources of aesthetic and organic nutrition for the gardener become more and more congested and stymied. And eventually it is time for either Round-Up or the bulldozer.

Now I am fervently in favor of both discovering nature's laws and pursuing higher moral purpose. But it seems to me that both enterprises suffer devastating consequences when the personal, the emotional, the subjective are factored out of them. Or, more accurately, it is the individual's participation in this factoring out that develops devastating consequences. We do not need less immediate experience of ourselves, but vastly more.

The conviction that our inner voices are unreliable, even illusory, insures that awakening from the abstract dream of virtual reality will be difficult at best, impossible for many. This double bind is a perfect example of what William Blake referred to as "the mind-forged manacles"—the beliefs we have adopted which prevent us from acting in our own, and our mutual, best interests. The escape from these conceptual manacles is through the immediate experience of the body and the direct engagement with its needs and capabilities. Until each of us for ourselves reopens the book, the text, the scripture of the body to the presence and attentiveness of the mind, there will continue to be many answers and many solutions that will refuse to appear.

Connective Tissue: A Cell Biologist's Insights

Every Nook and Cranny

In Chapter Three we saw that connective tissue penetrates, wraps, and gives structural shape to all our tissues. From the subcutaneous sheath of the entire body to the microscopic septa enclosing individual bundles of a few cells, collagen and its fluid ground substance are part of the immediate environment of all our organs and processes. This tough but pliable network is one of the primary structuring, ordering, and organizing elements of our body.

Surely this sponge-like mesh must have something directly to do with the preservation and the gradual developmental change of the body templates that directs the distribution of new cells and molecules in their overall seven-year cycles. In the first place, as we saw with relation to bone formation, it is the containing shapes of connective tissue that serve as the mold for tissue growth and arrangement. Secondly, we noted how connective tissue webbing responds to patterns of use, becoming thicker and tougher in areas that are commonly stressed, thinner and more pliant in places that are not. And third, connective tissue is one of our major repair mechanisms, deposited as scar tissue by fibroblasts that have migrated to the injury.

But how is it that a non-living substance can find its way into the appropriate nooks and crannies, orient its long molecular strands in the proper alignment, and bond into just the right shapes to support growth, stress, and healing? DNA codes in the fibroblasts' nuclei direct the production of collagen proteins, but where does the information come from that uses those proteins to continually recreate the evolving shape of the body in all its details?

I quoted George E. Snyder's observation that "The connective tissues not only bind the various parts of the body, but, in a broader sense, connect the numerous branches of medicine." More recent research and speculation seem to confirm just how prescient that statement was, and to indicate on how many levels those connective tissue connections may be taking place.

The Piezoelectric Effect

One of the properties of connective tissue is that it behaves like a liquid crystal, becoming more gel-like and solid when it is cooler and more sol-like and fluid when warmer—the *thixotropic* effect. This potential for warming and softening is one of the principle therapeutic effects of manipulation, and can allow for considerable restructuring and reconstituting of connective tissue.

There is another more recently appreciated property of connective tissue structures that greatly enhances this thixotrophy, and also adds a wide variety of energetic—as opposed to simply structural—functions of the connective webbing as well. Collagen belongs to a class of crystalline arrays of molecules that is called *piezoelectric*. Piezo- is a Greek root meaning to press or squeeze, and piezoelectric crystals generate spontaneous electricity when they are distorted by pressure. The entire connective tissue matrix is an electrical generator producing fields of current wherever pressure or movement is taking place, and this energy production has a great deal to do with the warmth and thixotropic responsiveness of fascia.

And even further, collagen structures are also *semiconductors* of the currents they are generating, carrying the electrical energy (and any sort of information that could be encoded in those electrical streams) throughout the body. It is also significant that collagen is a *semi*conductor, and not a pure conductor. Semiconductors are the various electrical mediums we use to transform electricity into other kinds of energy and information. Heating coils transform it into warmth, bulb filaments transform it into light, phonograph needles (another piezoelectric crystal) transform it into pulses that can be amplified as sound by yet other semiconductors.

Computer chips are also essentially semiconductor devices that route currents through their mazes to generate, encode, and combine large quantities of information in their sophisticated networks. The connective tissue web, then, does not merely generate electrical energy; it converts this energy into various forms, one of which is information. "In other words, a macromolecular array such as the connective tissue system could be a delicately tuned and extremely sensitive detector, amplifier, and processor of electromagnetic signals," says James Oschman.

"There has been a lot of interest in the possible biological significance of the piezoelectric phenomenon." He goes on:

> Research over the years has demonstrated that every movement made by the body generates electric fields due to the compression or stretching of bones, tendons, muscles, etc. In addition, electric fields occur as a consequence of nerve conduction and the activation of muscle contraction. Secretion by the various glands is also accompanied by generation of electrical fields. It is now widely thought that these fields spread through the tissues, providing signals that inform the cells of the movement, loads, or other activities occurring elsewhere in the body. The cells, in turn, are thought to use this information

to adjust their activities in maintaining and nourishing the surrounding tissues. For example, this mechanism accounts for the fact that movement and exercise maintain the skeleton, while long periods of bed rest or space travel in zero gravity lead to loss of bone mass.[1]

Moreover, many parallel arrays of other kinds of proteins in my body—like contractile proteins in muscles—are also semiconducting mediums, and can transfer and transform their currents back and forth with connective tissue. None of the body's tissues are electrically or magnetically shielded, so these currents and signals leak into virtually all of these tissues. All of the body's cells are "listening in" on this piezoelectric, semiconducted flow, and every movement and every compression augments it.

> Walking creates rhythmic compressions on bones and cartilage, and cycles of tension in tendons. These processes lead to pulsating fields that spread through the body. Each impact of the foot on the surface creates a veritable symphony of electromagnetic fields that tell a precise story of the force vectors developed with the body fabric. Walking on grass will produce a totally different melody than walking on pavement.

And since many of the body's proteins are able to initiate and receive these currents and signals, it is likely that there is a wide variety of kinds of information coursing through the network. There is even the strong suggestion that this conducting network is the basis for the therapeutic effects of acupuncture and acupressure: The meridians may well be channels of least resistance for specific organ signals, the points may be semiconductor nodes that process and refine that flow, and cross-over points may be the switch mechanisms through which energy in one meridian jumps directly to another.

The Connective Tissue Within Cells

The contents of cells was previously believed to be a viscous soup of "protoplasm," with the various organelles suspended in this more or less homogenous goo. However, improved magnification and imaging techniques have revealed that this is not the case. The old concept

> has given way to a more dynamic picture in which strands and arrays of fibrous elements form and reform an exquisitely ordered flexible moving network called the cytoskeleton. One of the most exciting developments during the recent period has been the recognition that muscle-like contractile proteins are present in virtually all cells, including those of plants.
>
> We now recognize that the "structureless" cytoplasmic ground substance in fact contains various filamentous proteins, tubulin, actin, myosin, intermediate filaments, and microtubules. These are able to form polymers, depolymerize, cross-link, intertwine, anchor, bind, elongate, and contract. This gives rise to cell movements, shape changes, cytoplasmic streaming, pigment migrations, pinocytosis, movement of organelles, phagocytosis, secretion, mitosis, the myriad of activities that constitute life.

That is, like the fascia network as a whole, the cytoskeleton within each cell is far more than merely structural material. Lapping myosin and actin strands create movement, other strands act as conveyor-chains that rapidly transfer specific molecules to specific sites within the cell, while still other kinds of strands link organelles together in an integrated energy/information network.

Connecting the Connectives

Further study has revealed something even more intriguing about this cytoskeleton matrix: Through a series of linking proteins, it is continuous outside the cell with a similar molecular net in the surrounding ground substance, and is ultimately linked directly to the local supporting collagen structures. And it is continuous inside the cell with a cytoskeletal structure within the cell nucleus. And all of these connecting strands of proteins are semiconducting pathways, conveyor-chains, and information lines.

> Other studies have shown that there are specific proteins, known as glycoproteins, that extend across the cell surface, from the cell interior to the exterior. It has been demonstrated that these proteins are connected to the filamentous network within the cell. On the outside of the cell, they form branching structures with charged groups (called sialic acid) at the ends of the branches. These in turn are joined by means of ionic calcium bridges to the negatively charged ends of the molecules comprising the extracellular ground substance, or gel. An additional set of proteins, known as fibronectins, binds the cell surface to extracellular collagenous fibers. The latter, along with the gel of the connective tissue, ultimately interconnect the larger fiber arrays of the myofascial system, including tendons, muscles, and bones.
>
> Hence, we now have a detailed description of an interconnected webwork that extends from organs, bones, tendons and muscles at the macroscopic level, down to the cell surface and through it, into the cytoplasm, where it contacts the various organelles and the genetic control center in the nucleus. Our classical concept of connective tissue must be extended to include the webwork within each cell.

Connective Tissue and the Templates of Renewal

As mentioned earlier, it was once believed that the structure of the adult body was relatively fixed and permanent. It was assumed that only a small fraction of dietary intake might be used to repair and replace structures that undergo wear and tear. However, studies using radioactively tagged dietary compounds revealed that in fact a very large fraction of them were rapidly integrated into virtually all tissues.

> The large molecules in the body, such as fats and proteins, are constantly being assembled and disassembled. The various bonds or linkages that hold them together are constantly being opened, enabling their building blocks (fatty acids and amino acids, respectively) to be liberated. The fragments then

enter what has been called the metabolic "pool," which is also fed by absorption of the same materials from the digestive tract. This pool is located within the sponge-like interstices of the meshwork of the connective tissue and cytoplasmic ground substance. Some of the molecules in this pool are degraded and excreted, while other are reassembled into the large molecules.

Now nature is generally extremely economical with the use of materials and energy, so why would all this extra activity be taking place to dismantle and reform so many molecules and fine structures that appear to be functioning as well as ever? The rate of turnover seems far beyond that necessary for merely repair—some molecules have a half-life of only nineteen minutes. Little explanation has been proffered for this surprising phenomenon, "But," Oschman says, "from our perspective there is an obvious explanation: Nature has allowed for rapid adaptive change in the structure of an organism in response to changes in the ways the organism interacts with its environment."

> An extreme example is the body-builder, who through the stimulus of constant exertion, brings about a dramatic alteration in body form. Not only do the muscles increase in size and strength, but the other components of the myofascial system—bones, tendons, ligaments, blood vessels—must increase as well. Likewise, when one stops repeating a particular activity, the body gradually reverts to its previous structure.

Here we are in the midst of our "template question": how are these orderly and responsive changes orchestrated, while at the same time preserving the details of a stable identity? Although specific clinical details are relatively sparse as yet, the electromagnetically and biochemically active connective web—from fascia to cell nucleus—would appear to be an obvious candidate. Indeed, it is now assumed by some biologists that this communication between various tissues and cells is, at least in part, generated and mediated by electrical fields produced by the piezoelectric effect coursing through this network.

There are many unanswered questions in all of this, to be sure. As Oschman observes, "The nature of the forces that determine the form of an organism is perhaps the central unsolved problem in biology." But it is already clear, he goes on to say, that

> we are far more than a machine consisting of mechanical parts held together with threads and fibers. The rapid and subtle flows, forces and movement that characterize life and set it apart from non-life originate within the molecules and molecular frameworks of which we are composed. At this level, the distinction between matter and energy becomes hazy, as does the distinction between different forces such as gravity, elasticity, heat, motion, electricity, magnetism, and chemical bonding. The mechanical properties of a molecule, such as its strength and flexibility, become inseparable from its electronic and other submolecular characteristics. It is here that biology, the science of life, becomes one with physics, the science of matter and motion.

There is another probable function of the structural molecular strands within the cell. In Chapter 2, we saw that cell membranes are populated with large proteins embedded in the phospholipid bilayer of the membrane. We noted at that time that these large molecules were "gateway" proteins that expanded their spiral pores to let nutrition in and toxic waste out, and contracted their pores to seal in life processes and seal out environmental toxins. These are, of course, crucial discriminating functions of the mem-"brain."

It now appears to be the case that the filament structures in the cell guide these gateway proteins towards the membrane, and fix their position there. In addition to spiral openings and closings for nutrition and toxins, there are a large variety of different kinds of reactive proteins embedded in every cell's membrane that stimulate an astonishing variety of cell functions, organ responses, and systems coordination. These "receptor" molecules are bonded by many kinds of proteins circulating in the intercellular fluid, and they trigger processes that were not even imagined until recently.

These receptors are also jockeyed into position and moored into place by filaments in the cytoskeleton, and in this role the filaments connect not only to the outside environment of the cell but also to the ground-breaking work of another contemporary biologist, Candace Pert.

Feeling and Healing: The Neuropeptide System

Receptors and Reception

Candace Pert's research has launched the ongoing discovery of a wide variety of molecules embedded in cell membranes whose functions are very different from those of the food/toxin, intake/output gateways. These molecules—called receptors—do not import and export material; nothing physical enters or leaves the cell through them. Rather, they take in *information,* information that dramatically alters the internal activities of the cell and its functional relationship to the rest of the body.

Pert has called these receptors "tiny eyes, or ears, or taste buds." Understanding their actions propels us into a much deeper appreciation of cellular intelligence, and profoundly alters our notions of the ways in which nerves and other cells send, receive, and process information that coordinates their interactions.

It was previously believed that there were two basic neuroactive chemicals: Norepinephrine was excitatory, and acetylcholine was inhibitory, making any given cell either more or less likely to fire action potentials, the electrical nerve signals that travel along axons and fire the next neuron down the line. These chemicals were understood to work exclusively at synapses. These principles suggested an "on/off" binary language, operating throughout the circuitry connections of a large and complex wiring network, something very like a hypercomplex computer. This in turn suggested a computational model for understanding the operations of consciousness, which has in its turn suggested the possibility of programming computers to possess an "artificial intelligence" that could rival—indeed perhaps utterly outstrip—our own. This is virtual reality at its logical extreme, with no other resources than the binary language of formal computational logic to somehow imitate the most complex manifestation of life.

An understanding of the roles of neuropeptides overwhelms this relatively simplistic computational model of consciousness and behavior. In the first place, there are not only two neuroactive chemicals, one excitatory and one inhibitory. Nearly ninety have been documented, and Dr. Pert believes they will find closer to 300 when the search is complete. And the action of each upon the responses of single cells and upon overall behavior are different. Furthermore, these substances are not active only at synapses; their bonding receptors can be *anywhere* on a nerve's membrane, and a single nerve may have millions of receptors at any given time. They are not transferred across synaptic clefts, but are circulating freely in the blood stream and all other bodily fluids. And, perhaps most surprisingly, neuropeptides are not produced and received only by neurons, but by *many* of the body's tissues, including the gut tube, the muscles, the glands, the lungs, and the various kinds of cells of the immune system. And they stimulate many more responses than that of action potentials. "The receptor," says Pert,

> having received a message, transmits it from the surface of the cell deep into the cell's interior, where the message can change the state of the cell dramatically. A chain reaction of biochemical events is initiated as tiny machines roar into action and, directed by the message of the [peptide], begin any number of activities—manufacturing new proteins, making decisions about cell division, opening or closing ion channels, adding or subtracting energetic chemical groups like the phosphates—to name just a few. In short, the life of the cell, what it is up to at any moment, is determined by which receptors are on its surface, and whether those receptors are occupied by [peptides] or not. On a more global scale, these minute physiological phenomena at the cellular level can translate to large changes in behavior, physical activity, even mood.[2]

The Biochemistry of Stress and Relaxation

There is some precedent to help us to understand the wide array of effects these molecular messengers may have upon our mood, behavior, and development. Hormones and the complex reactions they create have been studied for a long time, and the multi-leveled impact of the release of these messenger molecules into the bloodstream is in many respects well documented.

The body's organized response to a dangerous threat, the "fight or flight" response, was discussed briefly in Chapter Ten. When a threat is perceived, one hormone is released by an area in the hypothalamus of the brain; it triggers the release of a second from the pituitary gland; this second hormone is delivered to the adrenal glands, which in turn secrete a variety of chemicals that shift the activity of many target organs and tissues into an emergency mode:

- The whole central nervous system is aroused—the "rush."
- The sympathetic nervous system is strongly stimulated, while the para-sympathetic—which rests and restores the organism—is shut down.
- General muscle tone increases.
- Reflexes become hair-triggered—"facilitated."
- Respiration deepens and quickens.
- Blood lactate levels rise.
- Blood clotting mechanisms are heightened.
- Heart rate increases, blood pressure rises.
- Blood is directed away from internal organs, towards skeletal muscles.
- The spleen dumps extra blood cells into the blood stream.
- Digestion stops; or, more accurately, digestion of food in the gut tube stops, and we begin digesting ourselves—converting our more readily-available protein stores to glucose to immediately fuel the emergency activity.
- The kidneys retain water in anticipation of sweating or blood loss.
- Sensory perception becomes heightened, attention becomes vigilant.

When they are an orchestrated response to an actual threat, these complex interactions prepare the organism for the extra surge of energy and resources to deal with the situation quickly and effectively, either by fighting or fleeing, and by efficiently dealing with any injuries. When the threatening situation has passed, the body's hormonal chemistry then quickly reverts to normal levels for normal activity.

However, the research of Dr. Hans Selye on the physiological effects of chronic stress has shown that when these emergency reactions are repeated too frequently, or are sustained over long periods of time, they can have catastrophic consequences. The heightened stimulation eventually causes general exhaustion, and the shunting of blood from the internal organs adds to this depletion. Sustained high blood pressure has many pernicious effects. The digestion of our own protein stores first strips us of reserves, then begins consuming our own flesh. The heightened vigilance over time becomes aggressive and anxious, defensive behavior that creates a sense of urgency all its own, thus keeping the physiological machinery in high gear. Selye documented acute emotional distress, gastric lesions, tumor growth, wasting away, many kinds of organ dysfunction, and a variety of causes of eventual death as a result of these factors.

And if the hormonal chemistry of danger can do all these temporarily useful things, and when extended chronically can do all these destructive things over the long run, what about the chemistry of "unstress?" Might there be a way to counter this debilitation and death? Dr. Herbert Benson followed up Selye's research with a study of what he called "the Relaxation Response," the opposite of stress. Significantly, Benson chose to use the obvious calming and effects of Transcendental Meditation—a commonly known and readily learnable practice—in order to document the physiological profiles of relaxation.

What Benson discovered, in some cases with subjects who had as little as two weeks practice in TM, was a dramatic reversal of all the chemical and physical reactions to temporary stress, and a diminishing of the pathologies that had been set in motion by chronic stress. Specifically, he was able to show that in subjects who practiced TM, the following mirror-image changes occurred:

- A very different quality of central nervous system arousal occurs, characterized by a diffusely focused alertness rather than acute vigilance. Specifically, the brain enters an alpha-wave state.
- Sympathetic nervous system activity diminishes, while the restorative activities of the parasympathetic are increased.
- General muscle tone decreases.
- Reflexes are intact, but not in a hair-trigger state of facilitation.
- Respiration slows.
- Heart rate lowers, as does blood pressure.
- Blood clotting factor returns to normal.
- There is a drop in blood lactate levels.
- Blood is directed away from the skeletal muscles and towards internal organs.
- Normal digestion of food resumes, replacing rather than depleting the body's protein.
- Kidneys increase their urine output.
- Sensory perception normalizes, and attention relaxes its vigilance.

One thing that is very important to realize here is that cultivating this state of calm helps to make the organism more able, not less, to rally its defense mechanisms against the next threat. Animals that are exhausted by chronic stress lose the physiological ability to respond quickly and in an effectively orchestrated fashion to the next emergency and the next shot of adrenaline; they have worn themselves and their dosage out. On the other hand, animals that are emotionally and physically calm react strongly to the surge of chemistry for "fight or flight," perform at a high level while the threat is current, and then quickly restore to a resting state as soon as the crisis is resolved. They are then able to react just as quickly and effectively to the next emergency, and the next.

These examples of the actions of hormones, then, provide us with a previous model of molecules released from specific sites, circulating in the body's fluids, making contact with cell membrane receptors in distant tissues, and triggering shifts in an organ's operation as a response. Like the hormones, the neuropeptides are "information molecules," tiny bytes of data operating independently of the synaptic circuitry system, basic units of a language with which cells of many kinds "talk" and "listen" to each other, and mutually orchestrate their roles in the life and the vitality of the organism.

Molecules of Emotion

One of the most fascinating—and most significant—things about the neuropeptides is that above and beyond local responses in particular cells and tissues, they are in a more general way *mood specific*. The bonding of particular peptides upon particular cells initiates the experience of a particular feeling state. And the experience of a particular feeling state can release a particular neuropeptide to bond with receptors on specific cells. If a laboratory animal is incited to a strong state or emotion—rage, say—and the neuropeptides bonding to neurons in its brain's emotional centers are extracted and injected into a second animal, that second animal will immediately exhibit rage. And so on with other distinct feeling states: sadness, disgust, anger, anticipation, joy, acceptance, fear, surprise all appear to have their respective peptide, and all generate multilevels of response.

These shifts in feeling states are accompanied by simultaneous shifts in physiological profiles as well. An example offered by Pert is that of the peptide *angiotensin*: Once released into circulation, angiotensin travels to and bonds to several different target tissues, each with their angiotensin receptors distributed on cell membranes to get the message. Bonding to certain neurons in the brain, the peptide initiates the conscious feeling experience of *thirst*. Bonding to cells in the lungs' vascular system, it acts to lower the water vapor content of the exhale, preserving water during a time of thirst. And bonding to cells in the kidneys, it acts to lower the rate of urine production, further conserving water.

So a single molecular form—a single bit of information—acts at several tissue sites in order to connect a physical need with a feeling state, and that feeling state with appropriate physiological shifts, and eventually with an overall shift in behavior as the animal seeks water and drinks.

What is also fascinating is that the effects of angiotensin and the other peptides are not by any means exclusively human. All mammals, in fact all species so far assayed, have exactly the same peptide molecules and bonding molecules. They are present in creatures that do not even have nervous systems, and indeed are the messengers that even single cell populations use to communicate with one another and organize the collective actions of the colony. Perhaps this is why so many people achieve such a deep and evidently mutual emotional connection with a pet, why animals can "smell fear," and why some individuals have such immediate rapport with creatures of all kinds.

It appears that evolution has carefully preserved this basic communication system throughout the development of species, adding to its repertoire of agents and effects as more complexity opened up more possibilities. This must surely mean that their activity is basic to survival, that they are and always have been crucial elements in the complex development and coordination of life. They are the bridges between our emotions, our internal responses, and our overt behavior, the informing elements that tie these parts of ourselves to one another and that use their interactions to successfully regulate our lives.

How, then, can we imagine it is possible to factor our emotional responses out of our analysis of how our organisms operate, when even Darwin himself speculated that they were key to the survival of the fittest?[3] Once again, I believe we must be prepared to admit that subjective feeling states cannot be banished by so-called "objectivity," and that we must find a way to think more objectively about them. "When I use the term *emotion*," says Pert,

> I am speaking in the broadest of terms, to include not only the familiar human experiences of anger, fear, and sadness, as well as joy, contentment, and courage, but also basic sensations such as pleasure and pain, as well as the "drive states" studied by the experimental psychologists, such as hunger and thirst. In addition to measurable and observable emotions and states, I also refer to an assortment of other intangible, subjective experiences that are probably unique to humans, such as spiritual inspiration, awe, bliss, and other states of consciousness that we all have experienced but that have been, up until now, physiologically unexplained.

Brain or Body?

Another of the either/or debates that have been around for a long time is the question of whether what we experience as emotions, drives, and feeling states originate in the brain or in the various other tissues and systems of the body.

William James concluded that emotions are purely visceral events, originating in the body and are not cognitive. We perceive events and have bodily sensations. Then after the fact of perception and sensations, we associate the two in our memories and imaginations and then label the sensations as one emotion or another. There probably is not, he felt, an area or function in the brain for emotional expression—not even, in fact, any such separate entity as emotion, but simply perception and bodily response. The pounding heart, the sweaty palms, the tight stomach, the tensed muscles *are* the emotions.

Subsequent experimentation showed that this theory could not be the case. Direct electrical stimulation to parts of the brain do initiate bodily changes that can be felt, but these effects follow much too slowly to account for the onset of an emotional feeling in response to the stimulation. Also, the artificial stimulation of visceral changes typical to specific emotions, such as the strong intestinal contractions accompanying panic, do not produce the emotional feelings or the other physical manifestations of panic.

As one might expect, the answer to the puzzle is not a matter of deciding which position is correct, but of understanding how the domains of brain and body interact to produce both visceral states and the experience of emotions. These interactions are always taking place upon a two-way street, with simultaneous and accumulative responses developing in both directions. Pert quotes Elmer Green, a Mayo clinic physician:

> Every change in the physiological state is accompanied by an appropriate change in the mental emotional state, conscious or unconscious, and conversely, every change in the mental emotional state, conscious or unconscious, is accompanied by an appropriate change in the physiological state.

We now know that there are structures in the brain that appear to be involved specifically in emotional experiences and reactions. The limbic system occupies a deep central region in the brain, just on top of the brain stem. It includes the thalamus, hypothalamus, hippocampus, and parts of the basal ganglia. The list of associated structures and functions grows as an understanding of its inputs and effects expands. What is clear at this point is that these structures are situated at the crossroads of many channels of information and response in the nervous system. It is the area through which all sensory information coming up through the spinal cord enters the brain, and through which all motor commands flow back downward. It is also the area through which information from all the special sense organs in the cranium enter the brain. And it is directly connected, both with dense nerve networks and with rich capillary saturation, to the pituitary gland, and through it to the autonomic visceral system.

As it turns out, neuron groups situated within this limbic region have the densest collection of peptide bonding sites in the brain. Each of these neurons can manufacture, send, and receive every one of the neuropeptides now known. Each neuron can display millions of peptide receptors on its membrane at any given time, and can change the populations of specific types of receptors displayed either according to previous stimuli or current operational needs. Concentrations of peptide receptors are especially dense in areas where sensory information from all sense organs enter the brain, and where motor connections are distributed to both the skeletal and autonomic muscular systems.

This strategic concentration of receptors adds an immeasurable dimension to our experience of our perceptions and to our responses. This is where impressions about the world and about ourselves are imbued with how we *feel* about them, where situations take on personal significance, where raw data is given meaning and our world becomes a human one.

Things happen to us. We react. We observe the consequences. We draw conclusions about the goodness or badness of certain things, depending upon our experience with them. When confirmed by repeated experiences, these conclusions become beliefs about the nature of those things and about their significance to us personally. These beliefs then become organizing assumptions for all further experience, categories into which all apparently similar experiences are folded, and expectations about all future possibilities.

Facts are things that have happened. Laws are the ways in which we come to understand that things must happen if they are going to. Truth is what we make of all these objects and ideas, how we feel about them, what we believe about their relationship to us, and what compels our particular responses with regard to them. Our deepest convictions, those that unconsciously structure all of our experience and behavior, are products of this limbic function, and are utterly saturated with emotion.

There is also another major information channel that enters the brain in this central limbic region: the bloodstream. It is the vascular system that delivers peptides secreted by the brain to their target cells in distant tissues, and it is the blood that carries peptides secreted by other tissues to their targets, and back to the brain. Thus the limbic system is where chemical information in the blood and chemical information secreted by neurons are mixed with overall firing patterns in the brain—where action potential and chemistry meet. And in this meeting, weird and wonderful effects between mind and matter take place.

A cell or organ in the body is in a functional state that requires some sort of response to coordinate it with other systems. It synthesizes and secretes appropriate messenger peptides, which are carried by circulation to the limbic entrance to the brain. These peptides then bond to receptors on specific neurons, where they 1) alter the neurons' internal chemical activity and peptide secretion, and 2) alter those neurons' pattern of activity in the overall brain state. The original cell or organ is then "answered" both by chemicals from the brain delivered by the blood steam and by nerve impulses directly through the nervous system. Both responses are then factored together by the cell or organ to orchestrate its next response.

This mixture of chemical and action potential coding results in a remarkable transduction across their boundaries. Chemical messages (matter) are transformed into shifts in brain states (mind), and brain states are transformed into chemical messages. And both these transformations initiate an endless series of further transductions of one kind of information into another as they course through the organism. What we see here is not a question of mind versus matter, but a process of each being continually converted into the other, as physical events are being translated into feeling states and feeling states are being translated into physical responses of all kinds, conscious and unconscious. Pert observes:

> We can no longer think of the emotions as having less validity than physical, material substance, but instead must see them as cellular signals that are involved in the process of translating information into physical reality, literally transforming mind into matter. Emotions are at the nexus between matter and mind, going back and forth between the two and influencing both.

Emotions and Muscles

One of the major impacts of shifting emotions and their underlying peptide chemistry is upon our muscular performance. Feeling states can have as much to do with our strength, agility, and skill as does physical practice. Imagine the following mental experiment, which all of us have experienced in some form in our lives:

You have played piano for many years and are very good at it. You have your favorite Chopin nocturne, which you have practiced and enjoyed innumerable times in the quiet of your room, candelabra lit and night falling. Every nuance of tempo, volume, and tone are at your command, and can be spontaneously embellished at your will. Every note is a caress, every phrase a perfection.

Then on a holiday your mother drags you to the downstairs piano and bids you play your nocturne for a crowded room of people, about a third of whom you don't know particularly well, and another third whom you don't even particularly like. The opening phrase is halting. Wrong notes begin to creep in, and the tempo wobbles. Subtle effects become lurches, and the whole sense of the piece begins to disintegrate, until finally your heart is in your throat, your mind is a blank, and your fingers refuse to recall the next chord. Slamming the cover down, you retreat in humiliation.

What happened to your finely honed skill? Where did your highly organized "nocturne engram" disappear to? The answer is that nothing really happened to the skill, and the pattern of its performance in no way disappeared. It was simply unable to operate the same way in a profoundly different emotional state. Eyeing your surroundings, your mental response released a flood of anxiety and resentment peptides into your limbic system, where they dramatically altered the rhythm and the feel of all your perceptions and responses. Every command sent to a finger and every sensation returning to you from the keyboard had to find its way through the hazing of this emotional clatter, much of it getting confused and lost along the way.

Now this is an isolated-and somewhat dramatic an example. But the broader truth is that almost everything we do routinely with our muscles relies on patterned habits that operate smoothly with each repetition. And the list of things our muscles routinely do for us is very large indeed.

- They hold the joints of the skeleton in place and suspend all body parts from our frame—*stability*.
- They pattern the overall arrangement of that framework in space—*posture*.
- They create movements around the joints—*gesture*.
- They orchestrate these gestures to move us around—*locomotion*.
- They fill and empty our lungs—*primary respiration*.
- They provide the pumping mechanisms for all the body's fluids—*secondary respiration, intracellular circulation, lymphatic drainage*.
- They seek out nourishment, pick it up, bite, chew, swallow, move food through the gut tube, collect the waste products, urinate, defecate, and clean up afterwards.
- They are the means by which all the skills of survival and the arts of civilization are realized.
- They copulate, support gestation, and deliver the infant into the world.
- They start and stop the secretions of every gland.

- They aim and focus our special senses (eyes, ears, taste, smell, equilibrium).
- They mobilize and shape our sense of touch.
- They are sense organs themselves, contributing enormously both to our own body image and to our sense of the mass and extension—*the substance*—of the world.

Changes in feeling states, mediated through both the nerve impulses and the peptide reactions in the limbic system, create changes in all of these varieties of motor performance, just as surely as they conditioned the performance of the piano piece for better or for worse. Our emotions are constantly leaking into all of our muscular activities, and are either enhancing or debilitating our performances on every level.

Emotions and Memory

Note in our Chopin example that at the height of the emotional disturbance not only was motor performance compromised, but even *the memory of the correct notes* was jammed, unavailable for recall. Learning theorists now recognize that the feeling states that are being experienced at the same time learning is taking place are recorded along with the performance or the information, and are closely associated with future recall. They refer to this principle as *state-bound learning*, and it bears upon both our muscular and our intellectual competence in many ways.

For instance, experimental subjects are given a few shots of vodka. Twenty or thirty minutes later, they are given a list of random numbers to memorize. After they master the list with a high proficiency, they are then allowed to sober up. Asked again about the list, their memory falters and their proficiency plunges. Again given a few shots of vodka, they can successfully re-access the memory of the list and return to their former proficiency—without having actually re-consulted the list in the meantime.

This general principle seems to hold for all kinds of learning and colors both our learning and our recall in many ways. Songs we enjoy are easier to memorize, and their memory lasts longer. Experiences incurred in a state of heightened emotion—pleasant or unpleasant—print more deeply upon our psyche and become more ingrained in our beliefs about the world.

This may help to explain some puzzling human behaviors, especially some of the more obviously self-destructive ones. For example, some of our most important skills, knowledge, and beliefs are developed during our first two or three years—basic motor control, locomotion, language, initial human relations, and so on. The emotional tones of our family, and our own emotional responses, become closely associated with those early lessons, and the efficiency of their recall hinges upon the degree to which the emotional state is recalled and evoked as well.

If my early emotional life was not a happy one, this state-bound learning effect can lead to a serious dilemma. I cannot effectively recall much important information and many important skills if I cannot resurrect to some degree the emotions associated with their learning. So I must either sacrifice some degree of ready recall, or I must come up with ways to perpetuate those emotional tones to facilitate my memories. So I unconsciously continue to set up relationships and situations that will continually evoke those feelings and keep them current. I learn to use unconsciously my adult resources to reconstruct my infantile emotional pain so that I can more easily remember crucial things I learned then.

And with regard to deeper trauma, this effect might be a piece of the puzzle of repressed memory as well. If the feelings associated with an event are too painful for me to tolerate in my present emotionally weakened state (weakened by the trauma itself), then I am less likely to be able to recall the event, even if that recollection could resolve a great deal of confusion and conflict for me.

Antonio Damasio is a contemporary neurologist who has studied these limbic effects extensively, and in his book *Descartes' Error* he describes what he calls our "emotional I.Q."—the relationship between our capacity to entertain and process emotions, our capacity to learn, and our capacity to recall. Like Pert, he sees the life of the emotions and their peptide chemistry as being keys not only to our feelings, but to our intelligence as well.

Emotions and the Immune System

Anecdotal and common sense connections between emotional states and susceptibility to diseases have been around for a long time. One of the conclusions of Selye's and Benson's work was that particular feeling states have a great deal to do with the immune system's robustness or weakness.

As early as the 1920s and 1930s Russian scientists had demonstrated that the immune response could be either suppressed or enhanced by classical Pavlovian conditioning. Lab animals were given two paired cues—a loud trumpet blast followed by an injection of bacteria to stimulate their immune system. After repeated conditioning trials, the animals had learned to activate their immune systems purely in response to the associated trumpet blast, without the following bacterial injections. Experiments in America in the 1970s confirmed the reverse case—that when animals were conditioned with an immune-suppressing drug sweetened with saccharine, their immune systems were immediately suppressed by the taste of saccharine alone.

Pert's discoveries have extended this model of classical conditioning of the immune system enormously. Just like neurons, all of the white blood cells of the immune system manufacture, secrete, and receive the full range of neuropeptides. And like neurons, their membranes are studded with millions of receptors at any given time, and they can alter their populations according to past stimulation or current events. They are fully equipped to communicate back and forth with the brain from anywhere they may travel in the body.

This gives them such a close informational tie to the nervous system that it invites us to view white blood cells in an entirely new way. They are, in a sense, *wandering neurons*, information generators and receptors that float freely in and out of every nook and cranny of the body. This extends the sensitivity of the nervous system tremendously, making it responsive to microscopic elements in the body that are much too fine to be picked up by nerve ends. It has long been recognized that one of the primary jobs of the immune system is to discriminate between "me" and "not me" within my tissues. This remarkable sensitization of the white blood cells takes this discrimination down to the level of stray molecules.

Furthermore, these immune cells can use their peptide and receptor patterns to communicate with all other peptide-active cells and tissues as well. In this way they are active information links—receiver and broadcast stations—between the brain and the tissues that are both mobile and microscopically local. Among other important functions, this chemical linkage is very likely what makes it possible for immune cells to *chemotax* so quickly and accurately. Chemotaxy is the means by which a white blood cell follows the "scent" of a chemical trail to a specific site where it is needed. Local tissues experiencing invasion or trauma of some kind release a peptide "distress" signal, which diffuses outward and remains most dense closest to the local area. The white blood cells then pick up this message and begin quickly maneuvering themselves towards the increasing density of the messengers. In this way they locate the trouble very efficiently.

Given this tight informational linkage, many more effects are imaginable between the brain, the tissues, the white blood cells, and foreign invaders. For instance, this might well be the mechanism that makes it possible to classically condition immune responses. Once the brain has associated a trumpet blast with a pathogen, it can activate white blood cells even in the absence of that pathogen, operating on the stimulus of the trumpet blast alone.

This is to say that immune activity can be directly enhanced or suppressed by different feeling states. This finally is hard clinical evidence of the connection between emotions and disease. When chronically perpetuated, some feelings and beliefs predispose us to immune deficiency; others predispose us to robust resistance. Emotions directly impact the immune system.

This insight has in fact given rise to an entirely new branch of cell biology, dubbed by Pert and her partners *psychoneuroimmunology*—the clinical study of how brain firing patterns (psyche) impact individual neurons' peptide production (neuro), which in turn direct the activity of the immune system. "Can anger or other 'negative' emotions cause cancer?" asks Pert.

> In addition to the recent studies by various researchers like David Speigel of Stanford who have convincingly shown that being able to express emotions like anger and grief can improve survival rates in cancer patients, we now have a theoretical model to explain why this might be so. Since emotional

expression is always tied to a specific flow of peptides in the body, the chronic suppression of emotions results in a massive disturbance of the psychosomatic network. Many psychologists have interpreted depression as suppressed anger; Freud, tellingly, described depression as *anger redirected against oneself*. Now we know something about what this looks like at a cellular level.

"Is it possible," she continues,

> we could learn to consciously intervene to make sure our natural killer cells keep doing their job? Could being in touch with our emotions facilitate the flow of the peptides that direct these killer cells at any given moment? Is emotional health important to physical health? And if so, what is emotional health? These are the sort of questions we have to start addressing if we take the links between body and mind seriously.
>
> Let me begin to answer by saying that I believe *all* emotions are healthy, because emotions are what unite the mind and the body. Anger, fear, and sadness, the so-called negative emotions, are as healthy as peace, courage, and joy. To repress these emotions and not let them flow freely is to set up a disintegrity in the system, causing it to act at cross-proposes rather than as a unified whole. The stress this creates, which takes the form of blockages and insufficient flow of peptide signals to maintain function at the cellular level, is what sets up the weakened conditions that can lead to disease. All honest emotions are positive emotions.
>
> Health is not just a matter of thinking "happy thoughts." Sometimes the biggest impetus to healing can come from jump-starting the immune system with a burst of long-suppressed anger.... The key is to express it and then let it go, so that it doesn't fester or build, or escalate out of control.

The molecules of emotions, it seems, are the informational "connective tissue" that unites and coordinates all of the organism's cells and systems, "weaving them into a single web that reacts to both internal and external environmental changes with complex, subtly orchestrated responses."

Soup and Circuitry

The stunning range of activities orchestrated by these fluid-borne neuropeptides forces us to completely rethink any wiring-diagram model or computational model of the functioning of the brain. Consciousness is not a matter of "hardware" or "software." Rather, it is more dependent on what we might call "wetware." Pert suggests that the vast majority of the channeling and processing of information in the nervous system itself is organized and directed by the distribution of peptides and receptors, and not by synaptic firing patterns, since these firing patterns are obliged to play themselves out in a system whose biases are set by the molecules of emotion. And as for the dimensions of consciousness arising from the rich connections made by the peptides with the rest of the body's tissues, the older computer models do not even offer a clue.

It is as though the "computer" were immersed in a bath that is saturated with chemicals that can alter the patterns of action of every individual chip in innumerable ways, and as though every chip could decide independently which chemicals to output to other chips, and which to uptake themselves at any given moment. No digital programmer could possible cope with these kinds of variables. This is a model that invites us to look at mental states not as coded binary read-outs, but as the delicate, complex, and subtle spicing of a soup, a soup whose recipe is continually changing even while it is continually *my* soup, its spicing based on my accumulated experience and my beliefs.

What is also clear is that "mind" is a great deal more than what "brain" does. The brain is only one of the nodes, albeit a very powerful one, that interconnects a multileveled system of information distribution and behavioral organization. James Oschman's research suggests that arrays of liquid crystal protein forms of many kinds may conduct energy and information, and that the connective tissue network is a major example of this. Cell membranes exercise many kinds of discrimination that directly effect their survival, and the survival of the organism. Immune cells function in many ways as an extension of the nervous system. And with the neuropeptides, we see that even molecular forms are bits of in*form*ation that in*form* all of the tissues of the body and materially alter their cooperative functions.

"So what we have been talking about all along is information," says Pert.

> I like to speculate that what the mind is is the flow of information as it moves among the cells, organs, and systems of the body. And since one of the qualities of information flow is that it can be unconscious, occurring below the level of awareness, we see it in operation at the autonomic, or involuntary, level of our physiology. The mind as we experience it is immaterial, yet it has a physical substrate, which is both the body and the brain. It may also be said to have a non-material, nonphysical substrate that has to do with the flow of that information. The mind, then, is that which holds the network together, often acting below our consciousness, linking and coordinating the major systems and their organs and cells in an intelligently orchestrated symphony of life. Thus, we might refer to the whole system as a psychosomatic information network, linking psyche, which comprises all that is of an ostensibly non-material nature, such as mind, emotion, and soul, to *soma*, which is the material world of molecules, cells, and organs.... With information added to the process, we see that there is an intelligence running things. It's not a matter of energy acting on matter to create behavior, but of intelligence in the form of information running all the systems and creating behavior.

Now all this is nothing more nor less than the rudiments of a clinical description of "the wisdom of the body." Our conscious experience of this wisdom is our own Witness. And our working relationship with this Witness, and its working relationship to the development of our minds and bodies, are powerfully obscured by the beliefs and behaviors entailed in a pure, mythical "objectivity."

The Implications for Bodywork

"What makes this model so different," says Pert, "is that it can explain how it is possible for our conscious mind to enter the network and play a deliberate part." What also makes it so different is that it can explain how a wide variety of effective psychosomatic therapies operate within the organism.

Following her initial discovery of the neuropeptides in the 1970s Candace Pert first speculated that therapeutic applications would involve the laboratory synthesis of peptides and their pharmaceutical application. What she has more recently come to understand, largely through healing experiences of her own, is that what the neuropeptide system really implies therapeutically is that each one of us is *already* our own ambulatory pharmacopoeia. All that is necessary is for us to learn to open and close the peptides' cellular vials. And the most direct way of doing that turns out to be the exploration of the world of *feelings*—feelings that are caused by peptide shifts, feelings that create peptide shifts, and a feel for the physiological effects that follow.

Getting in conscious touch with tissues and looking for the feelings that are healing are what effective bodywork has always been about. Both Oschman's energy/information model of the connective tissue web and Pert's energy/infor-mation model of the neuropeptide system seem to me to offer large leaps of insight into how these physical and emotional interventions accomplish their therapeutic effects, and to provide much clearer frameworks for their further improvement. These models are huge strides towards a broader knowledge of and consensus for truths about our sickness and our health that have been known by many therapists for a long time.

Among other insights, the neuropeptide theory is another helpful way of un-derstanding what kinds of templates are at work as we change over our bodies' substances, shifting tissues and functions to continually adapt and yet always managing to remain ourselves. One of those major templates is clearly the mo-lecular orchestration—the prevailing peptide homeostasis—of our emotional life and its impact on tissue response, beliefs, and behaviors. The system is designed to be highly variable and adaptive, using the kaleidoscopic language of emotions to generate all kinds of self-awareness and opportunities for self-regulation. On the other hand, we can suppress large portions of those feelings, which will seriously disrupt that awareness and regulation. Or we can fixate on patterns of behavior that favor the dominance of one feeling state over all others, which drastically limits our available responses and strategies for adapting.

I believe that I have witnessed this system at work in my own Trager practice. The basis of the Trager Approach is the projection and transference from one individual to another of a deep feeling state that Dr. Trager called "Hook-Up." He practiced Transcendental Mediation for many years and said that the feelings were indistinguishable to him, except for the fact that in Hook-Up he was at his table and actively engaged with someone else—hooked-up not only with his own consciousness, but with someone else's at the same time. And he insisted that unless the practitioner cultivated this state for him or herself, there was no hope of stimulating it in someone else—"You can only give away what you honestly have."

I think the "giving away" happens like this: The practitioner calls up within himself the feeling of Hook-Up. Once established (as a neuropeptide cocktail), this feeling informs every gesture, every contact he makes, in the same way that a dancer's feeling state informs every move of the choreography. This feeling quality is experienced by the client in his tissues through this contact. That feeling quality first enters his system as *sensations*. Now sensations can and do act as analogs to *feelings*—jarring touch, threatening touch, soothing touch, hooked-up touch. The shift from a sensation quality to an emotional feeling quality is the transduction of a train of nerve impulses into a reinforcing neuropeptide release and distribution. As the process continues, contacts imparted by the practitioner stimulate more and more peptide release that corresponds to the emotional quality projected by the quality of touch. Eventually the client ends up in a feeling state—and a neuropeptide bath—that is more like the practitioner's. And remember, that feeling state is closely allied to Benson's Relaxation Response. It is a state in which the organism can much more easily access self-awareness, self-regulation, and repair of all kinds.

This is using touch to speak the ancient language of the molecules of emotion. It is a non-verbal language—more than that, it is a profoundly *pre*-verbal language that all organisms for all of the history of life have used to commune with themselves and to successfully adapt to their circumstances. It is speaking within us all the time. All that is necessary is that we slow down and listen, and reacquaint ourselves with the meanings of its silent intonations.

It is the improvement of this flow of information, and the reconnection with the personal Witness who listens to it all, that are the genuine healers. All that the therapist can really do is to re-introduce these things to the consciousness of the client. It is within each of us that our own templates are evolving, and it is only there that any of us can sort out the differences between a muddle of accumulating confused reactions and a conscious series of truly adaptive shifts in patterns that leads to cycles of renewal, not degeneration.

Escaping Virtual Reality

Living in realities divorced from the needs of our bodies, the false conundrum of nature versus nurture, the mythical distinction between subjective and objective—these are the situations and ideas that hold us in bondage while we witness degenerative cycles relentlessly overwhelm us. They all entail the core beliefs that we are passive agents in the midst of our own lives, buffeted about by genes and the environment, and that we can only be ministered to in our distress by experts who seek to reduce the complexity of human life to the rules of formal logic.

It is this very belief in our helplessness in a deterministic universe which dooms us to degeneration. The truth is that the complexities of human consciousness open up choices so vast that they positively terrify us. Something in us senses that the more choices we entertain, the more ways we will encounter to louse things up, and this introduces an extremely conservative constraint on our willingness to change. This uneasiness has become so collectively powerful that we would evidently prefer to adopt a belief in our powerlessness than face the burden of the responsibility of choice.

And the accumulative momentum of our physical patterns, our knee-jerk reactions, and our habits of mind do tend to take on an automatic repetitiveness that is powerful enough to effectively reinforce this belief. We may be intermittently aware somewhere in the backs of our minds that we have choices, but what we are faced with on a practical daily basis is the compelling nature of our jumble of habits and addictions—addictions to postures, to limited repertoires of movement, to our own stressful neuropeptide cocktails, to attitudes, to beliefs.

Habits are not easy to overcome, even when their pursuit is obviously detrimental. New behaviors take time and persistence to establish, and they seldom feel comfortable or natural at first. You have to behave in a reality before you can perceive in that reality. For personal change, for pattern change, for template change it is necessary to *act*, to voluntarily, willfully engage in the process of self-observation and self-regulation, and to stay with it over time—it takes several miles of ocean to turn a battle ship. And the belief that we are helpless in our situation is very often the decisive factor in our failure to cultivate new habits that can lead to positive change.

I believe the information that we are not helpless is the primary element that effective bodywork imparts: We are the embodiment of what we do, and we can change what we do once we know that we can, and once we have some rudimentary internal tools with which to begin. Awakening and empowering the Witness is powerful medicine. But it is much more than that; it is the return of our own fates to our own hands. The ability of the human will to initiate a new step in a new direction is our tiny lever of freedom, and once that lever is actively utilized it opens up vast extensions of consciousness and altogether new physical possibilities.

Says Pert, "Information theory releases us from the trap of reductionism and its tenets of positivism, determinism, and objectivism." Coupled with effective therapeutic guidance it can do far more than that. The path to enlightenment is within an organism that is flowing with unimpeded information the way it was designed to do, and Nature's most profound possibility for us is the ability to act so as to enhance that flow. And She has provided for us the instruction materials for the process—our bodies, our feelings, and a mind that can genuinely listen to them. If we learn to listen, we can be free.

What is now proved was once only imagined.
　　　　　　　　—WILLIAM BLAKE

Deane Juhan
Mill Valley, California
July 20, 1998

Epilogue:
Job's Body

As a matter of fact, I must start from the idea that I had a body, in other words, that I was continually under the threat of a two-fold danger, external and internal. But even there I spoke in that way only for convenience of expression. For the internal danger, such as a cerebral hemorrhage, is external, being of the body. And having a body constitutes the principle danger that threatens the mind.
—Marcel Proust, *Remembrance of Things Past*

Contemporary Job

Today we are able to appreciate the fantastic complexities of our bodies with a degree of concrete detail that was previously unknown. The physical and biological sciences have discovered—and continue to discover—the host of parts and interactions that work together to make up our living organisms, and with each new discovery our sense of awe for nature's wisdom and the perfection of the body deepens further. Modern physiology presents us with an expanding view of the miraculousness of our life process as a whole, and with clearer and clearer images of the many separate elements of that process.

And yet we also continue to discover that this miracle, the body, can be our greatest enemy—"the principle danger that threatens the mind." It often reacts in ways that appear capricious, unconscious, and cruel. The same laws and elements that govern it may also sicken and even destroy it with utterly no compunctions. The very cells and systems that are designed to sustain and protect us can—seemingly with perverse minds of their own—turn viciously against us and wreak havoc with our existence. There is an unsuspected evil in our garden, a dark side to our miracle, a tormentor in the midst of our very selves.

Nothing, in fact, can torment us as intensely and inescapably as our own tissues.

The very language that we frequently use to describe this predicament is indicative of the problematical position most of us have taken with regard to our bodies: "I"—a collection of conscious sensibilities—am threatened by my "body"—an aggregate of physical components. Nowhere does this Cartesian schizophrenia, this separation of mind from matter, create more catastrophic confusion than it does in our relationships to our bodies. Our experiences of pleasure, pain, health, or disease—indeed, our experience of life itself—are entirely subjective, internal impressions. And yet the only terms that can be allowed in identifying or dealing with these impressions in any practical way are objective, materially verifiable ones.

That is to say, we live largely in the subjective, but we will accept as practical and reliable knowledge about ourselves only such propositions as can be demonstrated to be rigorously objective. This fundamental dislocation of logic immediately forces us into insurmountable dilemmas. My body, in this view, is "external," apart from and categorically different than my mental sensibilities which "inhabit" it; my body is an object that is alien to the properties of my mind. And yet I cannot even observe it and learn to control it like I can any number of *other* objects because, paradoxically, I am too close to it: My own impressions about it are merely sensory and subjective and are therefore, we are told, almost certainly misleading or irrelevant. Any idea that may be formed by my sensibilities concerning the very object that is most inextricably me cannot be trusted and must await objective verification by experts before I can safely act upon it. "Know thyself" is no longer a meaningful philosophical dictum; on the contrary, it is an epistemological impossibility.

And the idea that is objectively verifiable, according to the current experts, is that our genes—coded chemical sequences coiled inside our cell's nuclei—are ultimately responsible for everything about us: our structure, the success or failure of our organic functions, and even the most significant features of our learned behavior. Some lip-service is paid to "nurturing" and "environmental factors," but for all intents and purposes the role of the genes is rather god-like with its dictatorial powers. The only means of fundamentally altering the shape or quality of life are either by manipulating these chemical codes (a project that is almost unimaginably complex, and which is proving to be full of serious pitfalls even as it is scarcely begun) or by designing procedures which directly alter the mechanical parts or the chemical balances previously developed by those codes. And these are technically formidable tasks which the average individual cannot hope to achieve.

So it would appear that we are in a nearly helpless situation when we are confronted with the ills of our bodies. The laws of nature in general being what they are, and our particular genetic inheritance being what it is, things could not be other with us than they are. We are caught in a network of material determinism which our own subjective impressions cannot adequately perceive and which our own efforts are powerless to alter.

Like Job, we live in our prosperity by divine dispensation—the free inheritance of the developments which through the ages have established the mechanisms necessary for healthy, thriving organisms. And like Job, we are always the potential victims of the darker side of this same dispensation: The very systems that support us can turn against us, simple causes and effects can turn into vicious circles, habits that we have learned innocently can destroy us. A fraction too little or too much of any one of the elements which make up our lives, and those lives become filled with pain and fear. And like Job, when these catastrophes befall us our first impulse is to appeal the the experts, who often turn out to be miserable comforters indeed.

Self-Responsibility

Some crucial factors in our situation are forgotten or paid scant attention in this view of the ultimate power of the gene, and in this process of objectifying and distancing the body so that it may be examined and treated according to the rules of modern science. One thing that seems to have been forgotten to a surprising degree is the lifelong plasticity of most of our bodies' tissues, and the manner in which our whole style of living molds this plasticity in innumerable ways that could never be anticipated by the coding of our DNA.

Whenever we examine a human being, we are observing both the concrete realization of a predetermined genetic blueprint and the collective results of upbringing, training, diet, habit, and attitudes which have been superimposed upon that basic blueprint. Our genetic materials are one thing. What we do with them is quite another. And what we are aware of doing with them is another thing again.

The amount and distribution of our muscle tissue—the most abundant tissue in our bodies—is almost entirely dependent upon our personal activity, and not upon any fixed genetic plan. The strength, flexibility, and function of all our connective tissues also depends largely upon our diet and upon our characteristic ranges of movement. Throughout our lives our bones are directed in their growth and are adjusted in their shapes and densities in response to the specific stresses we put upon them. Our pituitary/adrenal axis, and the functioning of the glands and organs that are engaged by the pituitary/adrenal secretions, are conditioned by our life experiences and the attitudes which those experiences foster. Our digestion, respiration, and circulation are all radically affected by what we are doing and what we are thinking or feeling moment to moment. Even the sprouting of neural axons and the habituation or facilitation that continually occurs at their synapses appears to depend to a considerable degree upon the specific qualities of experiences and responses that are going on inside us.

And, of course, over-arching all of these particular tissue conditions is the more or less successful orchestration of all their interacting functions—complex coordinations on conscious and unconscious levels that depend almost entirely upon our active engagement with objects around us and events within us in order to develop effectively. We have observed, for example, that newborn animals in the laboratory do not learn to defecate or urinate without timely and adequate tactile stimulation, and they die. Indeed, many of our organic functions can go seriously awry at any time in our lives without enough of the sensory stimulation that is needed to initiate and organize their responses. Even our image of our bodies is not at all clear without the voluntary movements that generate the streams of information in our joint receptors, our skin, our muscle spindles, and our Golgi tendon organs—information which is used by the mind to construct a coherent picture of our pieces and parts in space. And on the other hand, the degrees of imaging and self-control that focussed awareness and voluntary movement can produce are astonishing—down to the twitching of a single motor unit, the opening or closing of specific blood vessels, or even the deliberate altering of the salinity of our urine.

Once we have lost sight of these continual and life-long changes that are occurring in all of our tissues and in our over-all patterns of coordinated behavior, we inevitably lose sight of another most significant fact: We are personally responsible for the tendencies of those changes and for the results they create in our lives. Each of us must always decide whether to act or not, whether to focus our awareness or not, whether to continue a tendency or not. Nor is this responsibility something that can in any way be avoided or even postponed. By refusing to consciously confront issues and consequences, I only narrow my range of possible choices; by ignoring my options I make certain choices by default; and by passively accepting the notion that I have no real choices to make, perhaps I make the most irrevocable choice of all.

The truth of the matter is that in the relationships between mental states and physical states there simply is no hard and fast line between subjectivity and objectivity. My body is an object that inescapably conditions my perceptions, my thoughts, and my feelings. And these mental events in turn condition every cell, every organ, and every function of my physiology, for better or for worse. Subjective and objective are not two distinct ways I have of viewing reality; they are two sides of a continuous feedback loop which together make up that reality. How completely I sense my body and how I feel about it has everything to do with the particular course of events going on within it. Attitudes, postures, patterns of behavior, and physiological functions are inextricably fused together in our organisms, and it is primarily my conscious awareness of their interrelationships which gives me some measure of control over my well-being. Once I have lost this awareness and the conscious decisions that go with it, I have relinquished the conditioning and control of my organic processes; without continual self-reflection, my behavior increasingly becomes the expression of my most deeply ingrained traditional values and my most insistent unconscious compulsions—both of which are just as capable of leading us into circles of self-destruction as they are of reliably steering us through the dangers we encounter within and without.

If, then, we rely only upon our basic genetic materials, the traditional values with which we were raised, and our dominant compulsions, it seems clear that our chances for continued strength and health are limited at best. This sort of passivity leaves us groping in the dark, trusting to luck, so that when we get into a muddle we neither know how we got there nor how we might get out. And in the process our basic genetic materials themselves will almost certainly become eroded—not just by accidents and disease, but even more thoroughly by counter-productive attitudes, postures, conditioned reflexes, habits, vicious circles. These are the traps that await us when we abdicate our responsibility for our own condition, and they usually follow that abdication like night follows day.

Two things are necessary in order for me to break these vicious circles that erode my constitution and my health: Self-awareness and the will to participate actively in my own development. I need self-awareness to be able to accurately assess my situation as it stands—the locations of my parts and the qualities of their functions; and beyond that, I need to persevere over time in this self-awareness so that I can gain an understanding of my various systems' plasticity and the ways in which they alter according to my experiences and my responses. In addition to this expanding self-awareness, I also need the exertion of my will—the will to focus my attention upon symptoms I would often rather avoid, and the will to actively intervene in the personal habits and tendencies I perceive to be self-destructive.

Both of these things—awareness and will—are personal, active principles. I must voluntarily open myself up to the messages my body is sending me, I must consciously seek out the information that will clarify and complete my body image. And I must mentally and physically engage myself in the internal events which my awareness reveals if I am to have any hope of positively affecting their course. Active, personal involvement is absolutely crucial to this process of self-discovery and self-development; the only facts about myself that are altogether real to me are those that have come to me through my own attentiveness, and the only knowledge that is of any practical value is the knowledge I can actually use in the formation of my patterns of behavior. For these endeavors the mind does not need ideas about the body that can be read or told. It needs physical movement and concrete sensations, the raw data that directly informs, that activates the reflexes, and that organizes patterns.

The Role of Bodywork

Physical sensations—particularly the large variety of tactile sensations—are the foundations of self-awareness. Whenever I contact an object, two streams of information are opened up: one that gives me impressions about the object, and another that gives me impressions about the body part that is doing the contacting and its relationship to the rest of me. Likewise, whenever I make a movement I also initiate a double stream of sensory impressions: one that register the shape of the movement as a whole, and another that register the size, shape, and specific qualities of all the body parts that are rubbed and pressed and distorted against one another in the course of the movement. And unless I am either touching an external object or moving my internal parts against one another, I do not generate any concrete sensory impressions about myself for the mind to entertain.

Descartes' fundamental principle—"I think, therefore I am"—is in this regard a false beginning upon which to base our reflections. It begs innumerable questions because it fails to address the sources of consciousness that are in fact primary in the formation of my sense of existence and identity. Far more to the point in our actual modes of self-perception would be the assertion that "I *feel*, and therefore I am." *Thinking* about those primary sensory realities comes later, and always introduces a host of variables and complications.

And what is most significant about these physical sensations in the context of bodywork is that tactile stimulation does not only announce the presence and delineate the condition of our various tissues; it is also crucial to the successful *organization* of the functions of those tissues. Without an adequate amount of continual sensory impressions, the mind very quickly loses track of the body's myriad activities and so loses control over their regulation and coordination. This organizing role of sensory impressions is so central to the managing of our complex organisms that psychosis and death are the inevitable sequels to sustained sensory isolation. Sensation is information, information without which the mind simply cannot make accurate and timely judgements about the legions of factors that continually affect our well being. In situations where adequate physical contact and movement are impossible, or even problematical, the sensory input of skilled bodywork can be invaluable to the mind as a source of the necessary data.

The flow of sensory information useful to the mind is the operative principle in effective bodywork. Self-awareness is obviously dependent upon this flow; and the constructive exertion of our will power relies completely upon it as well. Until we have some concrete knowledge about how things are situated, and until we know in what ways and to what degrees they can change, we are generally not motivated to initiate those changes; nor do we know how to make meaningful, helpful changes even if we want to. We are literally feeling our way along the course of our lives, and until we have *felt* something more complete, more harmonious, we don't know how to *be* more complete and harmonious. "It is important to remember," we have observed a neurologist remarking, "that in its simplest form the nervous system is merely a mechanism by which a muscular movement can be initiated by *some change in the peripheral sensation,* say, an object touching the skin." Sensory input is a primary initiator and organizer of all levels of behavior.

Then bodywork is essentially *education.* And as is the case in most forms of education, it is important that it be pleasurable. The mind tends to recoil from pain and ignore unpleasantness when it can, and for this reason we usually retain facts better when they come to us associated with feelings of pleasure. Pleasure engages our relaxed attentiveness. And besides, the possibilities of the body as a source of pleasure are precisely what we normally forget when we are ill or injured or deeply confused. This does not mean that the titillating of the nerve ends ís the *point* of bodywork, or that its pleasures are nothing more than a temporary hedonistic indulgence. Clear sensory information is the *content* of effective bodywork; pleasure is simply the medium which is most apt to help us focus upon those contents openly enough and long enough to integrate them to the degree that they become genuinely useful to the mind in organizing its perceptions and responses. Any old sort of physical manipulation will not do for this process of education, any more than any old sort of instruction will prepare us to find and appreciate the delights in Shakespeare or mathematics.

Hence it is the particular *quality* of sensations, and not just their quantity, that is of central importance to effective bodywork. When our consciousness is distracted by disturbing physical symptoms or when it is trying to disassociate itself from a source of pain, it can be of utmost importance to remind ourselves that most of the body is still intact and continues to be a possible source of pleasure and of the insights that can help lead us out of the discomforts we experience. What we most need to recall when our body dismays us with one debilitation or another is that it is still our most intimate friend, and often our only source of the information most useful for our recovery, not "the principle danger that threatens the mind."

This is not by any means to suggest that the quality of pleasurableness is the only quality significant to effective bodywork. Touch can be superficial or penetrating, general or quite precise; it can evoke particular feelings or be quite neutral; it can mimic to a high degree the sensations that would accompany unrestricted and pain-free movements, or it can be merely an incoherent jumble of pressures and stretches. It can provoke in us altogether new concepts about our bodies, or it can just riffle over familiar territory. After all, any number of people can stroke us pleasantly without prompting us to refer to their touching as "bodywork."

It is in these distinctions that the intent, the training, and the experience of the practitioner become crucial. How to reach the depths without causing discomfort, how to focus pressure to produce particular sensations in particular places,

how to manipulate a limb in such a way as to convey a sense of optimal movement rather than random or restricted movement, how to bring into the sensory
foreground areas long since forgotten, how to alleviate compensations without
worsening the original injury—these, and not simply pleasure per se, are the
specific qualities that can contribute to increased awareness, enhanced organization, improved health.

Practiced touch can not only deliver a collection of sensory impressions, but it
can also impart those impressions in such a way as to convey a smoother style, a
larger repertoire, a greater flexibility, and a finer appropriateness to our movements. It can help us to learn—in ways that our upbringing did not—a whole new
manner of sensing and behaving. It can help us to learn to more accurately assess
our condition, to identify and resolve stress, to reverse vicious circles, to move
towards health rather than towards increasing involvement with our infirmities.
It can help us to establish the new sensory engrams and master the new conditioned responses that are necessary for successfully breaking out of our
ingrained patterns and our compulsions.

Let us not suppose that bodywork can ever be anything like a "cure-all." There
are many physical and mental misfortunes that do in fact result from genetic
anomalies, many kinds of severe accidental traumata, many external
mechanisms of disease that invade us. For these we need to continually improve
the devices and procedures of modern medicine.

But there is also a vast array of ills that cannot improve until our conscious
relationship to our bodies improves, until we learn to sense more exactly what we
are made of and how we undergo internal change. True, there are amputations
we cannot replace, damages we cannot fix, diseases we cannot cure; but there are
ways in which we might more successfully *manage* their results and make life
more productive and more comfortable in spite of them. There are recoveries
that can be more complete, degenerations that can be slowed down, pitfalls in the
courses of chronic illnesses that can be avoided. And there are intractable suffer-
ings that can be mitigated in important ways even when we cannot give the
promise, or even the hope, of a final cure.

In short, we can expect bodywork to be helpful in any situation where
heightened self-awareness and improved control over conditioned responses
might be of constructive use. These kinds of situations include a variety of human
ills that is wide indeed. Bodywork will not replace the resources of the modern
physician, nor should it seek to do so. But there is certainly no reason for it not to
be a part of these resources. It can offer the patient a kind of information about
himself and a depth of insight into his personal involvement with his problem
that no lecturing, no prescription, and no surgery could ever impart.

Awareness is the only medicine that accomplishes its healing without disrupt-
ing the natural functions of the organism. In the face of his adversities Job cried,
"Is not my help within me?" The answer is yes. But it is up to each of us to seek
this help out, recognize it, develop it, and learn to use it successfully. We

inhabit—we are—the most complex and marvelous manifestation of nature that we know. The more we inform ourselves about those marvels and complexities, the more success we can expect in utilizing our inheritance. And the more we distance ourselves from the organic intimacies that comprise us, the more problems we can be sure of encountering.

> Before the bar of nature and fate, unconsciousness is never accepted as an excuse; on the contrary, there are severe penalties for it.

Bodywork is one of the readiest and most effective antidotes to this unconsciousness.

Notes

Introduction

[1] Kamenetz, H. L., "History of Massage," In *Manipulation, Traction, and Massage,* 2nd ed., edited by Joseph B. Rogoff, Williams and Wilkins, Baltimore, 1980, pp. 37-8. I am indebted to Dr. Kamenetz for my brief history of massage, which he treats at much greater length in this chapter. Quoted here by permission.

[2] Rolf, Ida P., *Rolfing,* Dennis-Landman, Santa Monica, California, 1977, p. 180. Quoted here by permission.

[3] Ibid., p. 153.

[4] Quoted by Juhan, D. in "The Trager Approach: Psychophysical Integration and Mentastics," *The Bodywork Book,* edited by Neville Drury, Prism Alpha Press, Dorset, England, 1984, pp. 36-7.

[5] Schilder, P., *The Image and Appearance of the Human Body,* International Universities Press, Inc., New York, 1950, p. 70. Quoted here by permission.

[6] Ibid., p. 87.

[7] Quoted from an anonymous review of Anna Halprin's "Movement Ritual," *Somatics* magazine, 1980, p. 24. Quoted here by permission.

1. Job's Dilemma

[1] *Job,* Chapter 19, verses 26 and 27, *The Book of God and Man,* translated by Robert Gordis, University of Chicago Press, London and Chicago, 1965.

[2] Dali, S., *The Secret Life of Salvador Dali,* translated by Haakon M. Chevalier, Dial Press, New York, 1942, p. 3. Quoted here by permission.

[3] Ibid., p. 2.

[4] Todd, M. E., *The Thinking Body,* Dance Horizons Inc., Brooklyn, NY, 1979, p. 8. Quoted here by permission.

[5] Bohm, D., and Weber, R., "Nature as Creativity," in *ReVision* journal, vol. 5, no. 2, Fall 1982, pp. 36-7.

[6] Merleau-Ponty, M., *Phenomenologie de la perception,* 1945, p. 231, excerpt translated by Father Aelrid Squire in Chapter 5 of his *Asking the Fathers,* S.P.C.K., London, 1973.

[7] Sheldrake, R. and Weber, R. V., "Morphogenic Fields: Nature's Habits?," in *ReVision* journal, vol. 5, no. 2, Fall 1982, p. 30. Quoted here by permission.

[8] Todd, p. 24.

2. Skin

[1] Montagu, A., *Touching: The Human Significance of the Skin,* Harper and Row, New York, 1971, p. 4. Much of the thrust of my arguments in this chapter, and many of my references, are indebted to this marvelous book by Mr. Montagu. When utilizing ideas or articles to which Mr. Montagu refers, I will note the original source when possible; otherwise I will refer the reader to the appropriate pages in his book. Material used by permission of Mr. Montagu.

[2] *The Encyclopaedia Britannica,* 15th ed., Chicago, 1976, 16:842.

[3] Montagu, p. 4 *Britannica,* 16:840.

[4] ———, pp. 4-5, *Britannica*, 16:839.

[5] *Britannica*, 16:844.

[6] Ibid., 16:839.

[7] Ibid., 16:840-1.

[8] Ross, R., "Wound Healing," in *Scientific American*, June 1969.

[9] *Britannica*, 16:547.

[10] Montagu, p. 1.

[11] Russell, B., *The ABC of Relativity*, Harper Brothers, New York, 1925 (in Mongagu, p. 6).

[12] Montagu, p. 102.

[13] Ibid., p. 80.

[14] *Britannica*, 16:549.

[15] Guyton, A.C., *Textbook of Medical Physiology*, 6th ed., W.B. Saunders Co., Philadelphia, 1981. p. 597.

[16] Schilder, P., *The Image and Appearance of the Human Body*, International Universities Press, New York, 1950, p. 124.

[17] Montagu, p. 230.

[18] Guyton, p. 578.

[19] Sperry, R.W., "The Growth of Nerve Circuits," *Scientific American*, Nov. 1959. Quoted here by permission.

[20] Ibid.

[21] Gardner, L.I., "Deprivation Dwarfism," *Scientific American*, July 1972. Quoted here by permission.

[22] Ibid.

[23] Ibid.

[24] Ibid.

[25] Ibid.

[26] Rosenzweig, M.R., Bennett, and Diamond, E.L., and Diamond, M.C., "Brain Changes in Response to Experience," *Scientific American*, Feb. 1972. Quoted here by permission.

[27] Rosenblatt, J.S. and Lerhman, D.S., "Maternal Behavior in the Laboratory Rat," in H.L. Rheingold (ed.), *Maternal Behavior in Mammals*, Wiley, New York, 1963, p. 14 (in Montagu, pp. 18-9).

[28] Roth, L.L., "Effects of Young and Social Isolation on Maternal Behavior in the Virgin Rat," *American Zoologist*, vol. 7, 1967, p. 800 (in Montagu, p. 33).

[29] Montagu, p. 16.

[30] Ibid., p. 14.

[31] Levine, S., "Stimulation in Infancy," *Scientific American*, May 1960.

[32] Ibid.

[33] Ibid.

[34] Rosenzweig, Bennett, and Diamond.

[35] Montagu, pp. 187-8.

[36] Rosenzweig, Bennett, and Diamond.

[37] Montagu, pp. 236, 317-8.

[38] Harlow, H., "Love in Infant Monkeys," *Scientific American,* June 1959. Quoted here by permission.

[39] Ibid.

[40] Montagu, p. 30.

[41] Ibid., p. 170.

[42] Ibid., p. 206.

[43] Forer, B.R., "The Taboo Against Touching in Psychotherapy," *Psychotherapy: Theory, Research, and Practice,* Vol. 6, 1969, p. 230 (in Montagu, p. 223). Quoted here by permission.

[44] Prescott, J.H., "Body Pleasure and the Origins of Violence," *The Futurist,* April 1975, pp. 64-5; "Early Somatosensory Deprivation as an Ontogenetic Process in the Abnormal Development of the Brain and Behavior," in E.I. Goldsmith and J. Moor-Janowski (eds.), *Medical Primatology,* S. Karger, New York, 1971, pp. 1-20 (in Montagu, pp. 177-8). Quoted here by permission.

[45] Mead, M., "Sex and Temperament in Three Primitive Societies," William Morrow, New York, 1935, pp. 40-1 (in Montagu, pp. 256-7). Quoted here by permission.

[46] Montagu, p. 257.

[47] Ibid., p. 258.

[48] Ibid., p. 257.

[49] Ibid.

3. Connective Tissue

[1] Gross, J., "Collagen," *Scientific American,* May, 1961.

[2] Verzar, F., "The Aging of Collagen," *Scientific American,* April, 1963.

[3] Gross.

[4] Snyder, G.E., "Fascia—Applied Anatomy and Physiology," in *The Journal of the American Osteopathic Association,* 68:675-685, March 1969, p. 677.

[5] *Britannica,* Micropaedia 6:817.

[6] Snyder, p. 66.

[7] Gross.

[8] Snyder, p. 70.

[9] Ibid.

[10] Ibid., p. 67.

[11] Ross, R. "Wound Healing," *Scientific American,* June, 1967.

[12] Little, K.E., "Toward More Effective Manipulative Management of Chronic Myofascial Strain and Stress Syndromes," in *The Journal of the American Osteopathic Association,* 68:675-685, p. 679.

[13] Taylor, R.B., "Bioenergetics of Man," in *Academy of Applied Osteopathic Association,* 68:675-685, p. 679.

[14] Little, p. 679.

[15] Doty, P., "Proteins," *Scientific American,* Sept. 1957.

[16] Verzar.

[17] Erlingheuser, R.F., "The Circulation of Cerebrospinal Fluid Through the Connective Tissue System," in the *Academy of Appled Osteopathy Yearbook,* 1959.

[18] Robbie, D.L., "Tensional Forces in the Human Body," in the *Orthopaedic Review,* vol. 6, no. 11, Nov. 1977.

[19] Snyder, p. 71.

[20] Ibid.

[21] Gardner, L.

[22] Snyder, p. 72.

[23] Ibid.

4. Bone

[1] Romanes, G. J., ed., *Cunningham's Textbook of Anatomy,* Oxford University Press, London, 1972, p. 75.

[2] Ibid., p. 78.

[3] Lockhart, R.D., Hamilton, G.F., and Fyfe, F.W., *Anatomy of the Human Body,* J. B. Lippincott Co., Philadelphia, 1969, p. 12. Quoted here by permission.

[4] Cunningham, p. 75.

[5] Vander, A. J., Sherman, J.H., and Luciano, D.S., *Human Physiology — The Mechanisms of Body Function,* 2nd ed., McGraw-Hill Co., New York, 1970, p. 217.

[6] This brief description of the formation of bone is presented in more detail in Cunningham, pp. 79-82.

[7] Lockhart, Hamilton, and Fyfe, p. 15.

[8] Ibid., p. 11.

[9] Romanes, p. 83.

[10] Lockhart, Hamilton, and Fyfe, p. 18.

[11] Romanes, p. 84.

[12] Lockhart, Hamilton, and Fyfe, p. 18.

[13] Romanes, p. 84.

[14] Ibid., p. 85.

[15] Guyton, p. 980.

[16] Romanes, p. 83.

[17] Vander, Sherman, and Luciano, pp. 346-7.

5. Muscle

[1] Lockhart, Hamilton, and Fyfe, p. 21.

[2] The following example, and the reflections upon its significance, are from the observations of Dr. Milton Trager, made during his internship at St. Francis Hospital in Honolulu, Hawaii in 1955.

[3] It is the skeletal, or striated muscles upon which we are presently focusing our attention. These are distinct in some important ways from the smooth muscles of the glands and organs, and the cardiac muscle of the heart.

[4] Little, K., "Toward More Effective Manipulative Management of Chronic Myofascial Strain and Stress Syndromes," in *The Journal of the American Osteopathic Association,* 68:675-685, March 1969.

[5] Cloud, P., "The Biosphere," *Scientific American,* Sept, 1983.

[6] This explanation has been considerably truncated and simplified. For fuller discussions of these events, the reader is referred to Guyton, A.C., *Textbook of Medical Physiology,* W.B. Saunders Co., Philadelphia, 1981, pp. 124-7; Vander, A.J., Sherman, H.H., and Luciano, D.S. *Human Physiology: The Mechanisms of Body Function,* pp. 193-204; Murray, J.M. and Weber, A., "The Cooperative Action of Proteins," *Scientific American,* Feb. 1974.

[7] This description of the Krebs cycle has been extremely condensed. For a much fuller and more detailed analysis of the steps involved in the reconstituting of ATP, the reader is referred to Guyton, A.C., pp. 842-5.

[8] Guyton, A.C., *Basic Human Physiology: Normal Function and Mechanisms of Disease,* W.B. Saunders Co., Philadelphia. Quoted here by permission.

[9] Lockhart, Hamilton, and Fyfe, p. 152.

[10] Little, K., p. 676.

[11] Vander, Sherman, and Luciano, pp. 412-13.

6. Nerve.

[1] Lockhart, Hamilton, and Fyfe, p. 269.

[2] Ibid., p. 264.

[3] Ibid., p. 366.

[4] Guyton, A.C., *Textbook of Medical Physiology,* p. 66.

[5] Lockhart, Hamilton, and Fyfe, p. 364.

[6] Vander, Sherman, and Luciano, 1975, p. 558. Quoted here by permission.

[7] This brief recapitulation of neural development from hydras to earthworms is condensed from Romanes, G.J., *Cunningham's Textbook of Anatomy,* 11th ed., Oxford Universtiy Press, London, 1978, pp. 581-3. Quoted here by permission.

[8] Lockhart, Hamilton, and Fyfe, p. 263.

[9] Romanes, p. 583.

[10] Ibid., p. 600.

[11] Lockhart, Hamilton, and Fyfe, p. 367.

[12] Ibid., p. 269.

[13] Vander, Sherman, and Luciano, p. 536.

7. Muscle as Sense Organ

[1] Schilder, P., *The Image and Appearance of the Human Body,* International Universities Press, Inc., New York, 1950, p. 101. Quoted here by permission.

[2] Ibid., pp. 112-3 [italics mine—D.J.].

[3] Ibid., p. 87.

[4] Ibid.

[5] Rutherford, D., "Auditory-Motor Learning and the Acquisition of Speech," *American Journal of Physical Medicine,* vol. 46, no. 1, February, 1967, p. 246. Quoted here by permission.

[6] Ibid., p. 245.

[7] Rankin, J. and Dempsey, J., "Respiratory Muscles," *American Journal of Physical Medicine, vol. 46, no. 1, February, 1967, p. 231.*

[8] Merton, P.A., "How We Control the Contraction of Our Muscles," *Scientific American,* May, 1972. Quoted here by permission.

[9] Guyton, A.C., *Basic Human Physiology: Normal Function and the Mechanism of Disease,* p. 461. Quoted here by permission.

[10] Ibid., p. 427 [italics mine—D.J.].

[11] Lockhart, Hamilton, and Fyfe, p. 378.

[12] Snider, R.S., "The Cerebellum," *Scientific American,* August 1958.

[13] Ibid.

[14] Lockhart, Hamilton, and Fyfe, p. 377.

[15] Snider, R.S.

[16] Ibid.

[17] Ibid.

8. The Sense of Effort

[1] Romanes, p. 581. Quoted here by permission.

[2] Merton, P.A.

[3] Ibid.

[4] Todd, pp. 31-2.

[5] Schilder, P., *The Image and Appearance of the Human Body,* International Universities Press, Inc. New York, 1970, p. 93. Quoted here by permission.

[6] Ibid., p. 92.

[7] Ibid.

[8] Ibid. [Italics mine—D.J.].

[9] Ibid. p. 78.

[10] Ibid. p. 77.

[11] Fong-Torres, B., "Shearing's Joy: Sighted Folks at Ease," in the *San Francisco Chronicle,* Sunday Datebook, Nov. 11, 1984, p. 42. Quoted here by permission.

[12] Rosen, C., "The New Sound of Franz Liszt," in *The New York Review of Books,* vol. 31, no. 6, April 12, 1984, p. 18. Quoted here by permission.

9. Sensory Engrams

[1] Merton, P.A.

[2] Guyton, A.C., *Textbook of Medical Physiology, p. 668.*

[3] Ibid.

[4] For a fuller description of the birth engram in this light, see Grof, S., *Realms of the Human Unconscious,* E.P. Dutton, New York, 1976.

10. Movements Toward Disease and Movements Toward Health

[1] Held, R., "Plasticity in Sensory-Motor Systems," *Scientific American,* Nov. 1965. Quoted here by permission.

[2] Merton, P.A. [Italics mine—D.J.]

[3] Guyton, A.C., Basic Human Physiology: Normal Function and Mechanisms of Disease, p. 494.

[4] Evarts, E., "Brain Mechanisms of Movement," Scientific American, Sept. 1979.

[5] Guyton, p. 472.

[6] Vander, Sherman, and Luciano, p. 539.

[7] Ibid., p. 547.

[8] William James, quoted by Evarts.

[9] *Dancemagazine,* June 1982, p. 80. [Italics mine—D.J.]. Quoted here by permission.

[10] "Profile: "The Winchester Kids," *The New Yorker,* May 9, 1977.

[11] Sacks, O., *A Leg To Stand On,* Summit Books, New York, 1984, p. 129.

[12] Ibid., pp. 135-6.

[13] Ibid., p. 135.

[14] Ibid., p. 143.

[15] Ibid., p. 144-5

[16] Ibid., p. 145-6

[17] Evarts.

[18] Nauta, W.J.H., and Feirtag, M., "The Organization of the Brain," *Scientific American,* Sept. 1979.

[19] Vander, Sherman, and Luciano, p. 166.

[20] Ibid.

[21] Wallace, R.K. and Benson, H., "The Physiology of Meditation," *Scientific American,* Feb. 1972. Quoted here by permission.

[22] Ibid.

[23] Romanes, p. 775.

[24] DiCara, L.V., "Learning in the Autonomic Nervous System," *Scientific American,* Jan. 1970. Quoted here by permission.

[25] Ibid.

[26] Ibid. [Italics mine—D.J.].

[27] Kandel, E.R., "Small Systems of Neurons," *Scientific American,* Sept. 1979. Quoted here by permission.

[28] Ibid.

[29] Ibid.

[30] Levine, S., "Stress and Behavior," *Scientific American,* Jan. 1971. Quoted here by permission.

[31] Constantinades, P.C. and Carey, N., "The Alarm Reaction," *Scientific American,* March 1949. Quoted here by permission.

[32] Levine.

[33] Ibid.

[34] Vander, A.J., *Readings From Scientific American: Human Physiology and the Environment in*

Health and Disease, W.H. Freeman Co., San Francisco, 1976, p. 134. Quoted by permission.

[35] Levine.

[36] Levine, S., "Stimulation in Infancy," *Scientific American*, May 1960. Quoted here by permission.

[37] ———, "Stress and Behavior."

[38] Ibid.

[39] Vander, Sherman, and Luciano, p. 501.

[40] Wallace, R.K. and Benson, H.

[45] Levine, S., "Stress and Behavior."

[46] Ibid.

[47] Weiss.

[48] Wallace, R.K. and Benson, H.

[49] Constantinides, P.C. and Carey, N.

[50] Ibid.

[51] Weiss.

[52] Constantinides, P.C. and Carey, N.

[53] Ibid.

[54] Ibid.

[55] *Britannica*, 16:550.

[56] Vander, Sherman, and Luciano, p. 509.

[57] Guyton, p. 406.

[58] Ibid., p. 405.

[59] Ibid., p. 408

[60] Vander, Sherman, and Luciano, p. 511.

[61] Ibid., p. 510

[62] Ibid.

[63] Guyton, p. 465.

[64] Vander, Sherman, and Luciano, p. 555 [Italics mine—D.J.].

11. Further Views at the Turn of the Century

[1] Oschman, James L., *Readings on the Scientific Basis of Bodywork,* Nature's Own Research Association, Dover, NH, 1993. All of the quotes and new information in this second edition, as well as Oschman's references to cited research, are included in this collection of articles, which is available by contacting N.O.R.A., box 5101, Dover, NH 03820.

[2] Pert, Candace, *Molecules of Emotion,* Scribner, New York, 1997. All of the included quotes and the information specifically relating to neuropeptides are from this book, which summarizes much of Pert's research.

[3] See Darwin's *Expression of Emotions in Man and Animals.*

Epilogue: Job's Body

[1] Jung, C.G., "An Answer to Job," The Viking Portable Jung, ed. Joseph Campbell, Viking Press, New York, 1971, p.637. Quoted here by permission.

Sources of Illustrations

Illustrations are listed sequentially by chapter. The Publisher gratefully acknowledges the artists, authors, and publishers for permission to use these illustrations from the following sources:

2. Skin

2-1: *The Encyclopaedia Brittanica,* vol. 4, H. H. Benton, Chicago, p. 860.

2-2: Vander, A. J., Sherman, J.H., and Luciano, D.S., *Human Physiology: the Mechanisms of Body Function,* McGraw-Hill Co., New York, 1970, p. 56.

2-3: Brachet, J., "The Living Cell," *Scientific American,* Sept. 1961.

2-4: Bretscher, M.S., "The Molecules of the Cell Membrane," *Scientific American,* Oct. 1985.

2-5: Ibid.

2-6: Vander, Sherman, and Luciano, p. 49.

2-7: Bretscher.

2-8: Ibid.

2-9: Montagna, W., "The Skin," *Scientific American,* Feb. 1965.

2-10: *Brittanica,* vol. 16, p. 841.

2-11: (A) Guyton, A.C., *Textbook of Medical Physiology,* W.B. Saunders Co., Philadelphia, 1981, p. 589.
(B) Nilsson, L., *Behold Man,* Little, Brown and Co., Boston, 1973, p. 225.

2-12: (A) Romanes, G. J., *Cunningham's Textbook of Anatomy,* Oxford University Press, London, 1978, p. 796.
(B) Kandel, E.R. and Schwartz, J.H., *Principles of Neural Science,* Elsevier/North-Holland, New York, 1981, p. 166.
(C) Schlossberg, L., *The Johns Hopkins Atlas of Human Functional Anatomy,* The Johns Hopkins University Press, Baltimore, 1977, p. 97.
(D) Romanes, p. 798.

2-13: Eldred, E., "Peripheral Receptors: Their Excitation and Relation to Reflex Patterns," *American Journal of Physical Medicine,* vol. 46, no. 1, Feb. 1967, p. 71.

2-14: Romanes, p. 801.

2-15: Grey, G.W., "The Organizer," *Scientific American,* Nov. 1957.

2-16: Romanes, p. 41.

2-17: Ibid., p. 28.

2-18: Guyton, p. 603.

2-19: Ibid., p. 578.

2-20: (A) Kandel and Schwartz, p. 171.
(B) Ibid., p. 309.

2-21: Vannini, V. and Pogliani, G., eds., *The Color Atlas of Human Anatomy,* translated and revised by Dr. Richard T. Jelly, Harmony Books, New York, 1980, p. 51.

2-22: Kandel and Schwartz, p. 517.

2-23: Cowan, W.M., "The Development of the Brain," *Scientific American,* Sept. 1979.

2-24: Netter, F.H., *The Ciba Collection of Medical Illustrations,* vol. 1, *Nervous System,* CIBA, New Jersey, 1983, p. 197.

3. Connective Tissue

3-1: Guyton, p. 42.

3-2: Ibid., p. 4.

3-3: Vander, Sherman, and Luciano, p. 40.

3-4: Vesalius, A., *The Illustrations From the Works of Andreas Vesalius of Brussels,* with annotations and translations, by J.B. de C. M. Saunders and Charles D. O'Malley, Dover Publications, New York, 1973, p. 87, p. 89.

3-5: (A) Romanes, p. 315.
      (B) Ibid., p. 268.

3-6: Singer, E., *Fasciae of the Human Body and Their Relationships to the Organs They Envelop,* Williams and Wilkins, Baltimore, 1935, p. 97.

3-7: Caplan, A.I., "Cartilage," *Scientific American,* Oct. 1984.

3-8: Gross, J., "Collagen," *Scientific American,* May, 1961.

3-9: Ross, R., "Wound Healing," *Scientific American,* June 1969.

3-10: Kessel, R.G. and Kardon, R.H., *Tissues and Organs: A Text-Atlas of Scanning Electron Microscopy,* W.H. Freeman and Co., San Francisco, 1979, p. 12.

3-11: Ibid., p. 11.

3-12: Ibid., p. 15.

3-13: Ibid.

3-14: Gross.

3-15: Ibid.

3-16: Ibid.

3-17: Ibid.

3-18: Ibid.

3-19: Doty, P., "Proteins," *Scientific American,* Sept. 1957.

3-20: Ibid.

3-21: Ibid.

3-22: Kessel and Kardon, p. 139.

3-23: Romanes, p. 210.

3-24: Ibid., p. 328.

3-25: Kessel and Kardon, p. 16.

3-26: Ibid., p. 41.

3-27: Romanes, p. 337.

3-28: Kessel and Kardon, p. 79.

3-29: Romanes, p. 885.

3-30: Ibid., p. 440.

3-31: Ibid., p. 916.

3-32: Juhan, D.

3-33: Robbie, D.L., "Tensional Forces in the Human Body," *Orthapaedic Review,* vol. 6, no. 11 Nov. 1977, p. 47.

3-34: Lockhart, R.D., Hamilton, G.F., and Fyfe, F.W., *Anatomy of the Human Body,* J.B. Lippincott Co., Philadelphia, 1972, p. 170.

3-35: Rolf, I., *Rolfing. The Integration of Human Structures,* Dennis-Landman, Santa Monica, 1977, p. 39.

4. Bone

4-1: Caplan.

4-2: Ibid.

4-3: Kessel and Kardon, p. 21.

4-4: Romanes, p. 79.

4-5: Kessel and Kardon, p. 29.

4-6: Ibid.

4-7: Ibid., p. 28.

4-8: Romanes, p. 80.

4-9: Kessel and Kardon, p. 24

4-10: Ibid., p. 25.

4-11: (A) Guyton, p. 979.
     (B) *Brittanica,* vol. 3, p. 20.

4-12: Kessel and Kardon, p. 24.

4-13: (A) Romanes, p. 97.
     (B) Ibid., p. 76.

4-14: *National Geographic,* May 1984, p. 604.

5. Muscle

5-1: Vesalius, p. 87, p. 89.

5-2: Romanes, p. 259.

5-3: Vander, Sherman, and Luciano, p. 195.

5-4: Kessel and Kardon, p. 139.

5-5: (A) Huxley, H.E., "The Mechanism of Muscular Contraction," *Scientific American,* Dec. 1965.
     (B)——, "The Contraction of Muscle," *Scientific American,* Nov. 1958.

5-6: Murray, J.M. and Weber, A., "The Cooperation of Muscle Proteins," *Scientific American,* Feb. 1974.

5-7: Huxley, H.E., "The Mechanism of Muscular Contraction."

5-8: Ibid.

5-9: Ibid.

5-10: Ibid.

5-11: (A) Guyton, p. 125.
     (B) Ibid., p. 127.

5-12: Murray and Weber.

5-13: Kessel and Kardon, p. 143.

5-14: Murray and Weber.

5-15: Huxley, H.E., "The Contraction of Muscle."

5-16: Vander, Sherman, and Luciano, p. 194.

5-17: Ibid., p. 193.

5-18: Huxley, H.E., "The Mechanism of Muscular Contraction."

5-19: Kessel and Kardon, p. 141.

5-20: Porter, K. and Franzini-Armstrong, C., "The Sarcoplasmic Reticulum," *Scientific American,* March 1965.

5-21: Vander, Sherman, and Luciano, p. 201.

5-22: Ibid., p. 73.

5-23: Ibid., p. 75.

5-24: Ibid., p. 82.

5-25: Chapman, C.B. and Mitchell, J.H., "The Physiology of Exercise," *Scientific American,* May 1965.

5-26: Guyton, p. 655.

5-27: Basmajian, J.V., "Control of Individual Motor Units," *American Journal of Physical Medicine,* vol. 46, no. 1, 1967, p. 481.

5-28: Hubel, D.H., "The Brain," *Scientific American,* Sept. 1979.

5-29: (A) Lester, H.A., "The Response to Acetylcholine," *Scientific American,* Feb. 1977.
(B) Guyton, p. 138.
(C) Vander, Sherman, and Luciano, p. 203.

5-30: Juhan, D.

6. Nerve

6-1: Cowan, M.W. J., "The Development of the Brain," *Scientific American,* Sept. 1979.

6-2: (A)Kandel and Schwartz, p. 144.
(B)Katz, B., "How Cells Communicate," *Scientific American,* Sept. 1961.

6-3: Guyton, p. 561.

6-4: Stevens, C.F., "The Neuron," *Scientific American,* Sept. 1979.

6-5: Kessel and Kardon, p. 77.

6-6: Stevens.

6-7: Ibid.

6-8: Vander, Sherman, and Luciano, p. 136.

6-9: Kandel and Schwartz, p. 34.

6-10: Guyton, p. 394.

6-11: ———, *Basic Human Physiology: Normal Function and Mechanisms of Disease,* W.B. Saunders Co., Philadelphia, 1971, p. 49.

6-12: Ibid.

6-13: ———, *Textbook of Medical Physiology,* p. 112.

6-14: Stevens.

6-15: Katz.

6-16: Guyton, Ibid., p. 569.

6-17: Iverson, L.L., "The Chemistry of the Brain," *Scientific American,* Sept. 1979.

6-18: Vander, Sherman, and Luciano, p. 168.

6-19: Katz.

6-20: Guyton, Ibid., p. 658.

6-21: Romanes, p. 734.

6-22: Guyton, Ibid., p. 640.

6-23: Hubel.

6-24: (A) Vander, Sherman, and Luciano, p. 545.
     (B) Ibid., p. 544.

7. Muscle as Sense Organ

7-1: Vander, Sherman, and Luciano, p. 254.

7-2: Smith, K.U. and Henry, J.P., "Cybernetic Foundations for Rehabilitation," *American Journal of Physical Medicine,* vol. 46, no. 1, Feb. 1967, p. 382.

7-3: Netter, p. 185.

7-4: Merton, P.A., "How We Control the Contraction of Our Muscles," *Scientific American,* May 1972.

7-5: Romanes, p. 802.

7-6: Vander, Sherman, and Luciano, p. 542.

7-7: Lippold, O., "Physiological Tremor," *Scientific American,* March 1971.

7-8: Vander, Sherman, and Luciano, p. 541.

7-9: Kandel and Schwartz, p. 19.

7-10: Ibid., p. 296.

7-11: Ibid., p. 299.

7-12: Merton.

7-13: Eldred, p. 83.

7-14: Kandel and Schwartz, p. 297.

7-15: Buchwald, J.S., "Exteroceptive Reflexes and Movement," *American Journal of Physical Medicine,* vol. 46, no. 1, Feb. 1967, p. 123.

7-16: Kandel and Schwartz, p. 302.

7-17: Guyton, Ibid., p. 562.

7-18: Eldred, E., "Functional Implications of Dynamic and Static Components of the Spindle Response to Stretch," *American Journal of Physical Medicine,* vol. 46, no. 1, Feb. 1967, p. 132.

7-19: Guyton, Ibid., p. 634.

7-20: Romanes, p. 708.

7-21: Netter, p. 186.

7-22: Vander, Sherman, and Luciano, p. 543.

7-23: Guyton, Ibid., p. 672.

7-24: Ibid., p. 648.

7-25: Eldred, Ibid., p. 130.

7-26: Ibid., p. 133.

7-27: Ibid., p. 134.

7-28: Snider, R.S., "The Cerebellum," *Scientific American,* Aug. 1958.

7-29: Netter, p. 189.

7-30: Snider.

7-31: Ibid.

7-32: Guyton, Ibid., p. 660.

7-33: Snider.

10. Movements Toward Disease and Movements Toward Health

10-1: Buchwald, J.S., "A Functional Concept of Motor Control," *American Journal of Physical Medicine,* vol 46, no. 1 Feb. 1967, p. 71.

10-2: Guyton, Ibid., p. 789.

10-3: Snyder, S.H., "The Molecular Basis of Communication Between Cells," *Scientific American,* Oct. 1985.

10-4: (A) Vander, Sherman, and Luciano, p. 164.
       (B) Guyton, Ibid., p. 711.
       (C) Ibid., p. 710.
       (D) Ibid., p. 720.

10-5: Levine, S., "Stimulation in Infancy," *Scientific American,* May 1960.

10-6: (A) Vander, Sherman and Luciano, p. 180.
       (B) Ibid., p. 182.
       (C) Ibid., p. 185.

10-7: Netter, p. 210.

10-8: Levine, S., "Stress and Behavior," *Scientific American,* Jan. 1971.

10-9: Snyder, S.H., "Opiate Receptors and Internal Opiates," *Scientific American,* March 1977.

10-10: Guyton, Ibid., p. 617.

Works Cited

Basmajian, J.V., "Control of Individual Motor Units," *American Journal of Physical Medicine,* vol. 46, no. 1, 1967.

Bohm, D. and Weber, R., "Nature as Creativity," in *ReVision* journal, vol. 5, no. 2, Fall 1982.

Brachet, J., "The Living Cell," *Scientific American,* Sept. 1961.

Bretscher, M.S., "The Molecules of the Cell Membrane," *Scientific American,* Oct. 1985.

Buchwald, J.S., "A Functional Concept of Motor Control," *American Journal of Physical Medicine,* vol. 46, no. 1, Feb. 1967.

——, "Exteroceptive Reflexes and Movement," *American Journal of Physical Medicine,* vol. 46, no. 1, Feb. 1967.

Caplan, A.I., "Cartilage," *Scientific American,* Oct. 1984.

Chapman, C.B. and Mitchell, J.H., "The Physiology of Exercise," *Scientific American,* May 1965.

Cloud, P., "The Biosphere," *Scientific American,* Sept. 1983.

Constantinades, P.C. and Carey, N., "The Alarm Reaction," *Scientific American,* March 1949.

Cowan, W.M., "The Development of the Brain," *Scientific American,* Sept. 1979.

Dali, S., *The Secret Life of Salvador Dali,* translated by Haakon M. Chevalier, Dial Press, New York, 1942.

DiCara, L.V., "Learning in the Autonomic Nervous System," *Scientific American,* Jan. 1970.

Doty, P., "Proteins," *Scientific American,* Sept. 1957.

Eldred, E., "Functional Implications of Dynamic and Static Components of the Spindle Response to Stretch," *American Journal of Physical Medicine, vol. 46, no. 1, Feb. 1967.*

——, "Peripheral Receptors: Their Excitation and Relation to Reflex Patterns," *American Journal of Physical Therapy,* vol. 46, no. 1, Feb. 1967.

Erlingheuser, R.F., "The Circulation of Cerebrospinal Fluid Through the Connective Tissue System," in the *Academy of Applied Osteopathy Yearbook, 1959.*

Evarts, E., "Brain Mechanisms of Movement," *Scientific American,* Sept. 1979.

Fong-Torres, B., "Shearing's Joy: Sighted Folks at Ease," in *The San Francisco Chronicle,* "Sunday Datebook," Nov. 11, 1984.

Forer, B.R., "The Taboo Against Touching in Psychotherapy," *Psychotherapy, Theory, Research, and Practice,* vol. 6, 1969.

Gardner, L.I., "Deprivation Dwarfism," *Scientific American,* July 1972.

Grey, G.W., "The Organizer," *Scientific American,* Nov. 1957.

Grof, S., *Realms of Human Unconscious,* E.P. Dutton, New York, 1976.

Gross, J., "Collagen," *Scientific American,* May, 1961.

Guyton, A.C., *Basic Human Physiology: Normal Function and Mechanisms of Disease,* W.B. Saunders Co., Philadelphia, 1971.

——, *Textbook of Medical Physiology,* W.B. Saunders Co., Philadelphia, 1981.

Harlow, H., "Love in Infant Monkeys," *Scientific American,* June 1959.

Held, R., "Plasticity in Sensory-Motor Systems," *Scientific American,* Nov. 1965.

Hubel, D.H., "The Brain," *Scientific American,* Sept. 1979.

Huxley, H.E., "The Contraction of Muscle," *Scientific American,* Nov. 1958.

——, "The Mechanism of Muscular Contraction," *Scientific American,* Dec. 1965.

Iverson, L.L., "The Chemistry of the Brain," *Scientific American,* Sept. 1979.

Job, Chapter 19, verses 26 and 27, *The Book of God and Man,* translated by Robert Gordis, Chicago University Press, London and Chicago, 1965.

Juhan, D., "The Trager Approach: Psychophysical Integration and Mentastics," *The Bodywork Book,* edited by Neville Drury, Prism Alpha Press, Dorset, England, 1984.

Jung, C.G., "An Answer to Job," *The Viking Portable Jung,* edited by Joseph Campbell, Viking Press, New York, 1971.

Kamenetz, H.L., "History of Massage," in *Manipulation, Traction, and Massage,* 2nd ed., edited by Joseph B. Rogoff, Williams and Wilkins, Baltimore, 1980.

Kandel, E.R. and Schwartz, J.H., *Principles of Neural Science,* Elsevier/North-Holland, New York, 1981.

——, "Small Systems of Neurons," *Scientific American,* Sept. 1979.

Katz, B., "How Cells Communicate," *Scientific American,* Sept. 1961.

Kessel R.G. and Kardon, R.H., *Tissues and Organs: A Text-Atlas of Scanning Electron Microscopy,* W.H. Freeman and Co., San Francisco, 1979.

Lester, H.A., "The Response to Acetylcholine," *Scientific American,* Feb. 1977.

Levine, S., "Stimulation in Infancy," *Scientific American,* May 1960.

——, "Stress and Behavior," *Scientific American,* Jan. 1971.

Lippold, O., "Physiological Tremor," *Scientific American,* March 1971.

Little, K.E., "Toward More Effective Manipulative Management of Chronic Myofascial Strain and Stress Syndromes," *The Journal of the American Osteopathic Association,* 68:675-685, March 1969.

Lockhart, R.D., Hamilton, G.F. and Fyfe, F.W., *Anatomy of the Human Body,* J.B. Lippincott Co., Philadelphia, 1972.

Mead, M., *Sex and Temperament in Three Primitive Societies,* William Morrow, New York, 1935.

Merleau-Ponty, M., *Phenomenologie de la Perception,* excerpt translated by Father Aelrid Squire in Chapter 5 of his *Asking the Fathers,* S.P.C.K., London, 1973.

Merton, P.A., "How We Control the Contraction of Our Muscles," *Scientific American,* May 1972.

Montagna, W., "The Skin," *Scientific American,* Feb. 1965.

Montagu, A., *Touching: The Human Significance of the Skin,* Harper and Row, New York, 1971.

Murray, J.M. and Weber, A., "The Cooperation of Muscle Proteins," *Scientific American,* Feb. 1974.

Nauta, W.J.H. and Feirtag, M., "The Organization of the Brain," *Scientific American,* Sept. 1979.

Netter, F.H., *The Ciba Collection of Medical Illustrations,* vol. 1, The Nervous System, CIBA, New Jersey, 1983.

Nilsson, L., *Behold Man*, Little, Brown and Co., Boston, 1973.

Porter, K. and Franzini-Armstrong, C., "The Sarcoplasmic Reticulum," *Scientific American*, March 1965.

Prescott, J.H., "Body Pleasure and the Origins of Violence," *The Futurist*, April 1975.

———, "Early Somatosensory Deprivation as an Ontogenetic Process in the Abnormal Development of the Brain and Behavior," in E.I. Goldsmith and J. Moor-Janowski (eds.), *Medical Primatology*, S. Karger, New York, 1971.

"Profile: The Winchester Kids," *The New Yorker*, May 9, 1977.

Rankin, J. and Dempsey, J., "Respiratory Muscles," *American Journal of Physical Medicine*, vol. 46, no. 1, February 1967.

Robbie, D.L., "Tensional Forces in the Human Body," *Orthopaedic Review*, vol. 6, no. 11, Nov. 1977.

Rolf, I., *Rolfing: The Integration of Human Structures*, Dennis-Landman, Santa Monica, 1977.

Romanes, G.J., *Cunningham's Textbook of Anatomy*, Oxford University Press, London, 1978.

Rosen, C., "The New Sound of Franz Liszt," in *The New York Review of Books*, vol. 31, no. 6, April 12, 1984.

Rosenblatt, J.S. and Lerhman, D.S., "Maternal Behavior in the Laboratory Rat," in H.L. Rheingold (ed.), *Maternal Behavior in Mammals*, Wiley, New York, 1963.

Rosenzweig, M.R., Bennett, E.L. and Diamond, M.C., "Brain Changes in Response to Experience," *Scientific American*, Feb. 1972.

Ross, R., "Wound Healing," *Scientific American*, June, 1967.

Roth, L.L., "Effects of Young and Social Isolation on Maternal Behavior in the Virgin Rat," *American Zoologist*, vol. 7, 1967.

Russel, B., *The ABC of Relativity*, Harper Brothers, New York, 1925.

Rutherford, D., "Auditory-Motor Learning and the Acquisition of Speech," *American Journal of Physical Medicine*, vol. 46, no. 1, February, 1967.

Sacks, O., *A Leg To Stand On*, Summit Books, New York, 1984.

Schilder, P., *The Image and Appearance of the Human Body*, International Universities Press, Inc., New York, 1950.

Schlossberg, L., *The Johns Hopkins Atlas of Human Functional Anatomy*, The Johns Hopkins University Press, Baltimore, 1977.

Sheldrake, R. and Weber, R., "Morphogenic Fields: Nature's Habits?", in *ReVision*, vol. 5, no. 2, Fall 1982.

Singer, E., *Fasciae of the Human Body and Their Relationships to the Organs They Envelop*, Williams and Wilkins Co., Baltimore, 1935.

Smith, K.U. and Henry, J.P., "Cybernetic Foundations for Rehabilitation," *American Journal of Physical Medicine*, vol. 46, no. 1, Feb. 1967.

Snider, R.S., "The Cerebellum," *Scientific American*, Aug. 1958.

Snyder, G.E., "Fascia—Applied Anatomy and Physiology" *The Journal of the American Osteopathic Association*, 68:675-685, March 1969.

Snyder, S.H., "The Molecular Basis of Communication Between Cells," *Scientific American*, Oct. 1985.

———, "Opiate Receptors and Internal Opiates," *Scientific American,* March 1977.

Somatics (magazine), 1980, anonymous review of Anna Halprin's *Movement Ritual.*

Sperry, R.W., "The Growth of Nerve Circuits," *Scientific American,* Nov. 1959.

Stevens, C.F., "The Neuron," *Scientific American,* Sept. 1979.

Taylor, R.B., "Bioenergetics of Man," in *Academy of Applied Osteopathy Yearbook,* 1958.

Todd, M.E., *The Thinking Body,* Dance Horizons Inc., Brooklyn, NY, 1979.

Vander, A.J., *Readings From Scientific American: Human Physiology and the Environment in Health and Disease,* W.H. Freeman Co., San Francisco, 1976.

———, Sherman, J.H. and Luciano, D.S., *Human Physiology: The Mechanisms of Body Function,* McGraw-Hill Co., New York, 1970.

Verzar, F., "The Aging of Collagen," *Scientific American,* April, 1963.

Vesalius, Andreas, *The Illustrations from the Works of Andreas Vesalius of Brussels,* with annotations and translations by J.B. deC. M. Saunders and Charles D. O'Malley, Dover Publications, New York, 1973.

Wallace, R.K. and Benson, H., "The Physiology of Meditation," *Scientific American,* Feb. 1972.

Weiss, J.M., "Psychological Factors in Stress and Disease," *Scientific American,* June, 1972.

Additional Works Cited

Oschman, James L., *Readings on the Scientific Basis of Bodywork,* Nature's Own Research Association, Dover, NH, 1993.

Pert, Candace, *Molecules of Emotion,* Scribner, New York, 1997.

Index

 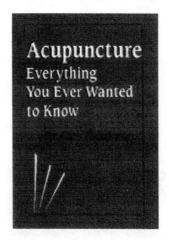

BodyStories

A Guide to Experimental Anatomy: Expanded Edition

Andrea Olsen

BodyStories engages the general reader as well as serious students of anatomy, dancers, artists, athletes, bodyworkers, massage therapists, teachers, and individuals with injuries or with a special interest in learning about their body. Thirty-one "days" of self-guided learning sessions heighten awareness of each bone and body system. Amusing and insightful personal stories enliven the text. Richly illustrated.

A perfect gift for anyone who wishes to understand anatomy in more depth through their own inner journeying. — **Bonnie Bainbridge Cohen**, Co-Director and Founder of The School of Body-Mind Centering

Andrea Olsen is an Associate Professor of Dance at Middlebury College in Vermont. As a dancer, she has choreographed over fifty works and toured internationally for twenty years. She lives in Thornes Market, Massachusetts.

Order No. P1060, $26.95 paper, ISBN 1-58177-023-5, 8 ½ x 11, 176 pages.

Acupuncture

Everything You Ever Wanted to Know

Dr. Gary F. Fleischman

A complete layperson's introduction to the principles and practices of acupuncture and oriental medicine, this easily readable Q & A book is your guide to the fastest growing, and perhaps most popular form of alternative medicine in America.

Gary F. Fleischman is a board-certified acupuncturist and medical doctor practicing in New Haven, Connecticut.

Order No. P9090, $13.95 paper, ISBN 1-886449-09-0, 6 x 9, 192 pages.